Quantum Bio-Informatics II

From Quantum Information to Bio-Informatics

QP–PQ: Quantum Probability and White Noise Analysis*

QP–PQ: Quantum Probability and White Noise Analysis

Vol. 24: Quantum Bio-Informatics II
From Quantum Information to Bio-Informatics
eds. L. Accardi, W. Freudenberg and M. Ohya

Vol. 23: Quantum Probability and Related Topics
eds. J. C. García, R. Quezada and S. B. Sontz

Vol. 22: Infinite Dimensional Stochastic Analysis
eds. A. N. Sengupta and P. Sundar

Vol. 21: Quantum Bio-Informatics
From Quantum Information to Bio-Informatics
eds. L. Accardi, W. Freudenberg and M. Ohya

Vol. 20: Quantum Probability and Infinite Dimensional Analysis
eds. L. Accardi, W. Freudenberg and M. Schürmann

Vol. 19: Quantum Information and Computing
eds. L. Accardi, M. Ohya and N. Watanabe

Vol. 18: Quantum Probability and Infinite-Dimensional Analysis
From Foundations to Applications
eds. M. Schürmann and U. Franz

Vol. 17: Fundamental Aspects of Quantum Physics
eds. L. Accardi and S. Tasaki

Vol. 16: Non-Commutativity, Infinite-Dimensionality, and Probability at the Crossroads
eds. N. Obata, T. Matsui and A. Hora

Vol. 15: Quantum Probability and Infinite-Dimensional Analysis
ed. W. Freudenberg

Vol. 14: Quantum Interacting Particle Systems
eds. L. Accardi and F. Fagnola

Vol. 13: Foundations of Probability and Physics
ed. A. Khrennikov

QP–PQ

Vol. 11: Quantum Probability Communications
eds. S. Attal and J. M. Lindsay

Vol. 10: Quantum Probability Communications
eds. R. L. Hudson and J. M. Lindsay

*For the complete list of the published titles in this series, please visit:
www.worldscibooks.com/series/qqpwna_series.shtml

QP–PQ
Quantum Probability and White Noise Analysis
Volume XXIV

Quantum Bio-Informatics II

From Quantum Information to Bio-Informatics

Tokyo University of Science, Japan 12 – 16 March 2008

Editors

L. Accardi
Università di Roma "Tor Vergata", Italy

W. Freudenberg
Brandenburgische Technische Universität Cottbus, Germany

M. Ohya
Tokyo University of Science, Japan

World Scientific

NEW JERSEY • LONDON • SINGAPORE • BEIJING • SHANGHAI • HONG KONG • TAIPEI • CHENNAI

Published by

World Scientific Publishing Co. Pte. Ltd.
5 Toh Tuck Link, Singapore 596224
USA office: 27 Warren Street, Suite 401-402, Hackensack, NJ 07601
UK office: 57 Shelton Street, Covent Garden, London WC2H 9HE

British Library Cataloguing-in-Publication Data
A catalogue record for this book is available from the British Library.

QP-PQ: Quantum Probability and White Noise Analysis — Vol. 24
QUANTUM BIO-INFORMATICS II
From Quantum Information to Bio-Informatics

ISBN-13 978-981-4273-74-9
ISBN-10 981-4273-74-0

Printed in Singapore by Mainland Press Pte Ltd

PREFACE

This volume is based on the second international conference of quantum bio-informatics held at the QBI Center of Tokyo University of Sciences.

The purpose of the conference is towards new stage making interdisciplinary bridges in mathematics, physics, information and life sciences, in particular, research for new paradigm for information science and life science on the basis of quantum theory. For this purpose we firstly return to the starting point of bio-informatics and quantum information, fields that are growing rapidly at present, and to seriously attempt mutual interaction between the two, with a view to enumerating and solving the many fundamental problems they entail. The papers submitted in this volume are all referred, whose contents are related to one of the following subjects:

(1) Mathematical foundation of quantum mechanics
(2) Quantum information
(3) Quantum algorithm and computation
(4) Quantum communication
(5) White noise analysis and quantum dynamics
(6) Chaos dynamics and adaptive dynamics
(7) Experimental studies of quantum computer
(8) Bio-Informatics
(9) Genome analysis

Luigi Accardi
Wolfgang Freudenberg
Masanori Ohya

CONTENTS

THE PROBLEM OF QUANTUM-LIKE REPRESENTATION IN ECONOMY COGNITIVE SCIENCE, AND GENETICS

L. ACCARDI and A. KHRENNIKOV *and M. OHYA

Mathematical Center Vito Volterra
Department of Mathematics
University of Rome II, Italy

Mathematical Center Vito Volterra
Department of Mathematics
University of Rome II, Italy

Department of Information Sciences,
Tokyo University of Science
Noda City, Chiba, 278-8510, Japan

We outline our programme to create quantum-like representations in economy, cognitive science, psychology, genetics,....

The basis of the *quantum-like paradigm* consists in understanding that the mathematical apparatus of quantum mechanics and especially quantum probability is not rigidly coupled with *quantum physics* but can have a wider class of applications.

Recall that differential and integral calculus were developed to serve classical Newtonian mechanics. However nowadays nobody is surprised that these tools are widely used everywhere – in engineering, biology, economy, ... In the same way, although the mathematical apparatus of quantum mechanics was developed to describe phenomena in the microworld, it could be applied to the solution of various problems outside physics.

One of the interesting open problems is to apply quantum probability e.g. to cognitive science or to financial markets. One of the main distinguishing features of quantum probability is the use of complex probability

*This work is supported by the profile of växjö university – "mathematical modelling and system collaboration".

amplitudes. In the abstract approach such amplitudes are represented by normalized vectors in a complex Hilbert space while the so called mixed states are represented by density matrices. Probabilities (which are compared with experimental relative frequencies) are given by Born's rule:

$$P_\psi(a = \alpha) = |\langle \psi, e_\alpha \rangle|^2,$$

where the observable a is represented by a self-adjoint operator $\widehat{a}$ and e_α is its eigenvector corresponding to the eigenvalue $\alpha : \widehat{a}e_\alpha = \alpha e_\alpha$ (Only the case of operators with purely discrete and nondegenerate spectra is considered here).

During the past 70 years the development of quantum mechanics has been characterized by intensive debates on the origin of quantum randomness and in particular on possibilities to reduce it to the classical ensemble randomness. For example, von Neumann was convinced that quantum randomness is irreducible, but Einstein had the opposite view to this problem: for him the discovery of quantum mechanics was merely a discovery of a special mathematical formalism (quantum formalism) for description of a special incomplete representation of information about microsystems.

This debate is directly related to the problem of creation of quantum-like representations outside of quantum physics. In the majority of applications, e.g., in economy or biology, the conventional models are based on macroscopic variables. For example, contemporary neurophysiology is based on a model with a neuron as the basic unit of information processing.

According to the Copenhagen interpretation of QM a pure quantum state (wave function) describes an *individual quantum system* not an ensamble of systems in the sense of classical probability. As a consequence of such an "individual interpretation" a concrete physical system can be prepared in a physical superposition of pure states.

In the majority of textbooks on QM we can read about e.g. an atom in superposition of different energy states, or an electron in superposition of spin-up and spin-down states, in the famous two slit experiment a photon is in a superposition of passing through both slits.

An attempt to apply the mathematical formalism of QM outside of the microworld in combination with the Copenhagen interpretation would create visible difficulties: it is not easy to imagine a macroscopic system e.g. in economics which is in a real, physical, superposition of two states. The authors of this paper are well aware about macroscopic quantum systems as well as of the attempts to use the Copenhagen interpretation even in this case – e.g. by Legget in superconductivity, by Zeilinger in the two slit experiments for macroscopic systems, by De Martini in experiments with "macroscopic Schrödinger cats" etc. ....

It is well known that such attempts to proceed with the Copenhagen interpretation for macroscopic quantum systems does not provide a clear physical picture of the phenomena. One of the possibilities is to use De Broglie's wave length for characterization of the wave features of a macroscopic system. Since it is very small for a large system, it is always possible to say that, although a macroscopic system has wave features, they are hardly observable.

This kind of compromise is hardly satisfactory as a solution to a conceptual problem.

Moreover, as already pointed out by Pauli in the early times of QM, any attempt to interpret the wave function as a physical wave clashes against the fact that, for most interesting physically systems, these wave functions are defined in a multi–dimensional mathematical space. Thus the supporters of the *wave–particle duality* face the paradox of believing in *a physical wave in a non physical space*.

As a consequence of the above mentioned difficulties with the interpretation of macroscopic quantum systems, a popular attitude today is to proceed beyond conventional models (e.g. in biology) which operate with states of macroscopic systems.

For example, in cognitive science a group of researchers (e.g., Penrose and Hameroff) developed the reductionist approach to the brain functioning. They moved beyond the conventional neuronal paradigm of cognitive science and tried to reduce processing of information in the brain to quantum micro processes – on the level of quantum particles composing the brain. Penrose repeated many times that a neuron (as a macroscopic system) could not be in a physical superposition of two states: firing and nonfiring.

As was already mentioned, the majority of attempts to apply the mathematical formalism of quantum mechanics outside physics were based on the reduction of the processes under consideration to some underlying quantum processes in the microworld. This reductionist approach was heavily based on the following argument: since everything in this world is composed of quantum particles, any kind of process might be (at least in principle) reduced to a quantum processes.

The unification dream is in principle correct, and it has played an important role in the development of natural sciences, in this spirit any attempt to apply quantum mechanics to e.g. cognitive science should be welcome. However, it is very difficult (if even possible at all) to establish a natural correspondence between conventional macroscopic models and underlying quantum models. There is a huge difference in scales of parameters in those models. Moreover even in quantum physics the *correspondence principle* is vaguely formulated and not totally justified and, on the other hand, even in classical physics, the unification dream is far from being accomplished in spite of the important successes of statistical mechanics in the reduction of thermodynamics to to mechanics. For example structures such as crystals, which are relatively simple with respect to biological structures, at the moment have not been deduced from first principles neither in classical nor in quantum physics.

We point out that it is possible to escape the above mentioned difficulties by rejection of the Copenhagen interpretation and association of a pure quantum state (wave function) not with an individual quantum system, but with an ensemble of identically prepared systems.

Such an interpretation is called the *statistical interpretation* of QM. It has been originally proposed by many authors among which Einstein, Popper, Margenau, De Broglie, Bohm, Ballentine, ... but only with the development of quantum probability it could overcome the traditional critiques which prevented, for over 50 years, the majority of physicists to accept this apparently natural interpretation. The main objection to it, to which the above mentioned authors never gave a satisfactory answer, was that the statistical interpretation is contradicted by the experimental data.

Concerning this objection the main point of quantum probability is that the experimental data contradict the use of the Kolmogorov model of probability and not the statistical interpretation. If one keeps to the statistical interpretation, then one can assume that the quantum probabilistic description need not be based on irreducible quantum randomness [1].

The quantum probabilistic calculus can be used for incomplete description of statistical data.[a]

One could not even exclude that in some cases a Kolmogorov model

[a]Here we should distinguish between the theoretical view of quantum probability as the study of *all* non Kolmogorovian models and the more restricted point of view that wants to limit the investigations to the original quantum model. Moreover quantum probability proves that the appeal to an *irreducible quantum randomness* is not necessary, but it does not prove that it is wrong. This is a personal belief that cannot be scientifically proved or disproved and it is a fact that a number of scientists, including renown pioneers of quantum probability like Hudson and Belavkin, still adhere to irreducible randomness.

can be found beyond the quantum probabilistic description. The crucial point is that the role of the presence of a "hidden Kolmogorovian model" is negligible if one has no access to data described by the latter (typically unobservable joint probabilities). In such cases the only reasonable possibility is to use the quantum probabilistic description or different non Kolmogorovian models.

Thus we propose to test the approach based on the accepting of Einstein's viewpoint: incompleteness of quantum mechanics. The natural question which is typically asked as first reaction to our proposal is the following:

What about the known no-go theorems?

We will not enter here into a debate on the complicated problem of the validity of no-go theorems (for this we refer to ([3])). In fact, the main problem of the no-go ideology is that it is directed against all possible prequantum models (the so called hidden variable models). [b] Supporters of no-go activity formulate new theorems excluding various classes of models with hidden variables, but one could never be sure that a natural model which does not contradict any known no-go theorem would be finally found. In particular, Accardi[2] pointed out to the possibility to produce non classical, i.e. non Kolmogorovian, statistics by using *classical adaptive local dynamical systems*, see e.g. Ohya[4], for modelling of the process of measurement. Such models are known as chameleon models: this animal adapts his color to color of surface. Chameleon realism differs essentially from Einstein realism – association of values of quantum observables directly with states of systems.

Einstein realism does not take into account the dynamics of the process of interaction of a system with the measurement device. In fact chameleon realism matches well with the ideas of the father of the Copenhagen interpretation N. Bohr who permanently pointed out that the whole experimental arrangement should be taken into account[5]. One might speculate that Bohr would prefer chameleon realism to such rather strange things as nonlocal realism or "quantum nonlocality".

[b]We do not agree with Bell's attempt to couple the so called "quantum nonlocality" with the problem of completeness of quantum mechanics. It has now been experimentally proved that "quantum nonlocality" is an absurd alternative to incompleteness. Unfortunately, in spite of the mathematical and experimental evidence, nowadays quantum nonlocality has become extremely popular in quantum information theory. Moreover, this idea diffuses outside quantum physics: it became fashionable to refer to quantum nonlocality in cognitive and social sciences and even in parapsychology.

We share Einstein's views only partially. We keep to the statistical interpretation of the quantum state and consequently incompleteness of QM, but we agree with Bohr in considering the values of some quantum observables as responses to interactions with apparata rather than objective properties of quantum systems.

We know that by using e.g. adaptive dynamical system one can obtain a local realistic model for quantum measurements. We could summarize our proposal as follows:

" Be not afraid to consider the quantum description as an incomplete one. Look for applications of quantum probability outside quantum physics! Use adaptive dynamical systems to describe interactions among "conventional variables" (e.g. neuronal states) which produce a quantum-like behavior".

As a comment the use of the notion quantum-like(QL) behavior, cf. Khrennikov [6, 7], we think that it would be useful to preserve the term "quantum" for quantum physics while, in other models which are still based on quantum or, more generally non Kolmogorovian, probabilistic description we should use the term "quantum-like". In particular, in this way we can distinguish our approach from a purely reductionist one. For example, the quantum brain model is a reductionist model of the brain functioning, but the quantum-like brain model is a model in which the wave function provides a (incomplete) probabilistic representation of information produced by the neurons[8] and not a model for the actual physical state of them.

The QL modelling immediately meets one complex problem: the creation of QL-representations (in complex Hilbert space) of classical probabilistic data. For example, looking for a QL model of image recognition, see e.g. Fichtner et al[9], it would be natural to represent an image by a wave function. Image processing may be modelled by using e.g. (in the simplest case) Schrödinger's equation. But we should solve the problem of initial conditions:

How does the brain represents the initial image by the wave function?

If one considers the brain as a kind of probabilistic machine, then this problem can be formulated as the *inverse Born problem:*

To construct a complex probability amplitude on the basis of probabilities.

An attempt to solve this problem was done in a series of works of Krennikov [6, 7]. There was created so called QL-representation algorithm. *Improvement of this algorithm, its generalization as well creation of new QL-*

representation algorithms is an important problem in the realization of the QL paradigm.

We now couple the QL paradigm with another important probabilistic paradigm. Nonclassical statistical data are not covered completely by the conventional quantum model. As one of the authors (Luigi Accardi) [2] pointed out the main distinguishing feature of quantum probability is its *non-Kolmogorovainity.*

It was emphasized that in the same way as in geometry (where starting with Gauss, Lobachevsky, Riemann, ..., various non-Euclidean geometries were developed and widely applied e.g. in relativity theory) in probability theory various non-Kolmogorov models may be developed to serve applications. The QM probabilistic model was one of the first non-Kolmogorovian models which had important applications. Thus one may expect development of other types of probabilistic models which would be neither Kolmogorovian nor quantum.

One of such models was presented – it is the model with so called hyperbolic interference [7]. It is based on representation of probabilities by amplitudes taking values in the algebra of so called hyperbolic numbers. This example motivates extension of the QL paradigm by attempting to develop and apply models in which probability amplitudes take values in various commutative and even noncommutative algebras. The corresponding generalizations of Born's rule should be presented, analogues of the QL-representation algorithms should be created. These are interesting and complex problems!

It is reasonable to extend our definition of a QL model by considering a multiplicity of probabilistic models differing from the conventional Kolmogorov model which is a purely mathematical construction expressing one possible probabilistic description of data. We would like to present a "physical definition" of a QL model coupled to the process of measurement. The main idea behind the QL paradigm is the possibility to represent incomplete probabilistic data about some class of systems. We formalize this idea in the following way.

Postulate H. (Weak Form of the Heisenberg Uncertainty Principle). There exist observables which cannot be measured simultaneously with arbitrary precision.

Definition. (QL Model). Any probabilistic model describing data obtained from observations satisfying the Weak Form of the Heisenberg Uncertainty Principle is called quantum-like.

To create a QL model, one should first find at least two observables which cannot be measured simultaneously. Corresponding probabilistic data should be incorporated into a model, typically by using some algebraic structure, e.g. complex or hyperbolic Hilbert space.

We are now analyzing several different families of empirical data in order to realize concretely the programme outlined in the present paper. These developments will be discussed elsewhere.

References

1. L. Accardi, Can mathematics help solving the interpretation problem of quantum theory? *Il Nuovo Cimento* **B 110**, 685–721 (1995).
2. L. Accardi, *Urne e Camaleoni: Dialogo sulla realta, le leggi del caso e la teoria quantistica,* Rome, 1997.
3. Accardi L.: Foundations of Quantum Mechanics: a quantum probabilistic approach, in The Nature of Quantum Paradoxes; eds. G.Tarozzi, A. van der Merwe Reidel (1988) 257–323. Preprint Dipartimento Di Matematica, Universita' di Roma Torvergata (1986)
4. M. Ohya, Addaptive dynamics and its applications to chaos and NPC problem. *Quantum Bio-Informatics*, Edited by L. Accardi, W. Freudenberg, M. Ohya, ser. *Quantum Prob. and White Noise An.* **21**, 2007, 181-216.
5. N. Bohr, *The philosophical writings of Niels Bohr*, 3 vols., Ox Bow Press, 1987.
6. A. Yu. Khrennikov, Contextual viewpoint to quantum stochastics. *J. Math. Phys.* **44**, N. 6, 2471- 2478 (2003).
7. A. Khrennikov, *Information Dynamics in Cognitive, Psychological and Anomalous Phenomena,* ser. *Fundamental Theories of Physics,* Kluwer Academic, 2004.
8. A. Khrennikov, Quantum-like brain: Interference of minds, *BioSystems* **84**, 225–241 (2006).
9. K. H. Fichtner, L. K.H. Fichtner, W. Freudenberg and M. Ohya, On the quantum model of the recognition process, in Quantum Bio-Informatics; eds. L. Accardi, W. Freudenberg and M. Ohya, QP-PQ series, **23** (2007) 64-84.

CHAOTIC BEHAVIOR OBSERVED IN LINEAR DYNAMICS

M. ASANO, T. YAMAMOTO, Y. TOGAWA

Department of Information Sciences, Faculty of Science and Technology,
Tokyo University of Science, Noda City, Chiba 278, Japan.

In this paper, we discuss chaos observed in linear dynamical systems. Mainly, we address log-linear dynamics, which is a dynamical system on open simplex, and show chaotic behaviors near the boundary on the simplex.

1. Introduction

In many dynamical systems, chaotic behaviors have been observed and discussed. Mechanism of generating chaos is usually very complicated, but in some cases, their mechanisms are explained simply. For example, the dynamics of Bernoulli shift is essentially determined by the linear map $f(x) = 2x$ ($x \in \mathsf{R}$), and any orbits drawn on R are non-chaotic. However, chaotic behaviors are observed on R/Z. Similarly, Anosov maps make chaotic behaviors on the torus $\mathsf{R}^2/\mathsf{Z}^2$. The maps are induced by linear maps $f(\mathsf{x}) = A\mathsf{x}$ ($\mathsf{x} \in \mathsf{R}^2$). (The matrix A has components of integers satisfying $\det(A) = \pm 1$.) The properties of quotient spaces R/Z and $\mathsf{R}^2/\mathsf{Z}^2$ induce chaotic behaviors, in spite of the dynamics determined by linear maps on R and R^2, respectively.

The dynamical system we discuss in this paper shows chaotic behaviors, and also has a simple background as a linear map. However, its mechanism for generating chaos is different from the ones in Bernoulli shift and Anosov maps. The space on which the dynamics is defined is simplex

$$\mathring{\Delta} = \{\mathsf{x} \in \mathsf{R}^{n+1} \mid x_0 + x_1 + x_2 + \cdots + x_n = 1,\ x_i > 0\}, \tag{1}$$

which is not a quotient space. Note that $\mathring{\Delta}$ includes no boundaries and it is homeomorphic to R^n. The closure of $\mathring{\Delta}$ is expressed by Δ. Let h be a homeomorphism from $\mathring{\Delta}$ to R^n. The dynamics on simplex is translated to a dynamics on R^n by h. In a sense, we observe the dynamics from two different viewpoints, that is, "inside of Δ" and "outside of Δ". The viewpoint "inside" is a mathematical viewpoint. The dynamics is defined on $\mathring{\Delta}$

without boundary and it is conjugate to the dynamics on R^n. On the other hand, when we observe the dynamics from the outside of Δ, we observe also the boundary $\Delta - \mathring{\Delta}$ which is not included in the domain. We address an interesting dynamics showing chaotic behaviors at the viewpoint from outside in spite of showing simple behaviors at the viewpoint from inside. This dynamics is called log-linear dynamics,[1–4] we discuss its property in detail in Sec. 3. First, in the next section, we mention the general form of dynamics on simplex and introduce some examples.

2. Dynamics on Simplex and Example

2.1. General Form of Dynamics on Simplex

In this section, we show the general form of dynamics on simplex $\mathring{\Delta}$. Let us consider the following differential equations on R^{n+1}.

$$\frac{dx_i}{dt} = x_i F_i(\mathbf{x}), \quad (i = 0, 1, ..., n), \tag{2}$$

where $F_i(\mathbf{x})$ stand for functions with variables of x_i, $i = 0, 1, .., n$. In this equation, for initial $x_i(0) > 0$, the $\mathbf{x}(t)$ stays on "first quadrant" $\{x|\ x_i > 0, i = 0, 1, .., n\}$. For the vector field $X_i(\mathbf{x}) = x_i F_i(\mathbf{x})$, we can describe the projected field on $\mathring{\Delta}$ as the form

$$\tilde{\mathsf{X}}(\mathbf{x}) = \mathsf{X}(\mathbf{x}) - c(\mathbf{x})\mathbf{x}. \tag{3}$$

Since, the field $\tilde{\mathsf{X}}(\mathbf{x})$ is orthogonal to the normal vector of simplex $\mathbf{u} = (1, 1, ..., 1)^T$, if $\mathbf{x} \in \mathring{\Delta}$, the function $c(\mathbf{x})$ is written as

$$c(\mathbf{x}) = \sum_{i=0}^{n} X_i = \sum_{i=0}^{n} x_i F_i(\mathbf{x}), \ \mathbf{x} \in \mathring{\Delta}. \tag{4}$$

Therefore, the dynamics on $\mathring{\Delta}$ is defined by the differential equations of

$$\frac{dx_i}{dt} = x_i (F_i(\mathbf{x}) - \sum_{i=0}^{n} x_i F_i(\mathbf{x})), \ \mathbf{x} \in \mathring{\Delta}. \tag{5}$$

The dynamical systems with such the form have been discussed in various areas. For an example, Replicator equation in game dynamics[5] is defined with $F_i(\mathbf{x}) = \sum_{j=0}^{n} a_{ij} x_j$. Also, Scarf-Hirota model explained in the next subsection are the interesting examples which have been discussed in economics.

2.2. Scarf-Hirota Model

Model of exchange economy

Scarf-Hirota model was proposed as a model of exchange economy.[6–8] Exchange economy models account for price movements of commodities. Suppose that there are n commodities and m consumers in a market. Let a_h be the initial endowment vector of h-th consumers and let p_i be the price of i-th commodity. The vector $\mathsf{p} = (p_1, \ldots, p_n)$ is called price vector, and then, $\mathsf{p} \cdot \mathsf{a}_h$ means h-th consumer's income. Here, we assume that price movements are decided by the following differential equation called price-scaled price adjustment process.

$$\frac{dp_i}{dt} = p_i E_i(\mathsf{p}), \qquad (i = 1, 2, \ldots, n). \tag{6}$$

$E_i(\mathsf{p})$ is called excess demand function, and $\mathsf{E}(\mathsf{p}) = (E_1, E_2, \ldots, E_n)$ is defined as follows. A function $u_h(\mathsf{x}_h)$ called utility function is given for each consumer, and $\mathsf{x}_h(\mathsf{p})$ is the vector maximizing the value of $u_h(\mathsf{x}_h)$ under the condition $\mathsf{p} \cdot \mathsf{x}_h = \mathsf{p} \cdot \mathsf{a}_h$, and then

$$\mathsf{E}(\mathsf{p}) \,, \sum_{h=1}^{m} (\mathsf{x}_h(\mathsf{p}) - \mathsf{a}_h)\,. \tag{7}$$

The excess demand $\mathsf{E}(\mathsf{p})$ satisfies the following conditions.

$$\mathsf{E}(\mathsf{p}) = \mathsf{E}(\lambda \mathsf{p}) \quad (\lambda \in \mathsf{R}), \tag{8}$$

$$\mathsf{p} \cdot \mathsf{E}(\mathsf{p}) = 0. \tag{9}$$

The equation (9) is well-known as Walras law in mathematical economics. Walras law implies that the price movements by Eq. (6) are described as dynamics on simplex, if $\sum_{i=1}^{n} p_i = 1$.

Scarf-Hirota model with three consumers and three commodities

Let us consider exchange economy model which consists of three consumers and three commodities ($n = m = 3$). In this case, Scarf-Hirota Model have the utility functions with the forms of

$$\begin{aligned} u_1(\mathsf{x}_1) &= \min[x_{12}, x_{13}] \\ u_2(\mathsf{x}_2) &= \min[x_{21}, x_{23}] \\ u_3(\mathsf{x}_3) &= \min[x_{31}, x_{32}]. \end{aligned} \tag{10}$$

Furthermore, the initial endowment vectors $\mathsf{a}_\mathsf{h} = (a_{\mathsf{h}1}, a_{\mathsf{h}2}, a_{\mathsf{h}3})$ satisfy

$$\sum_{\mathsf{h}=1}^{3} a_{\mathsf{hi}} = 1, \ \sum_{\mathsf{i}=1}^{3} a_{\mathsf{hi}} = 1, \tag{11}$$

for $i = 1,2,3$ and for $h = 1,2,3$. Then, the excess demand functions are written as

$$\begin{aligned} E_1(\mathsf{p}) &= \frac{\mathsf{p}\cdot\mathsf{a}_2}{p_3+p_1} + \frac{\mathsf{p}\cdot\mathsf{a}_3}{p_1+p_2} - 1, \\ E_2(\mathsf{p}) &= \frac{\mathsf{p}\cdot\mathsf{a}_3}{p_1+p_2} + \frac{\mathsf{p}\cdot\mathsf{a}_1}{p_2+p_3} - 1, \\ E_3(\mathsf{p}) &= \frac{\mathsf{p}\cdot\mathsf{a}_1}{p_2+p_3} + \frac{\mathsf{p}\cdot\mathsf{a}_2}{p_3+p_1} - 1. \end{aligned} \tag{12}$$

It is known that if $4a_{11}a_{22} > (1 - a_{12} - a_{21})^2$ is satisfied, the dynamics by price-scaled price adjustment process $\frac{\mathsf{dp_i}}{\mathsf{dt}} = p_\mathsf{i} E_\mathsf{i}(\mathsf{p})$ has a fixed point with asymptotic global stability. (See Fig. 1-(a).) When the endowment vectors are defined as $\mathsf{a}_1 = (0,1,0)$, $\mathsf{a}_2 = (0,0,1)$ and $\mathsf{a}_3 = (1,0,0)$, the dynamics does not have such a fixed point with asymptotic global stability. The orbits stay in the curves $\frac{1}{\mathsf{p}_1} + \frac{1}{\mathsf{p}_2} + \frac{1}{\mathsf{p}_3} = \text{const.}$ (See Fig. 1-(b).)

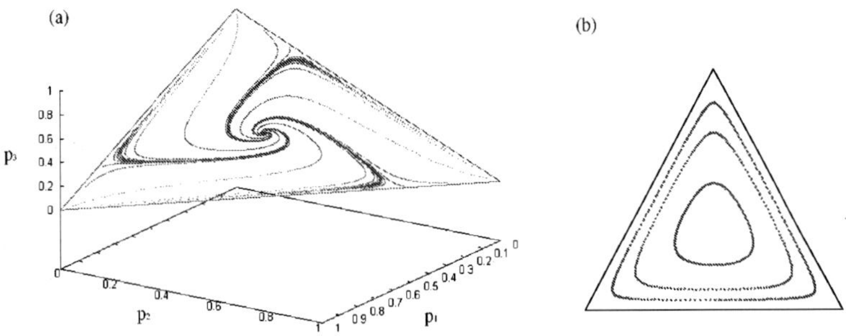

Fig. 1. (a) Orbits with asymptotic global stability in Sacrf-Hirota model. (b) Orbits in the special case with $\mathsf{a}_1 = (0,1,0)$, $\mathsf{a}_2 = (0,0,1)$ and $\mathsf{a}_3 = (1,0,0)$.

Utility functions for linear dynamics

Next, let us consider the following utility functions.

$$u_\mathsf{h}(\mathsf{x}_\mathsf{h}) = \sum_{\mathsf{i}=1}^{\mathsf{n}} b_{\mathsf{hi}} \log x_{\mathsf{hi}} \quad (h = 1, \ldots, m). \tag{13}$$

($\{b_{hi}\}$ are arbitrary parameters.) Then, the excess demand functions are

$$E_i(\mathsf{p}) = \frac{1}{p_i}\sum_{h=1}^{m}\frac{b_{hi}\mathsf{p}a_h}{\sum_{k=1}^{n} b_{hk}} - \sum_{h=1}^{m} a_{hi} \qquad (i = 1, 2, \ldots, n), \tag{14}$$

and $\frac{dp_i}{dt} = p_i E_i(\mathsf{p})$ is described as linear dynamics, that is,

$$\begin{aligned}\frac{dp_i}{dt} &= \sum_{j=1}^{n}\left(\sum_{h=1}^{m}\frac{b_{hi}a_{hj}}{\sum_{k=1}^{n} b_{hk}} - \delta_{ij}\sum_{h=1}^{m} a_{hi}\right)p_j \\ &= \sum_{j=1}^{n} M_{ij}p_j \qquad (i = 1, 2, \ldots, n).\end{aligned} \tag{15}$$

However, this dynamics is not a dynamics in simplex. There are cases that a trajectory go out the boundary of simplex. Such behavior is out of theme of our discussion.

3. Log-linear Dynamics and Chaotic Behaviors on Simplex

3.1. Log-linear Dynamics on Simplex

Log-linear dynamics on simplex $\mathring{\Delta}$ is given by the function of $F_i(\mathsf{x}) = \sum_{j=0}^{n} a_{ij}\log x_j$ in Eq. (5), and the form of differential equation is

$$\frac{dx_i}{dt} = x_i\Big(\sum_{j=0}^{n} a_{ij}\log x_j - \sum_{i=0}^{n} x_i \sum_{j=0}^{n} a_{ij}\log x_j\Big), \ \mathsf{x} \in \mathring{\Delta}. \tag{16}$$

For the parameters $\{a_{ij}\}$ in Eq. (16), we consider the matrix $A = \{a_{ij}\}$ with the condition of

$$A\mathsf{u} = \mathbf{0}, \tag{17}$$

where $\mathsf{u} = (1, 1, 1, \ldots 1)^{\mathsf{T}} (\in \mathsf{R}^{n+1})$. The simplex $\mathring{\Delta}$ is defined as

$$\mathring{\Delta} = \{\mathsf{x} \in \mathsf{R}^{n+1} \mid x_0 + x_1 + x_2 + \cdots + x_n = 1,\ x_i > 0\},$$

and it is homeomorphic to R^n. The log-linear dynamics with condition of Eq. (17) is related to a linear dynamics on R^n. It is explained as bellows. Let us consider the set of

$$\Gamma = \{\mathsf{y} \in \mathsf{R}^{n+1} \mid e^{y_0} + e^{y_1} + e^{y_2} + \cdots + e^{y_n} = 1, y_i < 0\}, \tag{18}$$

and define the map of $\mathring{\Delta}$ to Γ as

$$Log : \mathsf{x} \in \mathring{\Delta} \longrightarrow \log(\mathsf{x}) = \mathsf{y} \in \Gamma. \tag{19}$$

By this map, the dynamics of Eq. (16) is translated to a dynamics on Γ described as

$$\frac{dy_i}{dt} = (A\mathsf{y})_i - \sum_{k=0}^{n} e^{y_k}(A\mathsf{y})_k, \; \mathsf{y} \in \Gamma. \tag{20}$$

Next, we consider the set of

$$M = \{\mathsf{z} \in \mathsf{R}^{n+1} \mid z_0 + z_1 + z_2 + \cdots + z_n = 0\}, \tag{21}$$

which is homeomorphic to R^n, and the projection from Γ to M as

$$\Pi : \mathsf{y} \in \Gamma \longrightarrow \mathsf{y} - d(\mathsf{y})\mathsf{u} = \mathsf{z} \in M, \tag{22}$$

where

$$d(\mathsf{y}) = \frac{\sum_{i=0}^{n} y_i}{n+1}. \tag{23}$$

Noting the condition of Eq. (17), we can confirm that the map of $\Pi \circ \Gamma$ conjugates the log-linear dynamics and the dynamics of

$$\frac{d\mathsf{z}}{dt} = A\mathsf{z}, \; \mathsf{z} \in M, \tag{24}$$

which is the linear dynamics defined by the matrix A.

3.2. Discrete Log-linear Dynamics

Next, we represent the log-linear dynamics as a discrete dynamics on $\mathring{\Delta}$. An orbit $\mathsf{z}(t)$ satisfying the differential equation of Eq. (24) is described as

$$\mathsf{z}(t) = \exp(At)\mathsf{z}(0). \tag{25}$$

Let $\mathsf{z}_0, \mathsf{z}_1, \mathsf{z}_2, \mathsf{z}_3, \ldots$ be values of $\mathsf{z}(t)$ for $t = 0, 1, 2, 3, \ldots$. Then, the mapping of z_m to z_{m+1} is defined by $\mathsf{z}_{m+1} = B\mathsf{z}_m$, where the matrix $B = \{b_{ij}\}$ stands for $\exp(A)$. From the condition of Eq. (17), $B\mathsf{u} = \mathsf{u}$ is satisfied. We put this linear map, $g : M \longrightarrow M$, and give the map of discrete log-linear dynamics by

$$f \equiv (\Pi \circ Log)^{-1} \circ g \circ (\Pi \circ Log). \tag{26}$$

Since

$$\Pi^{-1} : \; \mathsf{z} \in M \longrightarrow \mathsf{z} - \log(\sum_{i=0}^{n} e^{z_i}) \in \Gamma,$$

$$Log^{-1} : \; \mathsf{y} \in \Gamma \longrightarrow e^{\mathsf{y}} \in \mathring{\Delta}, \tag{27}$$

i-th component of $f(\mathsf{x})$ is described as

$$f_{\mathsf{i}}(\mathsf{x}) = \frac{\prod_{\mathsf{j}=0}^{\mathsf{n}} x_{\mathsf{j}}^{\mathsf{b}_{\mathsf{ij}}}}{\sum_{\mathsf{k}=0}^{\mathsf{n}} \prod_{\mathsf{j}=0}^{\mathsf{n}} x_{\mathsf{j}}^{\mathsf{b}_{\mathsf{kj}}}}. \tag{28}$$

An orbit of $\mathsf{x}_{\mathsf{m}} = f^{\mathsf{m}}(\mathsf{x}_0) \in \mathring{\Delta}$ for an initial x_0 is linked to $\mathsf{z}_{\mathsf{m}} = B^{\mathsf{m}}\mathsf{z}_0 \in M$ with the relation of $\mathsf{z}_{\mathsf{m}} = \Pi(Log(\mathsf{x}_{\mathsf{m}}))$.

3.3. Chaotic Behavior

In this subsection, chaotic behaviors observed in the log-linear dynamics are shown. First, we give an example in the case of the discrete Log-linear dynamics. For the dimension $n+1$, we set $n = 2$, and give the matrix B as

$$B = (\mathsf{u}, \mathsf{v}_1, \mathsf{v}_2) \begin{pmatrix} 1 & 0 & 0 \\ 0 & k\cos\theta & -k\sin\theta \\ 0 & k\sin\theta & k\cos\theta \end{pmatrix} (\mathsf{u}, \mathsf{v}_1, \mathsf{v}_2)^{\mathsf{T}}, \tag{29}$$

where $\mathsf{u} = (1,1,1)$, and $\mathsf{v}_{\mathsf{i}=1,2}$ stand for vectors on M, which are orthogonal to u. The parameters of k and θ are real numbers, and $k > 1$ is postulated. For an initial $\mathsf{z}_0 = r_0\cos\phi_0\mathsf{v}_1 + r_0\sin\phi_0\mathsf{v}_2$, $\mathsf{z}_{\mathsf{m}} = B^{\mathsf{m}}\mathsf{z}_0$ is described as

$$\mathsf{z}_{\mathsf{m}} = k^{\mathsf{m}} r_0 \cos(m\theta + \phi_0)\mathsf{v}_1 + k^{\mathsf{m}} r_0 \sin(m\theta + \phi_0)\mathsf{v}_2. \tag{30}$$

It is a spiral orbit, which is non-chaotic for any values of k and θ. Despite this, the discrete log-linear dynamics defined with Eq. (28) has chaotic behaviors. As m becomes large, z_{m} moves away from the origin, and then, the corresponding point of $\mathsf{x}_{\mathsf{m}} = Log^{-1}(\Pi^{-1}(\mathsf{z}_{\mathsf{m}}))$ is posited on vicinity of a vertex of Δ, namely $P = (1,0,0)$ or $Q = (0,1,0)$ or $R = (0,0,1)$. (This property is discussed in detail in the previous paper.[4]) Thus, in limit of m, the discrete log-linear dynamics is recognized as a transition between vertices of P, Q, and R. For example, we obtain the array of results as

P	Q	Q	R	R	P	P	Q	R	R	P	P	Q	Q	R	R	P
P	Q	Q	R	R	P	P	Q	R	R	P	P	Q	Q	R	R	P
P	Q	Q	R	R	P	P	Q	Q	R	P	P	Q	Q	R	R	P
P	Q	Q	R	R	P	P	Q	Q	R	P	P	Q	Q	R	R	P
P	Q	Q	R	R	P	P	Q	Q	R	R	P	Q	Q	R	R	P
P	Q	Q	R	R	P	P	Q	Q	R	R	P	Q	Q	R	$\cdots$.	

Note that this sequence is non-periodic. Generally, such the chaotic behavior with non-periodicity is provided when the rotating angle θ is irrational.

As mentioned in Sec. 3.2, the map of discrete log-linear dynamics is derived from the differential equation of Eq. (16). The continuous dynamics defined by Eq. (16) also has chaotic behaviors in some cases. Let us consider the following case. The dimension is set by $n+1=5$, and the matrix A whose elements $\{a_{ij}\}$ decide the dynamics is given as

$$A = P\tilde{A}P^{-1}, \tag{31}$$

where

$$\tilde{A} = \begin{pmatrix} 1 & 0 & 0 & 0 & 0 \\ 0 & 0.2 & 2 & 0 & 0 \\ 0 & -2 & 0.2 & 0 & 0 \\ 0 & 0 & 0 & 0.2 & \sqrt{3} \\ 0 & 0 & 0 & -\sqrt{3} & 0.2 \end{pmatrix}, \tag{32}$$

and $P = (\mathsf{u}, \mathsf{p}_1, \mathsf{p}_2, \mathsf{q}_1, \mathsf{q}_2)$ with

$$\begin{aligned} \mathsf{p}_1 &= (1,-1,0,0,0)^{\mathsf{T}}, \ \mathsf{p}_2 = (0,0,1,0,-1)^{\mathsf{T}}, \\ \mathsf{q}_1 &= (0,1,-1,0,0)^{\mathsf{T}}, \ \mathsf{q}_2 = (0,0,0,1,-1)^{\mathsf{T}}. \end{aligned} \tag{33}$$

Then, the solution of differential equation $\frac{\mathsf{dz}}{\mathsf{dt}} = A\mathsf{z}$ ($\mathsf{z} \in M$) of Eq. (24) is described by

$$\mathsf{z}(t) = PR(t)P^{-1}\mathsf{z}(0), \tag{34}$$

with using the matrix of

$$R(t) = \begin{pmatrix} 1 & 0 & 0 & 0 & 0 \\ 0 & \mathrm{e}^{0.2\mathrm{t}}\cos(2\mathrm{t}) & \mathrm{e}^{0.2\mathrm{t}}\sin(2\mathrm{t}) & 0 & 0 \\ 0 & -\mathrm{e}^{0.2\mathrm{t}}\sin(2\mathrm{t}) & \mathrm{e}^{0.2\mathrm{t}}\cos(2\mathrm{t}) & 0 & 0 \\ 0 & 0 & 0 & \mathrm{e}^{0.2\mathrm{t}}\cos(\sqrt{3}\mathrm{t}) & \mathrm{e}^{0.2\mathrm{t}}\sin(\sqrt{3}\mathrm{t}) \\ 0 & 0 & 0 & -\mathrm{e}^{0.2\mathrm{t}}\sin(\sqrt{3}\mathrm{t}) & \mathrm{e}^{0.2\mathrm{t}}\cos(\sqrt{3}\mathrm{t}) \end{pmatrix} \tag{35}$$

We can see that the orbit of $\mathsf{z}(t)$ is related to the two kinds of spiral dynamics with angle rates, 2 and $\sqrt{3}$. As t becomes large, $\mathsf{z}(t)$ on M goes away from the origin, and $\mathsf{x}(t) = Log^{-1}(\Pi^{-1}(\mathsf{z}(t)))$ on $\mathring{\Delta}$ is posited on tremendously near by one of vertices of Δ, that is, $P_1 = (1,0,0,0,0)$, $P_2 = (0,1,0,0,0)$, $P_3 = (0,0,1,0,0)$, $P_4 = (0,0,0,1,0)$ and $P_5 = (0,0,0,0,1)$. Similar to the case of the discrete log-linear dynamics, the orbit of $\mathsf{x}(t)$ is regarded as a transition between the vertices, when t is large enough. This behavior is confirmed in calculations as shown in Fig. 2. The result of Fig. 2

shows the values of components of $\mathbf{x}(t)$ for $200 \leq t \leq 400$. The values of $x_i(t)$ have virtually 1 or 0. Importantly, appearance of $x_i(t) \cong 1$ occurs non-periodically for time t, and we see such the behavior chaotic.

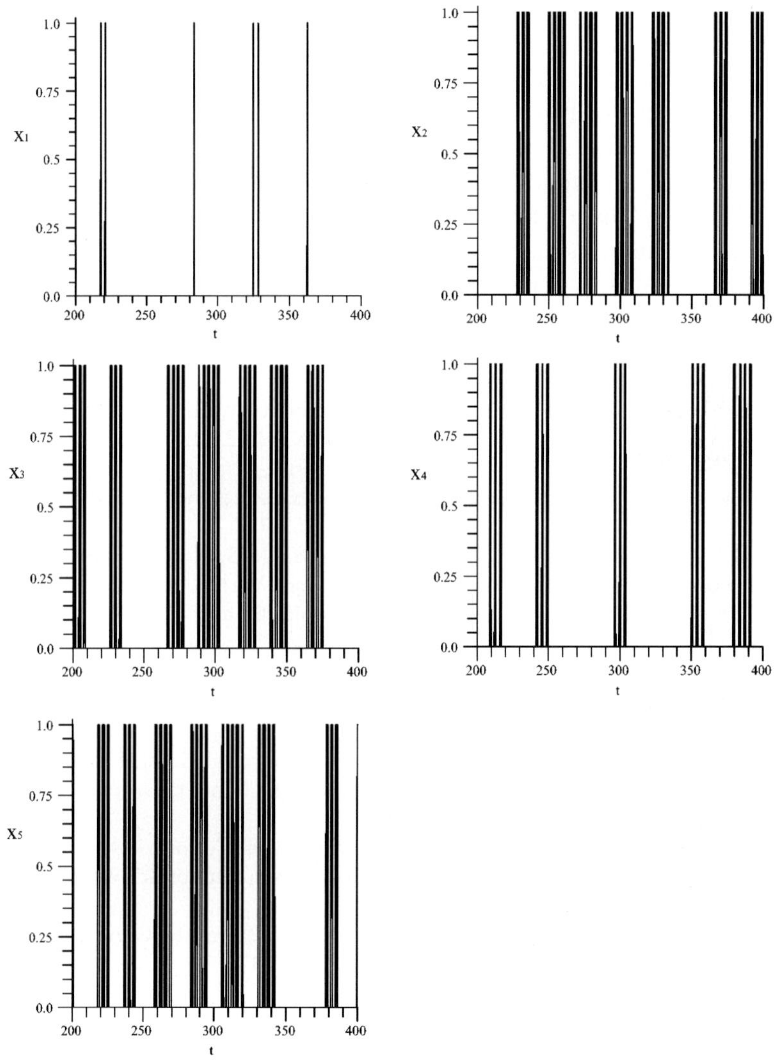

Fig. 2. Components of $x(t) = (x_1(t), x_2(t), x_3(t), x_4(t), x_5(t))$

4. Conclusion

In this paper, we show the chaotic behaviors in the log-linear dynamics. Certainly, as seen in Fig. 2, we recognize that a trajectory jumps from a node to a node non-periodically. It should be noted that such the recognition is caused from our scope of observation including the boundary of $\Delta - \mathring{\Delta}$ especially nodes of simplex, but mathematically, the dynamics which is defined exclusively on $\mathring{\Delta}$ is conjugate to a simple linear dynamics on R^n. Thus, the chaos we showed can be recognized just at the viewpoint from "outside of Δ", and it is important that such the relation between the observer and the dynamics influences observer's recognition for whether the system is chaos. The log-linear dynamics is just a simple example, however, behaviors in many phenomenons in nature may be chaotic or non-chaotic depending on our observations.

References

1. D. S. Dendrinos (1988) Theoretical developments in maps of discrete relative population dynamics, Sistemi Urbani, Vol. X, No.2/3: 211-235.
2. D. S. Dendrinos, M. Sonis (1986) Variational principles and conservation conditions in Volterra's ecology and in urban/regional relative dynamics, Journal of Regional Science, Vol. 26, No.2:359-377.
3. D. S. Dendrinos, M. Sonis (1990) Chaos and Socio-Spatial Dynamics, Applied Mathematical Sciences Series, Vol.86; Springer Verlag, New York.
4. M. Asano, M. Ohya, Y. Togawa, Entropic Chaos Degree of Rotations and Log-linear Dynamics, QP-PQ: Quantum Probability and White Noise Analysis-Vol XXI (2007)
5. Taylor, P., and L. Jonker, Evolutionarily stable strategies and game dynamics, Math. Biosciences, Vol.40, 145-156 (1978)
6. Anderson C.M., Plott C.R., Shimomura K.I., Granat S.,2004,"Global Instability in Experimental General Equilibrium: The Scarf Example." Journal of Economic Theory,115(2),:209-249.
7. Hirota M.,1981, "On the Stability of Competitive Equilibrium and Patterns of Initial Holdings, An Example", International Economic Review 22,:186-192.
8. Scarf H., 1960,"Some Examples of Global Instability of the Competitive Equilibrium", International Economic review 1,:157-171

COMPLETE M-LEVEL QUANTUM TELEPORTATION BASED ON KOSSAKOWSKI-OHYA SCHEME

M. ASANO, M. OHYA AND Y. TANAKA
Department of Information Sciences,
Faculty of Science and Technology,
Tokyo University of Science, Noda City, Chiba 278, Japan

In many models of perfect teleportation, maximum entangled states are usually used as resources. Kossakowski and Ohya showed that a linear teleportation map is defined mathematically even if a non-maximum entangled state is postulated. Based on K-O scheme, we design a perfect teleportation model by means of non-maximal entangled state.

1. Introduction

The idea of quantum teleportation is proposed by Bennett et al [1]. In Bennett's model, an EPR pair as $\frac{1}{\sqrt{2}}(|00\rangle + |11\rangle)$ is prepared between Alice and Bob. Alice's mission is to deliver an input state $\alpha|0\rangle + \beta|1\rangle$ to Bob. Alice interacts the input state with her half of the EPR pair, and then measures one of four possible states $\frac{1}{\sqrt{2}}(|00\rangle \pm |11\rangle)$ and $\frac{1}{\sqrt{2}}(|01\rangle \pm |10\rangle)$, that are called Bell states. Bob's mission is to recover the input state by applying a unitary transformation to the half of the EPR pair at his side. This operation is called unitary key.

Bennett's model is a simple example of complete teleportation. Recently, Kossakowski and Ohya defined a general form of teleportation process [2]. In their representation, an input state is written as an n-dimensional state vector in Hilbert space $\mathcal{H}_1$ and a state vector of an entangled state between Alice and Bob is defined in n^2-dimensional Hilbert space $\mathcal{H}_2 \otimes \mathcal{H}_3$. A set of n^2 orthogonal states on $\mathcal{H}_1 \otimes \mathcal{H}_2$ corresponds to n^2 possible results of a measurement. A process of teleportation is described as a map from an input state to an output state Bob gains after Alice's measurement. As mentioned in Sec. 2, if the map is a linear map, there exists an unitary key recovering the input state. In a complete teleportation model, all maps to outputs are linear. Conventional processes of teleportation, as seen in Bennett's model, have linear teleportation maps only when maximal entangled states are postulated as resources. However, the theory of Kossakowski and Ohya redefined a mathematical form of teleportation map such that its linearity is preserving regardless of an intensity of entangled state.

Basing on the theory of Kossakowski and Ohya, this paper discusses a specific type of teleportation that can be completed successfully even if a non-maximum entangled state is used. We call this type of teleportations m-level quantum teleportations. An input state in the m-level quantum teleportation is an n-dimensional state vector to be expanded by using m basis out of a set of n orthogonal basis, and it is called m-level state. (Note that the number of m is less than n.) An entangled state of resource is defined in n^2-dimensional Hilbert space. We define an ideal teleportation map for m-level state with aiming to conform to the map of Kossakawski and Ohya, and show an entangled state does not need to be maximal entangled one.

This paper is composed of three sections. First, in Sec. 2, we explain the theory of Kossakowski and Ohya briefly. In Sec. 3, the teleportation map of complete m-level teleportation is defined. In Sec. 4, the models to realize this map are proposed in cases of $m = 2$ and $n = 2^s$ $(s = 2, 3, ..)$.

2. Quantum Teleportation and Kossakowski-Ohya Scheme

A process of quantum teleportation consists of three steps, that is, a preparation of entangled state, an implementation of quantum measurement and a recovering of input state by a unitary key. In this section, the representation by Kossakowski and Ohya for these steps is explained. Furthermore, we review briefly Kossakowski-Ohya scheme of an ideal teleportation map that is proposed for models using resources of non-maximal entangled states.

An input state is written as an n-dimensional state vector $|x\rangle$ in Hilbert space $\mathcal{H}_1 = \mathbb{C}^n$. A vector of entangled state in n^2-dimensional Hilbert space $\mathcal{H}_2 \otimes \mathcal{H}_3 = \mathbb{C}^n \otimes \mathbb{C}^n$ is described as

$$|\Psi\rangle = \sum_{i=1}^{n} |i\rangle_2 \otimes f\,|i\rangle_3\,, \tag{1}$$

where $\{|i\rangle_{2,3}\}$ are orthogonal basis in $\mathcal{H}_{2,3}$, and f is an operator acting on $|i\rangle_3$. ($rank(f) = n$ is assumed.) Note that the reduced density operator $\rho_3 \equiv tr_2(|\Psi\rangle\langle\Psi|)$ is represented in the form of ff^*. The eigenvalues of ff^*, $\{\eta_i\}$ are closely related to intensity of entanglement. If $\sqrt{n}f$ is unitary, all

η_i are equal to $\frac{1}{n}$, and $|\Psi\rangle$ defined with such the f implies a vector of maximum entangled state.

Possible states in quantum measurements are described as state vectors in $\mathcal{H}_1 \otimes \mathcal{H}_2$ with the following form,

$$|\Phi_\alpha\rangle = \sum_{i=1}^{n} g_\alpha^* |i\rangle_1 \otimes |i\rangle_2 \ (\alpha = 1, ..n^2). \tag{2}$$

The set of operators $\{g_\alpha^*\}$, each of which acts on the base $|i\rangle_1$, satisfies $tr(g_\alpha g_\beta^*) = \delta_{\alpha,\beta}$ from the orthogonal condition $\langle\Phi_\alpha|\Phi_\beta\rangle = \delta_{\alpha,\beta}$. With using the operators f and g_α, the composite system $|x\rangle \otimes |\Psi\rangle$ is represented as

$$|x\rangle \otimes |\Psi\rangle = \sum_{\alpha=1}^{n^2} |\Phi_\alpha\rangle \otimes fg_\alpha |x\rangle . \tag{3}$$

Conventionally, an implementation of quantum measurement is described as a projection by $P_\alpha = |\Phi_\alpha\rangle \langle\Phi_\alpha|$. Thus, when Alice measures a result of a state vector $|\Phi_\alpha\rangle$ in $\mathcal{H}_1 \otimes \mathcal{H}_2$, the state vector Bob obtains in $\mathcal{H}_3$ is

$$|\tilde{x}_\alpha\rangle \equiv \frac{fg_\alpha |x\rangle}{\sqrt{\langle x| \kappa_\alpha |x\rangle}}, \tag{4}$$

where $(\sqrt{\langle x| \kappa_\alpha |x\rangle})^{-1}$ is the normalizing factor and then κ^α means $g_\alpha^* f^* f g_\alpha$. The map $T_\alpha : |x\rangle \langle x| \rightarrow fg_\alpha |x\rangle \langle x| g_\alpha^* f^*$ is written as

$$\begin{aligned} T_\alpha(\rho) &\equiv tr_{\mathcal{H}_T \otimes \mathcal{H}_A}(P_\alpha \otimes I)\rho \otimes \sigma(P_\alpha \otimes I) \\ &= fg_\alpha \rho g_\alpha^* f^*, \end{aligned} \tag{5}$$

where $\rho = |x\rangle \langle x|$ and $\sigma = |\Psi\rangle \langle\Psi|$. Note that the output state $|\tilde{x}_\alpha\rangle \langle\tilde{x}_\alpha| \equiv \tilde{\rho}_\alpha$ is equal to $T_\alpha(\rho)/trT_\alpha(\rho)$, so the teleportation map T_α is not trace preserving. Here, the teleportation map Λ_α^* connecting ρ and $\tilde{\rho}_\alpha$ is defined with the form of $\Lambda_\alpha^*(\rho) = \tilde{\rho}_\alpha$. Generally, this map is non-linear.

In the relation of $\Lambda_\alpha^*(\rho) = \tilde{\rho}_\alpha$, a unitary key to recover the input $|x\rangle$ from the output $|\tilde{x}_\alpha\rangle$ exists only if $fg_\alpha/\sqrt{\langle x| \kappa_\alpha |x\rangle}$ in Eq. (4) is unitary. It is when the map Λ_α^* is linear. For example, with setting $\sqrt{n}f$ and $\sqrt{n}g_\alpha$ to be unitary, the existence of key is assured. (Bennett's model is a one of such the specific cases.)

The idea by Kossakowski and Ohya is to make the teleportation linear map that the existence of unitary key is always assured. In order to do this, they considered the dual map $\tilde{T}_\alpha$ satisfying $trAT_\alpha(\rho) = tr\tilde{T}_\alpha(A)\rho$ for any linear operator A. Under the condition of $rank(f) = rank(g) = n$, $\tilde{T}_\alpha$ is normalized as

$$\kappa_\alpha^{-\frac{1}{2}} \tilde{T}_\alpha \kappa_\alpha^{-\frac{1}{2}} \equiv \tilde{\Gamma}_\alpha. \tag{6}$$

The dual map Γ_α of $\tilde{\Gamma}_\alpha$ is trace preserving. A mapped state is given by

$$\Gamma_\alpha(|x\rangle\langle x|) \equiv |\tilde{x}_\alpha^{KO}\rangle\langle \tilde{x}_\alpha^{KO}|,$$

where

$$|\tilde{x}_\alpha^{KO}\rangle \equiv f g_\alpha \kappa_\alpha^{-\frac{1}{2}} |x\rangle. \tag{7}$$

This state is different from the output of Eq. (4). Note that $f g_\alpha \kappa_\alpha^{-\frac{1}{2}}$ is unitary even if $\sqrt{n} f$ is not unitary. There always exists a unitary key $W_\alpha \equiv (f g_\alpha \kappa_\alpha^{-\frac{1}{2}})^*$ recovering $|x\rangle$ from $|\tilde{x}_\alpha^{KO}\rangle$, and in this sense, Γ_α is an ideal quantum teleportation linear map.

3. Teleportation Map of Complete m-level Teleportation

A specific type of teleportation we call m-level teleportation is proposed in this section. We consider the following form of state vector as an input.

$$|x(m)\rangle = \sum_{i=1}^{m} c_i |e_i\rangle, \tag{8}$$

where $\{|e_i\rangle\}$ are the basis made by the eigenstates of κ_α and the coefficients $c_{i=1,..,m}$ are not zero. Since the vector $|x(m)\rangle$ is defined as an n-dimensional vector, a possible value of m is from 1 to n in the above form. We name the state $\rho(m) \equiv |x(m)\rangle\langle x(m)|$ the m-level state, and call the teleportation of $\rho(m)$ the m-level quantum teleportation.

A complete m-level teleportation is characterized with a linear teleportation map. We make the teleportation map based on the K-O map Γ_α. Γ_α maps the m-level state $\rho(m)$ to $\Gamma_\alpha(\rho(m)) = f g_\alpha \kappa_\alpha^{-\frac{1}{2}} |x(m)\rangle\langle x(m)| \kappa_\alpha^{-\frac{1}{2}} g_\alpha^* f^* \equiv \rho_\alpha^{KO}(m)$, and the state vector $f g_\alpha \kappa_\alpha^{-\frac{1}{2}} |x(m)\rangle \equiv |\tilde{x}_\alpha^{KO}(m)\rangle$ is rewritten with the basis $\{|e_i\rangle\}$ as

$$|\tilde{x}_\alpha^{KO}(m)\rangle = \sum_{i=1}^{m} c_i \frac{f g_\alpha |e_i\rangle}{\sqrt{\langle e_i| \kappa_\alpha |e_i\rangle}} \equiv \sum_{i=1}^{m} c_i |\tilde{e}_{\alpha i}\rangle, \tag{9}$$

where $\{|\tilde{e}_{\alpha i}\rangle\}$ are the basis with the relation $|\tilde{e}_{\alpha i}\rangle = f g_\alpha \kappa_\alpha^{-\frac{1}{2}} |e_i\rangle$. Here, let us consider unitary operators V and U_α such that the two sets of basis

$\{|y_i\rangle \equiv V|e_i\rangle\}$ and $\{|\tilde{y}_{\alpha i}\rangle \equiv U_\alpha |\tilde{e}_{\alpha i}\rangle\}$ have the relations

$$\frac{fg_\alpha |y_i\rangle}{\sqrt{\langle y_i|\kappa_\alpha|y_i\rangle}} = |\tilde{y}_{\alpha i}\rangle .$$

Then, from the orthogonal condition of the basis $\{|\tilde{y}_{\alpha i}\rangle\}$,

$$\frac{\langle y_i|\kappa_\alpha|y_j\rangle}{\sqrt{\langle y_i|\kappa_\alpha|y_i\rangle\langle y_j|\kappa_\alpha|y_j\rangle}} = \delta_{i,j}, \tag{10}$$

is to be satisfied for $i,j = 1,..,m$. The V and U_α to satisfy this condition are decided depending on the form of κ_α and the value of m. In general, $\kappa_\alpha \neq \frac{1}{n}I$, and if $m = n$, these unitary operators must be $V = U_\alpha = I$ because only the basis $\{e_i\}$ of eigenstates of κ_α are allowed as $\{y_i\}$ in Eq. (10). If $m \neq n$, on the other hand, V and U_α not to be I can be found. By using such V and U_α, we give the states ρ'^*, $\tilde{\rho}'_\alpha(m) \equiv U_\alpha \rho_\alpha^{KO}(m) U_\alpha^*$. The state vectors of $\rho'(m)$ and $\tilde{\rho}'_\alpha(m)$ are written as

$$|x'(m)\rangle = \sum_{i=1}^{m} c_i |y_i\rangle , \tag{11}$$

and

$$|\tilde{x}'_\alpha(m)\rangle = \sum_{i=1}^{m} c_i \frac{fg_\alpha |y_i\rangle}{\sqrt{\langle y_i|\kappa_\alpha|y_i\rangle}} \equiv \sum_{i=1}^{m} c_i |\tilde{y}_{\alpha i}\rangle . \tag{12}$$

We define the linear map of complete m-level teleportation with $\Gamma'_\alpha(\rho'(m)) = \tilde{\rho}'_\alpha(m)$. Between K-O map Γ_α and Γ'_α, the relation $\Gamma_\alpha(\rho(m)) = U_\alpha \Gamma'^*_\alpha) U_\alpha^*$ is satisfied. In Sec. 4, we mention the teleportation model designed for the realization of this Γ'_α.

4. Model of Complete Two-level Quantum Teleportation

4.1. *Case of* $n = 4$

In this section, the model of complete m-level teleportation in $n = 4$ and $m = 2$ is designed. The entangled state between Alice and Bob is defined by the following state vector in $\mathcal{H}_2 \otimes \mathcal{H}_3 = \mathbb{C}^4 \otimes \mathbb{C}^4$.

$$|\Psi\rangle = \sum_{i=1}^{4} |e'_i\rangle \otimes f|e'_i\rangle = \sum_{i=1}^{4} \sqrt{\omega_i}\, |e'_i\rangle \otimes |e'_i\rangle , \tag{13}$$

where the basis $\{e'_i\}$ are the eigenstates of the operator f that has the form of $f^*f = \sum_{i=1}^{4} \omega_i |e'_i\rangle\langle e'_i|$. The eigenvalues of $\{\omega_i\}$ decide the intensity of entanglement, and here, we assume

$$\omega_1 + \omega_2 = \omega_3 + \omega_4 = \frac{1}{2}. \tag{14}$$

The group of entangled states satisfying this condition includes non-maximum entangled states not to be $\omega_1 = \omega_2 = \omega_3 = \omega_4 = \frac{1}{4}$. The states of possible results in quantum measurements $|\Phi_\alpha\rangle\langle\Phi_\alpha|$ $(\alpha = 1, ..16)$ are given with the state vectors

$$|\Phi_\alpha\rangle = \sum_{i=1}^{4} g_\alpha^* |e_i'\rangle \otimes |e_i'\rangle \,. \tag{15}$$

The operators $\{g_\alpha^*\}$ satisfy $tr(g_\alpha g_\beta^*) = \delta_{\alpha,\beta}$. The input state $\rho'(m = 2)$ is described with the state vector

$$|x'(2)\rangle = c_1 |y_1\rangle + c_2 |y_2\rangle \,, \tag{16}$$

in Hilbert space $\mathcal{H}_1 = \mathbb{C}^4$. As discussed in Sec. 3, in the complete m-level teleportation we consider, the basis $|y_{1,2}\rangle$ have to satisfy the equation of Eq. (10). Instead of Eq. (10), in our model, the two conditions,

$$\langle y_1| \kappa_\alpha |y_1\rangle = \langle y_2| \kappa_\alpha |y_2\rangle \,, \tag{17}$$

and

$$\langle y_1| \kappa_\alpha |y_2\rangle = \langle y_2| \kappa_\alpha |y_1\rangle = 0, \tag{18}$$

are postulated. It can be easily checked that under these condition, the state vector of output

$$\begin{aligned} |\tilde{x}'_\alpha(2)\rangle &= c_1 \frac{f g_\alpha |y_1\rangle}{\sqrt{\langle y_1| \kappa_\alpha |y_1\rangle}} + c_2 \frac{f g_\alpha |y_2\rangle}{\sqrt{\langle y_2| \kappa_\alpha |y_2\rangle}} \\ &= c_1 |\tilde{y}_{\alpha 1}\rangle + c_2 |\tilde{y}_{\alpha 2}\rangle \end{aligned}$$

is obtained through the quantum measurement $P_\alpha = |\Phi_\alpha\rangle\langle\Phi_\alpha|$. This state vector corresponds to Eq. (12), that is, the state vector of the mapped state $\tilde{\rho}'_\alpha(m = 2)$ derived by the complete m-level teleportation linear map Γ'_α.

Our model of the complete m-level teleportation is a feasible model, if there exist the basis $|y_{1,2}\rangle$ such that Eq. (17) and Eq. (18) are satisfied, and it depends on setting of the operator $\kappa_\alpha = g_\alpha^* f^* f g_\alpha$. Since the operator f is to be $\sum_i^4 \sqrt{\omega_i} |e_i'\rangle\langle e_i'|$ with $\omega_1 + \omega_2 = \omega_3 + \omega_4 = \frac{1}{2}$, the setting of g_α is important. Let us find the basis $|y_{1,2}\rangle$ in the following matrices of g_α

represented in the basis $\{|e_i'\rangle\}$.

$$
\begin{aligned}
2g_{\alpha=1,2,3,4} &= \begin{pmatrix} 1 & & & \\ & 1 & & \\ & & 1 & \\ & & & 1 \end{pmatrix}, \begin{pmatrix} 1 & & & \\ & -1 & & \\ & & 1 & \\ & & & -1 \end{pmatrix}, \\
&\quad \begin{pmatrix} 1 & & & \\ & 1 & & \\ & & -1 & \\ & & & -1 \end{pmatrix}, \begin{pmatrix} 1 & & & \\ & -1 & & \\ & & -1 & \\ & & & 1 \end{pmatrix} \\
2g_{\alpha=5,6,7,8} &= \begin{pmatrix} & 1 & & \\ 1 & & & \\ & & & 1 \\ & & 1 & \end{pmatrix}, \begin{pmatrix} & 1 & & \\ -1 & & & \\ & & & 1 \\ & & -1 & \end{pmatrix}, \\
&\quad \begin{pmatrix} & 1 & & \\ 1 & & & \\ & & & -1 \\ & & -1 & \end{pmatrix}, \begin{pmatrix} & 1 & & \\ -1 & & & \\ & & & -1 \\ & & 1 & \end{pmatrix} \\
2g_{\alpha=9,10,11,12} &= \begin{pmatrix} & & 1 & \\ & & & 1 \\ 1 & & & \\ & 1 & & \end{pmatrix}, \begin{pmatrix} & & 1 & \\ & & & -1 \\ 1 & & & \\ & -1 & & \end{pmatrix}, \\
&\quad \begin{pmatrix} & & 1 & \\ & & & 1 \\ -1 & & & \\ & -1 & & \end{pmatrix}, \begin{pmatrix} & & 1 & \\ & & & -1 \\ -1 & & & \\ & 1 & & \end{pmatrix} \\
2g_{\alpha=13,14,15,16} &= \begin{pmatrix} & & & 1 \\ & & 1 & \\ & 1 & & \\ 1 & & & \end{pmatrix}, \begin{pmatrix} & & & 1 \\ & & -1 & \\ & 1 & & \\ -1 & & & \end{pmatrix}, \\
&\quad \begin{pmatrix} & & & 1 \\ & & 1 & \\ & -1 & & \\ -1 & & & \end{pmatrix}, \begin{pmatrix} & & & 1 \\ & & -1 & \\ & -1 & & \\ 1 & & & \end{pmatrix}.
\end{aligned}
\tag{19}
$$

(Such the simple set of $\{g_\alpha\}$ specifies a set of maximum entangled states

$\{|\Psi_\alpha\rangle\langle\Psi_\alpha|\}$.) In this setting of g_α, the κ_α is written as

$$\kappa_\alpha = \frac{1}{4}\sum_{i=1}^{4} k_i^\alpha |e_i\rangle\langle e_i|. \tag{20}$$

The basis $\{|e_i\rangle\}$, the eigenvectors of κ_α, are equal to the basis $\{|e'_\alpha\rangle\}$. Here, we represent $4\kappa_\alpha$ as the vector of $\boldsymbol{k}_\alpha = (k_1^\alpha, k_2^\alpha, k_3^\alpha, k_4^\alpha)$. For $\alpha = 1, .., 16$,

$$\begin{aligned}
\boldsymbol{k}_{\alpha=1,2,3,4} &= (\omega_1, \omega_2, \omega_3, \omega_4) \equiv \boldsymbol{u}_1, \\
\boldsymbol{k}_{\alpha=5,6,7,8} &= (\omega_2, \omega_1, \omega_4, \omega_3) \equiv \boldsymbol{u}_2, \\
\boldsymbol{k}_{\alpha=9,10,11,12} &= (\omega_3, \omega_4, \omega_1, \omega_2) \equiv \boldsymbol{u}_3, \\
\boldsymbol{k}_{\alpha=13,14,15,16} &= (\omega_4, \omega_3, \omega_2, \omega_1) \equiv \boldsymbol{u}_4,
\end{aligned} \tag{21}$$

and these are categorized to four kinds of vectors $\boldsymbol{u}_{1,2,3,4}$. The arrays of eigenvalues of all κ_α can be represented with the matrix

$$\begin{aligned}
K_4 &\equiv (\boldsymbol{u}_1^T, \boldsymbol{u}_2^T, \boldsymbol{u}_3^T, \boldsymbol{u}_4^T) \\
&= \begin{pmatrix} \omega_1 & \omega_2 & \omega_3 & \omega_4 \\ \omega_2 & \omega_1 & \omega_4 & \omega_3 \\ \omega_3 & \omega_4 & \omega_1 & \omega_2 \\ \omega_4 & \omega_3 & \omega_2 & \omega_1 \end{pmatrix}.
\end{aligned} \tag{22}$$

When $|y_i\rangle$ is represented as $\sum_{j=1}^4 a_{ij}|e_j\rangle$, the conditions of Eq. (17) and Eq. (18) are written as

$$\sum_{j=1}^{4}(|a_{1j}|^2 - |a_{2j}|^2)k_j^\alpha, \quad \sum_{j=1}^{4}(a_{1j}^* a_{2j})k_j^\alpha.$$

By using the 4-dimensional vectors $\boldsymbol{p}$ and $\boldsymbol{q}$ defined as

$$\begin{aligned}
\boldsymbol{p} &= (p_1, p_2, p_3, p_4), \ p_j = |a_{1j}|^2 - |a_{2j}|^2, \\
\boldsymbol{q} &= (q_1, q_2, q_3, q_4), \ q_j = a_{1j}^* a_{2j},
\end{aligned} \tag{23}$$

these conditions are rewritten as

$$\boldsymbol{p}K_4 = \boldsymbol{0}, \ \boldsymbol{q}K_4 = \boldsymbol{0}. \tag{24}$$

It is clear that the case of $\boldsymbol{p} = \boldsymbol{q} = \boldsymbol{0}$ does not exist. However, from $\omega_1 + \omega_2 = \omega_3 + \omega_4 = \frac{1}{2}$, we can find the suitable $\boldsymbol{p}$ and $\boldsymbol{q}$. For example, the case of $p = \boldsymbol{0}$ and $q = \frac{1}{2}(1, 1, -1, -1)$ is suitable, and then $|y_1\rangle = \frac{1}{2}(1, 1, 1, 1)$ and $|y_2\rangle = \frac{1}{2}(1, 1, -1, -1)$ are derived.

4.2. *Expansion to Model of* $n = 2^s$

The model of complete two-level teleportation with $n = 4$ can be expanded to the model with $n = 2^s$. Let us demonstrate the case of $n = 2^3 = 8$. In this model, the vector of entangled state is written as $|\Psi\rangle = \sum_i^8 |e_i\rangle \otimes f |e_i\rangle$, where $f = \sum_{i=1}^8 \sqrt{\omega_i} |e_i\rangle \langle e_i|$. Note that

$$\sum_{i=1}^{4} \omega_i = \sum_{j=5}^{8} \omega_j = \frac{1}{2}, \tag{25}$$

is postulated. With using the matrix of

$$K_4 = \begin{pmatrix} \omega_1 & \omega_2 & \omega_3 & \omega_4 \\ \omega_2 & \omega_1 & \omega_4 & \omega_3 \\ \omega_3 & \omega_4 & \omega_1 & \omega_2 \\ \omega_4 & \omega_3 & \omega_2 & \omega_1 \end{pmatrix}, \tag{26}$$

of Eq. (22) and the matrix of

$$\tilde{K}_4 \equiv \begin{pmatrix} \omega_5 & \omega_6 & \omega_7 & \omega_8 \\ \omega_6 & \omega_5 & \omega_8 & \omega_7 \\ \omega_7 & \omega_8 & \omega_5 & \omega_6 \\ \omega_8 & \omega_7 & \omega_6 & \omega_5 \end{pmatrix}, \tag{27}$$

with components $\omega_{5,6,7,8}$, we make the 8×8 matrix of

$$\begin{aligned} K_8 &\equiv \begin{pmatrix} K_4 & \tilde{K}_4 \\ \tilde{K}_4 & K_4 \end{pmatrix} \\ &= \begin{pmatrix} \omega_1 & \omega_2 & \omega_3 & \omega_4 & \omega_5 & \omega_6 & \omega_7 & \omega_8 \\ \omega_2 & \omega_1 & \omega_4 & \omega_3 & \omega_6 & \omega_5 & \omega_8 & \omega_7 \\ \omega_3 & \omega_4 & \omega_1 & \omega_2 & \omega_7 & \omega_8 & \omega_5 & \omega_6 \\ \omega_4 & \omega_3 & \omega_2 & \omega_1 & \omega_8 & \omega_7 & \omega_6 & \omega_5 \\ \omega_5 & \omega_6 & \omega_7 & \omega_8 & \omega_1 & \omega_2 & \omega_3 & \omega_4 \\ \omega_6 & \omega_5 & \omega_8 & \omega_7 & \omega_2 & \omega_1 & \omega_4 & \omega_3 \\ \omega_7 & \omega_8 & \omega_5 & \omega_6 & \omega_3 & \omega_4 & \omega_1 & \omega_2 \\ \omega_8 & \omega_7 & \omega_6 & \omega_5 & \omega_4 & \omega_3 & \omega_2 & \omega_1 \end{pmatrix} \\ &\equiv (\boldsymbol{u}_1^T, \boldsymbol{u}_2^T, \boldsymbol{u}_3^T, \boldsymbol{u}_4^T, \boldsymbol{u}_5^T, \boldsymbol{u}_6^T, \boldsymbol{u}_7^T, \boldsymbol{u}_8^T). \end{aligned} \tag{28}$$

In the model of $n = 8$, the above 8 vectors $\boldsymbol{u}_1, \boldsymbol{u}_2, .., \boldsymbol{u}_8$ correspond to arrays of eigenvalues of κ_α written as $\frac{1}{8}\sum_{i=1}^8 k_i^\alpha |e_i\rangle \langle e_i|$. The vectors of

$\boldsymbol{k}_\alpha = (k_1^\alpha, k_2^\alpha, .., k_8^\alpha)$ are given as

$$\begin{aligned} &\boldsymbol{k}_{\alpha=1,..,8} = \boldsymbol{u}_1, \ \boldsymbol{k}_{\alpha=9,..,16} = \boldsymbol{u}_2, \\ &\boldsymbol{k}_{\alpha=17,..,24} = \boldsymbol{u}_3, \ \boldsymbol{k}_{\alpha=25,..,32} = \boldsymbol{u}_4, \\ &\boldsymbol{k}_{\alpha=33,..,41} = \boldsymbol{u}_5, \ \boldsymbol{k}_{\alpha=42,..,49} = \boldsymbol{u}_6, \\ &\boldsymbol{k}_{\alpha=50,..,57} = \boldsymbol{u}_7, \ \boldsymbol{k}_{\alpha=58,..,64} = \boldsymbol{u}_8. \end{aligned} \tag{29}$$

Since $\kappa_\alpha = g_\alpha^* f^* f g_\alpha$ and $f^* f = \sum_{i=1}^{8} \omega_i |e_i\rangle \langle e_i|$, we can simply find the appropriate set of $\{g_{\alpha=1,..,8^2}\}$, that is, we can define the possible states in quantum measurement $|\Phi_\alpha\rangle \langle \Phi_\alpha|$. For example, the operators $g_{\alpha=1,..,8}$ for $\kappa_{\alpha=1,..,8}$ defined with $\boldsymbol{k}_{\alpha=1,..,8} = \boldsymbol{u}_1 = (\omega_1, \omega_2, \omega_3, \omega_4, \omega_5, \omega_6, \omega_7, \omega_8)$ are represented as 8×8 diagonal matrices in the basis $\{|e_i\rangle\}$. These diagonal components are

$$\begin{aligned} diag(g_{\alpha=1}) &= \begin{pmatrix} + + + + + + + + \end{pmatrix}, \\ diag(g_{\alpha=2}) &= \begin{pmatrix} + - + - + - + - \end{pmatrix}, \\ diag(g_{\alpha=3}) &= \begin{pmatrix} + + - - + + - - \end{pmatrix}, \\ diag(g_{\alpha=4}) &= \begin{pmatrix} + - - + + - - + \end{pmatrix}, \\ diag(g_{\alpha=5}) &= \begin{pmatrix} + + + + - - - - \end{pmatrix}, \\ diag(g_{\alpha=6}) &= \begin{pmatrix} + - + - - + - + \end{pmatrix}, \\ diag(g_{\alpha=7}) &= \begin{pmatrix} + + - - - - + + \end{pmatrix}, \\ diag(g_{\alpha=8}) &= \begin{pmatrix} + - - + - + + - \end{pmatrix}, \end{aligned} \tag{30}$$

where + and − stand for $\frac{1}{\sqrt{8}}$ and $-\frac{1}{\sqrt{8}}$. It can be checked that the orthogonal conditions of $tr(g_\alpha^* g_\beta) = \delta_{\alpha,\beta}$ are satisfied.

So that this model is feasible, the basis $|y_i\rangle = \sum_{j=1}^{8} a_{ij} |e_j\rangle$ $(i = 1, 2)$ used for the input $|x(2)\rangle = c_1 |y_1\rangle + c_2 |y_2\rangle$ have to satisfy the conditions,

$$\boldsymbol{p} K_8 = \boldsymbol{0}, \ \boldsymbol{q} K_8 = \boldsymbol{0}, \tag{31}$$

where $\boldsymbol{p}$ and $\boldsymbol{q}$ are 8-dimensional vectors with components of $p_j = |a_{1j}|^2 -$

$|a_{2j}|^2$ and $q_j = a_{1j}^* a_{2j}$ $(j = 1,..,8)$. Such the basis can be found from $\sum_{i=1}^{4} \omega_i = \sum_{j=5}^{8} \omega_j = \frac{1}{2}$.

In general, the model of $n = 2^s$ uses a non-maximum entangled state with

$$\sum_{i=1}^{2^{s-1}} \omega_i = \sum_{j=2^{s-1}+1}^{2^s} \omega_j = \frac{1}{2}. \tag{32}$$

As seen in the case of expansion to $n = 8$, the matrix K_{2^s}, which gives the arrays of eigenvalues of κ_α, can be made from the matrix $K_{2^{s-1}}$ for the model of $n = 2^{s-1}$ as

$$K_{2^s} = \begin{pmatrix} K_{2^{s-1}} & \tilde{K}_{2^{s-1}} \\ \tilde{K}_{2^{s-1}} & K_{2^{s-1}} \end{pmatrix}. \tag{33}$$

This matrix enables us to define the measured states $|\Phi_\alpha\rangle \langle\Phi_\alpha|$ and to find appropriate basis $|y_i\rangle = \sum_{j=1}^{2^s} a_{ij} |e_j\rangle$ $(i = 1, 2)$ of the input $|x(2)\rangle = c_1 |y_1\rangle + c_2 |y_2\rangle$.

5. Conclusion

In this research, we designed a model of complete quantum teleportation that is feasible even if non-maximum entangled state is used. As mentioned Sec. 3, the model we call m-level teleportation is defined by a linear teleportation map homeomorphic to K-O map that describes a most ideal teleportation process. Actually, the entangled state we postulated as a resource in a two-level teleportation model, is not entangled maximally, and we consider it will be enough realizable experimentally.

References

1. C. H. Bennett, G. Brassard, C. Crepeau, R. Jozsa, A. Peres and W. K. Wootters, Teleporting an unknown quantum state via dual classical and EPR channels, Phys. Rev. Lett. 70, 1895 (1993)
2. A. Kossakowski, M. Ohya, New scheme of quantum teleportation, Infinite Dimensional Analysis, Quantum Probability and Related Topics Vol. 10, No. 3 411 (2007)
3. P.T.Cochrane, G.J.Milburn,and W.J.Munro, "Teleportation using coupled oscillator states",Phys. Rev.A62, 062307 (2000)
4. M.S.Kim,W.Son,V.Bužek, and P.L.Kinght, "Entanglement by a beam splitter: Nonclassicality as a prerequisite for entanglement", Phys. Rev.A65, 032323 (2002)
5. H. Yonezawa, T. Aoki, and A. Furusawa, Demonstration of a quantum teleportation network for continuous variables, Nature Vol. 431, 430-433 (2004)

TOWARDS QUANTUM CYBERNETICS: OPTIMAL FEEDBACK CONTROL IN QUANTUM BIO INFORMATICS

V P BELAVKIN
School of Mathematics, University of Nottingham,
E-mail:vpb@maths.nott.ac.uk

A brief account of the quantum information dynamics and dynamical programming methods for the purpose of optimal control in quantum cybernetics with convex constraints and concave cost and bequest functions of the quantum state is given. Consideration is given to both open loop and feedback control schemes corresponding respectively to deterministic and stochastic semi-Markov dynamics of stable or unstable systems. For the quantum feedback control scheme with continuous observations we exploit the separation theorem of filtering and control aspects for quantum stochastic micro-dynamics of the total system. This allows to start with the Belavkin quantum filtering equation and derive the generalized Hamilton-Jacobi-Bellman equation using standard arguments of classical control theory. This is equivalent to a Hamilton-Jacobi equation with an extra linear dissipative term if the control is restricted to only Hamiltonian terms in the filtering equation. A controlled qubit is considered as an example throughout the development of the formalism. Finally, we discuss optimum observation strategies to obtain a pure quantum qubit state from a mixed one.
Keywords: *Quantum Filtering, Quantum Control, Quantum Cybernetics*

1. Introduction

Cybernetics steamed from Greek 'controller' or 'governor' was defined by Norbert Wiener, in his book of that title, as the study of control and communication in the animal and the machine. A more philosophical definition, suggested in 1956 by Louis Couffignal, one of the pioneers of cybernetics, characterizes cybernetics as "the art of ensuring the efficacy of action". So far cybernetics was restricted mostly to classical self-organizing systems such as mechanical regulators, electrical networks, biological organisms, neurosytems and social systems described by the classical laws of physics, probability and information. It is base on the mathematical systems theory.

Quantum Cybernetics (QC) will be dealing with such self-organizing mechanical, electrical, biological, neuro- and social systems described by the *quantum laws* of physics, probability and information. Like the classical cybernetics is essentially the classical systems theory closely related with optimal feedback control theory, QC can be described as the quantum theory of optimally observed and feedback controlled open systems. Thus, the main ingredients of QC is the quantum optimal filtering and quantum feedback control theory based on the quantum stochastic innovation dynamics which was developed by the author since [1] in a serious cited in the recent review paper [2].

Here we give a brief account of quantum systems theory and optimal control theory in quantum open systems with and without observation. Such systems can be described by mathematical theory of conditionally Markov quantum processes governed by the Belavkin Master Equation. This is illustrated by the simplest such system, the unstable controlled quantum bit with continuous diffusive observation modelled by a single Wiener innovation process.

2. Some facts and notations

Let $\mathbb{A} \subseteq \mathfrak{B}(\mathfrak{h})$ be an operator subalgebra with left and right modular involutions $\sharp,\flat$ and the Hermitian conjugation $\dagger$ isometric with respect to a standard pairing

$$\left\langle \mathtt{A}^{\dagger}, \mathtt{B}\right\rangle := \varphi\left(\mathtt{A}^{\sharp}\mathtt{B}\right) \equiv (\mathtt{A}|\mathtt{B}) = \left(\mathtt{B}^{\dagger}|\mathtt{A}^{\dagger}\right) := \varphi\left(\mathtt{B}\mathtt{A}^{\flat}\right) \equiv \left\langle \mathtt{B}, \mathtt{A}^{\dagger}\right\rangle$$

given by a reference weight φ on $\mathbb{A}$. Then the predual space $\mathbb{A}_{\star}$ can be realized by the densities $\varrho \vdash \bar{\mathbb{A}}$ with respect to φ defined as generalized elements affiliated with the weak closure $\bar{\mathbb{A}}$ of $\mathbb{A}$, i.e. as in general unbounded sesquilinear forms commuting with $\mathbb{A}'$ [3].

Let $\mathcal{S} \subset \mathbb{A}_{\star}$ denote the state space realized by positive mass one densities $\varrho^{\dagger} = \varrho \geq 0$, $(\varrho|\mathtt{I}) = 1$ with the tangent space $\mathcal{T}_0 = \left\{v = v^{\dagger} \in \mathbb{A}_{\star} : (v|\mathtt{I}) = 0\right\}$ and the cotangent space $\mathcal{T}_0^{\star} = \bar{\mathbb{A}}^{\mathrm{h}}/\mathbb{R}\mathtt{I}$. Every state $\varrho \in \mathcal{S}$ can be parametrized as $\varrho(\mathtt{q}) = \varrho_0 - \mathtt{q}$ by a *tangent element* $\mathtt{q} \in \mathcal{T}_0$. Cotangent elements $\mathtt{p} \in \mathcal{T}_0^{\star}$ are the equivalence classes

$$\mathtt{p}(\mathtt{X}) = \left\{\mathtt{A} \in \bar{\mathbb{A}} : \mathtt{A} = \mathtt{X} + \lambda\mathtt{I}, \text{ for some } \lambda \in \mathbb{R}\right\}.$$

2.1. *Quantum state geometry*

For the semifinite quantum systems with the trace $\varphi = \mathrm{tr}$ we have $\sharp = \dagger = \flat$. In the simple case

$$\begin{aligned}
\mathbb{A} &= \mathfrak{B}(\mathfrak{h}) && \text{- the Banach operator algebra on } \mathfrak{h}; \\
\mathbb{A}_\star &= \mathfrak{T}(\mathfrak{h}) && \text{- the predual trace-class space on } \mathfrak{h}; \\
\mathcal{S} &= \mathfrak{S}(\mathfrak{h}) && \text{- the positive trace-one operators on } \mathfrak{h}; \\
\mathcal{T}_0 &= \mathfrak{T}_0(\mathfrak{h}) && \text{- the tangent zero-trace space on } \mathfrak{h}; \\
\mathcal{T}_0^\star &= \bar{\mathbb{A}}/\mathbb{R}\mathtt{I} && \text{- the cotangent space with } 0 = \{\mathtt{A} \propto \mathtt{I}\}
\end{aligned}$$

with respect to the pairing $\langle \varrho, \mathtt{A}\rangle := \mathrm{tr}\{\varrho\mathtt{A}\} \equiv (\varrho|\mathtt{A})$ of $\mathtt{A} \in \bar{\mathbb{A}}$ with $\varrho = \varrho^\dagger$.

Example: A single quantum bit is described by the matrix algebra of (2×2)-matrices $\mathtt{A} = \alpha\mathtt{I} + \sigma_{\vec{a}}$, where $\sigma_{\vec{a}} \in \mathcal{T}_0$ is decomposed into Pauli matrices

$$\sigma_{\vec{a}} = a_x\sigma_x + a_y\sigma_y + a_z\sigma_z \equiv \vec{a} \cdot \vec{\sigma}$$

with the normalized trace $\varphi(\mathtt{A}) = \mathrm{tr}\{\mathtt{A}\} = \alpha$ defining the standard pairing $\langle \varrho, \mathtt{A}\rangle = \alpha - \vec{q} \cdot \vec{a}$. Here the quantum bit states $\varrho = 1 - \sigma_{\vec{q}}$ are given by the state coordinate $\mathsf{q} = \sigma_{\vec{q}}$ parametrized by real vector $\vec{q} \in \mathbb{R}^3$ from the unit ball $|\vec{q}| \leq 1$ with respect to the central state $\varrho_0 = \mathtt{I}$.

2.2. *Derivations and Hessians*

A (nonlinear) functional $\varrho \mapsto \mathsf{F}[\varrho]$ on $\mathbb{A}_\star$ (or on $\mathcal{S} \subset \mathbb{A}_\star$) admits a Frechet derivative if

$$\lim_{h \to 0} \frac{1}{h} \{\mathsf{F}[\varrho + hv] - \mathsf{F}[\varrho]\} = \langle v, \nabla_\varrho \mathsf{F}[\varrho]\rangle,$$

for each $v \in \mathbb{A}_\star$ (for each $v \in \mathcal{T}_0$). Note that $\nabla_\varrho \mathsf{F}[\varrho] \in \mathcal{T}_0^\star$ if ∇_ϱ is defined only on $\mathcal{S}$.

A Hessian $\nabla_\varrho^{\otimes 2} \equiv \nabla_\varrho \otimes \nabla_\varrho$ with values in $\bar{\mathbb{A}}_{\mathrm{sym}}^{\otimes 2}$ (or in $\mathcal{T}_{0\mathrm{sym}}^{\star\otimes 2}$) is defined by

$$\lim_{h \to 0} \frac{1}{h} \langle v', \nabla_\varrho \mathsf{F}[\varrho + hv]\rangle - \nabla_\varrho \mathsf{F}[\varrho]\rangle = \left\langle v \otimes v', \nabla_\varrho^{\otimes 2} \mathsf{F}[\cdot]\right\rangle.$$

Example: Let $\mathsf{F}[\varrho] = f[\vec{q}]$ for $\varrho = 1 - \sigma_{\vec{q}}$ on $\mathcal{S}$ in the Pauli matrix basis. Then $\nabla_\varrho \mathsf{F}[\varrho] = \lambda\mathtt{I} - \vec{\sigma} \cdot \vec{\nabla} f(\vec{q}) \equiv -\vec{\sigma} \cdot \vec{\nabla} f(\vec{q})$,

$$\nabla_\varrho^{\otimes 2} \mathsf{F}[1 - \sigma_{\vec{q}}] = \left(\vec{\sigma} \cdot \vec{\nabla}\right)^{\otimes 2} f(\vec{q}).$$

2.3. *Affine and concave costs*

An affine functional $\mathsf{G}(u,\varrho) = \langle \varrho, G(u)\rangle$ of $\varrho \in \mathcal{S}$ given by a positive or bounded from below function $u \mapsto G(u) \vdash \mathbb{A}$ on a measurable space $\mathcal{U}$ is called *expected cost* of the control $u \in \mathcal{U}$. The *minimal expected cost*

$$\mathsf{S}[\varrho] = \inf\{\langle \varrho, G(u)\rangle : u \in \mathcal{U}\}$$

is not affine but concave finite functional on the convex set $\mathcal{S}$ in the sense

$$\mathsf{S}[\lambda\varrho_0 + (1-\lambda)\varrho_1] \geq \lambda \mathsf{S}[\varrho_0] + (1-\lambda)\mathsf{S}[\varrho_1]$$

for any $\lambda \in [0,1]$ and $\varrho_0, \varrho_1 \in \mathcal{S}$.

More generally, the concave functional $\mathsf{S}[\varrho]$ can have values in $\mathbb{R} \sqcup \{-\infty\}$ for an unbounded from below function $G(u)$.

Example: $\mathsf{S}[\varrho] = \inf_{\vec{u}\in\mathbb{R}^3}\{\langle \varrho, \sigma_{\vec{u}}\rangle + |\vec{u}|\} = O^-_{\mathcal{B}_1}(\vec{q})$, where $O^-_{\mathcal{B}_1}(\vec{q}) = \left\{\begin{smallmatrix} 0, & \vec{q}\in\mathcal{B}_1 \\ -\infty, & \vec{q}\notin\mathcal{B}_1\end{smallmatrix}\right. = -O^+_{\mathcal{B}_1}(\vec{q})$ is the max-plus indicator function of for the ball $\mathcal{B}_1 = \{\vec{q} \in \mathbb{R}^3 : |\vec{q}| \leq 1\}$.

2.4. *Legendre-Fenchel transform*

The above example for the affine G with $G(u) = |\vec{u}|\,\mathtt{I} + \sigma_{\vec{u}}$ is a special case of the Legendre-Fenchel transform

$$\mathsf{S}[\varrho] = \inf\{g(u) - \langle \mathsf{q}(\varrho), \mathsf{p}[G(u)]\rangle : u \in \mathcal{U}\}$$

of the convex function $g(u) = |\vec{u}|$. The result $\mathsf{S}[\varrho]$ is well defined as a concave functional for any $g : \mathcal{U} \mapsto \mathbb{R} \sqcup \{\infty\}$ not identically equal ∞. Moreover, every concave functional $\mathsf{S}[\varrho]$ on $\mathcal{S}$ can be obtained as a result of the Legendre transformation $\inf_{u\in\mathcal{T}_0^\star}\{\mathsf{G}(u,\varrho)\}$ for $\mathsf{G}(u,\varrho) = g(\mathtt{X}) - \langle \varrho, \mathtt{X}\rangle$ uniquely defined by a convex function $g(\mathtt{X}) = g(\mathtt{X} + \lambda\mathtt{I}) - \lambda$ on the representatives $\mathtt{X} = \mathtt{X}^\dagger$ for $u = \mathsf{p}(\mathtt{X})$. This g is found by the inverse Legendre transform

$$g(\mathtt{X}) = \sup\{\langle \varrho, \mathtt{X}\rangle + \mathsf{S}[\varrho] : \varrho \in \mathcal{S}\}$$

Thus, the relative entropies $\mathsf{S}[\varrho : \mu] = -\mathrm{tr}\left\{\varrho \ln \frac{\varrho}{\mu}\right\}$ of (a) $\ln\frac{\varrho}{\mu} = \ln\varrho - \ln\mu$ and (b) $\ln\frac{\varrho}{\mu} = \ln\left(\mu^{-1}\varrho\right)$ types are the transforms of some convex g_{a} and g_{b}, and

$$\mathsf{S}[\varrho : \mu] = \inf\left\{\langle \varrho, \mathtt{X}\rangle + \ln\left\langle \mu, \mathrm{e}^{-\mathtt{X}}\right\rangle : \mathtt{X} = \mathtt{X}^\dagger\right\}$$

is another (thermodynamic) relative entropy.

3. Quantum conditionally Markov dynamics

Continuously observed quantum systems are described by conditionally Markov stochastic dynamics over $\mathbb{A}$ with respect to an increasing family of operator algebras $\mathcal{B}^t \subseteq \mathcal{B}(H)$. It can be weakly defined by a hemigroup $\{\phi_r(t) | t > r \in \mathbb{R}_+\}$ of normal contracting completely positive maps $\phi_r(t) : \mathcal{B}^{t]} \to \mathcal{B}^{r]}$ on the W*-algebras $\mathcal{B}^{t]} \simeq \mathcal{B}^t \bar{\otimes} \mathbb{A}$ with $\mathcal{B}^t \simeq \mathcal{B}^r \bar{\otimes} \mathcal{B}_r^{t-r}$ such that

$$\phi_t(t+s, \mathrm{XY}) = \mathrm{Y}\phi_t(t+s, \mathrm{X}) \quad \forall \mathrm{Y} \in \mathcal{B}^t, \mathrm{X} \in \mathcal{B}_t^{s]},$$

where $\mathcal{B}_t^{s]} \simeq \mathcal{B}_t^s \bar{\otimes} \mathbb{A}$ for any $s > 0$. If $\mathcal{B}^t$ are commutative, generated by a family $\mathrm{v}^t = \{\mathrm{v}(r) | r < t\}$ of compatible observables $\mathrm{v}(t)$ representing an output classical stochastic process $v(t) = v(\omega, t)$, it induces on $\mathbb{A}$ a controlled hemigroup of sub-Markov maps $\tau_r^y(t) : \mathbb{A} \to \mathbb{A}$, where $y = \ell(v^r) \equiv y^r$ is a feedback control strategy defining the restriction $\phi_r(t) |\mathbb{A} = \tau_r^{\mathrm{y}}(t)$ as the function $\tau_r^{\bullet}(t)$ of commuting controlling operators $\mathrm{y} = \ell(\mathrm{v}^r) \equiv \mathrm{y}^r$.

3.1. *Quantum controlled generator*

The continuous feedback controlled sub-Markov dynamics $\{\tau_r^{\mathrm{y}}(t) | t > r \in \mathbb{R}_+\}$ is usually determined by its generator

$$\mathfrak{L}(y(t), \mathtt{X}) = \lim_{s \searrow 0} \frac{1}{s}\left(\tau_t^y(t+s, \mathtt{X}) - \mathtt{X}\right)$$

given by a completely dissipative map $\mathfrak{L}(u) : \mathbb{A} \to \mathbb{A}$ controlled by the values $y(t) = (u_1, \ldots, u_n)$ of some parameters $u_j \in \mathbb{R}$. For the simple quantum systems $\mathbb{A} = \mathcal{B}(\mathfrak{h})$ such dynamics is described by the controlled semi-Lindblad generators $\mathfrak{L}(u, \mathtt{X}) = \mathrm{i}[H(u), \mathtt{X}] + \mathfrak{L}^m(u, \mathtt{X})$ with the dissipation part

$$\mathfrak{L}^m(\mathtt{X}) = \mathfrak{K}^{\mathtt{L}}(\mathtt{X}) - \frac{1}{2}(\mathtt{MX} + \mathtt{XM}),$$
$$\mathfrak{K}^{\mathtt{L}}(\mathtt{X}) = \sum_j \mathtt{L}_j \mathtt{X} \mathtt{L}_j^\dagger, \quad \mathtt{M} \geq \sum_j \mathtt{L}_j \mathtt{L}_j^\dagger$$

prepared for a continuous indirect measurement, say, of the observables $\mathtt{L}_j + \mathtt{L}_j^\dagger$ affiliated to $\mathbb{A}$. Here we assumed that the dynamics is controlled only by the Hamiltonian $H(u) = H(u)^\dagger$ and $\mathtt{M} \vdash \mathbb{A}$ (affiliated to $\mathbb{A}$) determines the decay operator $\mathtt{D} = \mathtt{M} - \sum_j \mathtt{L}_j \mathtt{L}_j^\dagger = -\mathfrak{L}(\mathtt{I})$.

3.2. *Deterministic quantum Master equations*

The density operator ϱ_t of a quantum state evolved from an initial $\varrho_t = \varrho_0$ by resolving the *controlled Master equation* $\frac{\mathrm{d}}{\mathrm{d}t}\varrho = \mathfrak{L}_\star(u,\varrho)$ with the predual generator $\langle \mathfrak{L}_\star(\varrho), \mathtt{A}\rangle = \langle \varrho, \mathfrak{L}(\mathtt{A})\rangle$ is normalized on the decaying probability of the survival $p_t = \langle \varrho_t, \mathtt{I}\rangle$ if initially $\langle \varrho_0, \mathtt{I}\rangle = 1$. The renormalized $\varrho^t = \varrho_t/p_t$ describes the quantum state conditioned by the survival effect up to time t. Its evolution is described by the velocity $\upsilon(\varrho) = \frac{\mathrm{d}}{\mathrm{d}t}\mathtt{q}(\varrho) \in \mathcal{T}_0$ defining the *deterministic nonlinear filtering equation*

$$\frac{\mathrm{d}}{\mathrm{d}t}\varrho = \mathfrak{L}_\star(\varrho) - \langle \varrho, \mathfrak{L}(\mathtt{I})\rangle \varrho \equiv -\upsilon(\varrho).$$

Example: An unstable quantum bit is described by the Hamiltonian $H(\vec{u}) = \frac{1}{2}\sigma_{\vec{u}}$, $\mathtt{L} = \frac{1}{2}\lambda\sigma_z = \mathtt{L}^\dagger$, $\mathtt{M} = \mu\mathtt{I} + \sigma_{\vec{m}}$ with $\mu \geq |\vec{m}| + \lambda^2/4$. Then $\upsilon(\varrho) = \sigma_{\vec{u}\times\vec{q}} + \upsilon_m(\varrho)$ where

$$\upsilon_m(\varrho) = \left(\vec{m} - \frac{\lambda^2}{2}\vec{q}_z^{\perp} - (\vec{m}\cdot\vec{q})\,\vec{q}\right)\cdot\vec{\sigma} \equiv \vec{\upsilon}_m\cdot\vec{\sigma}.$$

3.3. *Output processes and continuous observation*

The state of an individual continuously measured quantum system does not coincide with the solution of the deterministic master equation but instead depends on the random measurement output ω in a causal manner. We allow the output to constitute the generalized trajectories $x_j(t,\omega)$ of a classical noise-like processes $\dot{\upsilon} = (x_j : j = 1,\ldots,m)$ with zero expectation but take a mathematically more convenient approach to work with usual left continuous trajectories of independent increment processes given by $\upsilon(t,\omega) = \int_0^t \omega(s)\,\mathrm{d}s$, and for simplicity will consider only the standard diffusive type such that $\mathrm{d}\upsilon_i\mathrm{d}\upsilon_k = \delta_{ik}\mathrm{d}t$. It is therefore natural to model this on the Wiener probability space $(\Omega,\mathfrak{A},\mathbb{Q})$ with the standard Gaussian probability measure $\mathbb{Q} = \mathbb{P}_w$ which can be induced from the vacuum state $\langle \mathrm{I},\mathrm{Y}\rangle_\emptyset = \langle \delta_\emptyset|\mathrm{Y}\delta_\emptyset\rangle$ as a reference on the product algebra $\mathcal{B} \sim L^\infty_{\mathfrak{A}}(\Omega,\mathbb{Q})$ represented in Fock space.

The conditionally Markov dynamics induces the output probability measure $\mathbb{P}^t_\upsilon(\mathrm{d}\omega) = P^t_\upsilon(\omega)\,\mathbb{Q}^t(\mathrm{d}\omega)$ defined by the output state

$$\left\langle \varrho^0, \phi^t_0(\mathrm{Y})\right\rangle = \int Y^t_\upsilon(\omega)\,\mathbb{P}^t(\mathrm{d}\omega) = \left\langle P^t_{\mathrm{v}}, \mathrm{Y}\right\rangle_\emptyset$$

for any $\mathrm{Y} = Y^t_{\mathrm{v}} \in \mathcal{B}^t$ given by a measurable functional Y^t_υ of $\upsilon = \upsilon^t$ and any initial state density $\varrho^0 \in \mathbb{A}_\star$. Here $P^t_{\mathrm{v}} \in \mathcal{B}^t_\star$ is given for each t by the density $P_v^t = \langle \varpi_v^t, \mathrm{I}\rangle$ with $\varpi_v^t = \phi_{\star v}^t\left(0, \varrho^0\right)$ evolved from $\varpi_v^0 = \varrho^0$ by the predual dynamics

$$\left\langle \phi_{\star \mathrm{v}}^{t-r}\left(r, \varpi_{\mathrm{v}}^r\right), \mathrm{X}\right\rangle = \left\langle \varpi_{\mathrm{v}}^r, \phi_r\left(t, \mathrm{X}\right)\right\rangle \;\; \forall \mathrm{X} \in \mathcal{B}^{t]}, \varpi_{\mathrm{v}}^r \in \mathcal{B}_\star^{r]}.$$

3.4. *Stochastic quantum Master equations*

The predual dynamical maps $\phi_{\star v}^t$ are usually determined as resolving for a *controlled stochastic Master equation.* This equation, derived for continuous nondemolition measurements of $\mathrm{L}_j + \mathrm{L}_j^\dagger$, $j = 1, \ldots, m$ in [4,5,6] reads for the general $\mathbb{A}_\star \in \varpi_v^t$ as

$$\mathrm{d}\varpi_v^t = \mathfrak{L}_\star\left(u, \varpi_v^t\right) + \sum_{j=1}^m \left(\mathrm{L}^j \varpi_v^t + \varpi_v^t \mathrm{L}^{j\dagger}\right) \mathrm{d}v_j,$$

where $\mathrm{L}^j = \mathrm{L}_j^\sharp$ ($= \mathrm{L}_j^\dagger$ in the case of trace $\varphi = \mathrm{tr}$). The conditional state $\varrho_v^t = \varpi_v^t / P_v^t$ satisfies the Belavkin *stochastic nonlinear filtering equation*

$$\mathrm{d}\varrho_v^t + \upsilon\left(\varrho_v^t\right)\mathrm{d}t = \sum_{j=1}^m \theta^j\left(u, \varrho_v^t\right)\left(\mathrm{d}v_j - \left\langle \varrho_v^t, \mathrm{L}_j + \mathrm{L}_j^\dagger\right\rangle \mathrm{d}t\right), \qquad (1)$$

where $\theta^j\left(u, \varrho\right) = \mathrm{L}^j \varrho_v + \varrho_v \mathrm{L}^{j\dagger} - \left\langle \varrho_v, \mathrm{L}_j + \mathrm{L}_j^\dagger\right\rangle \varrho_v$. Given $\varrho^{t_0} = \varrho$, the solution $\varrho_v^t\left(t_0, \varrho\right)$ defines an $\mathcal{S}$-valued Markov process, and we will have

$$\lim_{h \searrow 0} \frac{1}{h}\left\{\mathbb{E}\left[\mathsf{F}\left[\varrho_v^{t+h}\left(t, \varrho\right)\right] - \mathsf{F}\left[\varrho\right]\right]\right\} = D\left(u, \varrho\right)\mathsf{F}\left[\varrho\right],$$

where $D\left(t, \varrho\right)$ is the elliptic operator defined by $\varkappa\left(u, \varrho\right) = \sum_{j=1}^m \theta^j\left(\left(u, \varrho\right)\right)^{\otimes 2}$:

$$D\left(t, \varrho\right) \cdot + \left\langle \upsilon\left(u, \varrho\right), \nabla_\varrho \cdot\right\rangle = \frac{1}{2}\left\langle \varkappa\left(u, \varrho\right), \nabla_\varrho^{\otimes 2} \cdot\right\rangle.$$

4. Quantum dynamical programming

The *cost to go* of a feedback control law $y\left(t\right) = \ell\left(t, v^t\right)$ defining an adapted previsible process y_ω having $y_\omega\left(t_+\right) = \lim_{\varepsilon \searrow 0} y_\omega\left(t + \varepsilon\right)$ with respect to the innovation ω is

$$\mathsf{J}_\omega\left[\left\{y_\omega\left(\cdot\right)\right\}\right] = \int_t^\tau \mathsf{C}\left(y_\omega\left(r\right), \varrho_\omega^r\right)\mathrm{d}r + \mathsf{G}\left(y_\omega\left(\tau_+\right), \varrho_\omega^\tau\right).$$

Due to the statistical interpretation of quantum states,

$$\mathsf{C}\left(u, \varrho\right) = \left\langle \varrho, C\left(u\right)\right\rangle, \;\; \mathsf{G}\left(u, \varrho\right) = \left\langle \varrho, G\left(u\right)\right\rangle.$$

The optimal average cost on the interval $(t, \tau]$ to be

$$\mathsf{S}(t, \varrho) := \inf_{\{y_\omega(r)\}} \mathbb{E}\left[\mathsf{J}_\omega\left[y_\omega(\cdot); t, \varrho\right]\right],$$

with $\mathsf{S}(\tau, \varrho) = \inf\{\langle \varrho, G(u)\rangle : u \in \mathcal{U}\} \equiv \mathsf{S}[\varrho]$. It can be given by any concave functional $\mathsf{S} : \varrho \mapsto \mathbb{R} \sqcup \{-\infty\}$ of the terminal state $\varrho = \varrho_\omega^\tau$.

4.1. *Quantum Hamilton-Jacobi equation*

Here the control strategy $y = \ell$ will be non-random, $\ell = \{u_j(t)\}$ as will be any specific cost $\mathsf{J}[\{u\}]$. As for $\mathsf{S}(t, \varrho) = \inf \mathsf{J}[\{u\}; t, \varrho]$ at the times $t < t' \leq \tau$, one has

$$\mathsf{S}(t, \varrho) = \inf_{\{u\}} \left\{ \int_t^{t'} \mathsf{C}(u(r), \varrho^r)\, \mathrm{d}r + \mathsf{J}\left[u(\cdot); t', \varrho^{t'}\right] \right\}.$$

Suppose that $\{u^\circ(r, \varrho) : r > t\}$ is an optimal control when starting in state ϱ at the time t, and denote by $\{\varrho^r : r \in (t, \tau]\}$ the corresponding state trajectory starting at a state ϱ at t. Bellman's optimality principle observes that

$$-\frac{\partial}{\partial t}\mathsf{S}(t, \varrho) = \inf_{u \in \mathcal{U}} \left\{\mathsf{C}(u, \varrho) - \langle \upsilon(u, \varrho), \nabla_\varrho \mathsf{S}(t, \varrho)\rangle\right\},$$

The equation is then to be solved subject to

$$\mathsf{S}(\tau, \varrho) = \mathsf{G}(u^\circ(\tau_+, \varrho), \varrho) \equiv \mathsf{S}[\varrho].$$

4.2. *Pontryagin's maximum principle*

We may rewrite this as the *Hamilton-Jacobi* equation

$$-\frac{\partial}{\partial t}\mathsf{S}(t, \varrho(\mathsf{q})) + \mathsf{H}_\upsilon(\mathsf{q}, \mathsf{p}(\nabla_\varrho \mathsf{S}(t, \varrho))(\mathsf{q})) = 0$$

introducing the Pontryagin's Hamiltonian as the transform

$$\mathsf{H}_\upsilon(\mathsf{q}, \mathsf{p}) = \sup_{u \in \mathcal{U}} \{\langle \upsilon(u), \mathsf{p}\rangle - \mathsf{C}(u)\}(\mathsf{q}) = \sup_{u \in \mathcal{U}} \mathcal{K}_\upsilon(u, \mathsf{q}, \mathsf{p})$$

defined on $\mathcal{T}_0 \times \mathcal{T}_0^\star$ by the quadratic in q functional

$$\mathcal{K}_\upsilon(u, \mathsf{q}, \mathsf{p}(\mathtt{X})) := \langle \upsilon(u, \varrho(\mathsf{q})), \lambda \mathtt{I} + \mathtt{X}\rangle - \mathsf{C}(u, \varrho(\mathsf{q})),$$

which is affine in $\mathsf{p} \in \mathcal{T}_0^\star$ as independent of $\lambda \in \mathbb{R}$. This leads to the Hamiltonian boundary value problem

$$\begin{cases} \mathsf{q}^t - \nabla_{\mathsf{p}} \mathsf{H}_\upsilon(\mathsf{q}^t, \mathsf{p}^t) = 0, & \mathsf{q}^0 = 0 \\ \mathsf{p}^t + \nabla_{\mathsf{q}} \mathsf{H}_\upsilon(\mathsf{q}^t, \mathsf{p}^t) = 0, & \mathsf{p}^\tau = \mathsf{b} \end{cases}$$

as the *Hamilton-Pontryagin problem* with the solutions with $\mathsf{b} = \nabla_\varrho \mathsf{S}[\varrho]$ defining the minimal costs by the path integral in

$$\mathsf{S}(t_0, \varrho_0) = \int_{t_0}^{\tau} \left[\langle \mathsf{q}^t, \mathsf{p}^t\rangle - \mathsf{H}(\mathsf{q}^t, \mathsf{p}^t)\right] \mathrm{d}r + \mathsf{S}[\varrho(\mathsf{q}^\tau)].$$

4.3. *The filtered quantum Bellman equation*

If the filtering equation is in place of the Master equation, we have the diffusive Bellman equation

$$-\frac{\partial \mathsf{S}}{\partial t} = \inf_{u \in \mathcal{U}} \{\mathsf{C}(u, \varrho) + D(u, \varrho) \mathsf{S}(t, \varrho)\}, \quad \mathsf{S}(\tau, \varrho) = \mathsf{S}[\varrho].$$

This can be rewritten in the generalized HJ form as

$$\frac{\partial}{\partial t} \mathsf{S}(t, \varrho(\mathsf{q})) = \mathsf{H}_\upsilon^\varkappa\left(\mathsf{q}, \nabla_\varrho \mathsf{S}(t, \varrho(\mathsf{q})), \nabla_\varrho^{\otimes 2} \mathsf{S}(t, \varrho(\mathsf{q}))\right)$$

in terms of the generalized (Bellman) "Hamiltonian"

$$\mathsf{H}_\upsilon^\varkappa(\mathsf{q}, \mathsf{p}_1, \mathsf{p}_2) := \sup_{u \in \mathcal{U}} \left\{\mathcal{K}_\upsilon(u, \mathsf{q}, \mathsf{p}_1) - \frac{1}{2}\langle \varkappa(u, \varrho), \mathsf{p}_2\rangle\right\},$$

where $\mathsf{p}_2 \in \mathcal{T}_{0\mathrm{sym}}^{\star \otimes 2}$. If $\varkappa$ does not depend on u, this gives

$$-\frac{\partial \mathsf{S}}{\partial t}(\varrho) + \mathsf{H}_\upsilon(\mathsf{q}, \nabla \mathsf{S}(\varrho)) = \frac{1}{2}\left\langle \varkappa(\varrho), \nabla^{\otimes 2} \mathsf{S}(\varrho)\right\rangle,$$

since the optimal control strategy coincides in this case with the solution $u^{\mathrm{o}}(\mathsf{q}, \mathsf{p})$ of the corresponding nonstochastic problem for $\mathsf{q} = \varrho_0 - \varrho$ and $\mathsf{p} = \mathsf{p}(\nabla \mathsf{S}(t, \varrho))$.

4.4. *Linear–convex state costs*

Let $C(\boldsymbol{u})$, $G(\boldsymbol{u})$ and $H(\boldsymbol{u})$ be linear in $\boldsymbol{u} = (u^\alpha) \in \mathbb{R}^n$:

$$C(\boldsymbol{u}) = -u^\alpha \mathrm{B}_\alpha, \quad G(\boldsymbol{u}) = -u^\alpha \mathrm{F}_\alpha, \quad H(\boldsymbol{u}) = -u^\alpha \mathtt{Q}_\alpha$$

The additive real-valued functions $c(\boldsymbol{u})$ and $g(\boldsymbol{u})$ are assumed to be convex in $\boldsymbol{u}$, e.g. indicating a constraint $c(\boldsymbol{u}) = O_{\mathcal{B}_1}^{+}(\boldsymbol{u})$ for a convex compact $\mathcal{B}_1 \subset \mathbb{R}^n$, and similar for g. Using the affinity of

$$\langle \upsilon(\boldsymbol{u}), \mathrm{X}\rangle - \langle \varrho, C(\boldsymbol{u})\rangle = \langle \varrho, u^\alpha(\mathrm{B}_\alpha + \mathrm{i}[\mathtt{Q}_a, \mathrm{X}])\rangle + \langle \upsilon_m, \mathrm{X}\rangle$$

in $\boldsymbol{u}$, we can describe the Pontryagin Hamiltonian as

$$\mathsf{H}_\upsilon(\varrho_0 - \varrho, \mathsf{p}(\mathrm{X})) = \mathsf{E}(\boldsymbol{p}(\varrho, \mathrm{X})) + \langle \upsilon_m(\varrho), \mathrm{X}\rangle$$

by the usual Legendre transform $\mathsf{E}(\boldsymbol{p})$ of the constraint cost function $c(\boldsymbol{u})$. It is defined as the extremal value

$$\mathsf{E}(\boldsymbol{p}) = \boldsymbol{p}\cdot\boldsymbol{u}^{\mathrm{o}}(\boldsymbol{p}) - c(\boldsymbol{u}^{\mathrm{o}}(\boldsymbol{p})) := \sup_{\boldsymbol{u}}\{\boldsymbol{p}\cdot\boldsymbol{u} - c(\boldsymbol{u})\}$$

at $\boldsymbol{p} = (p_\alpha)$ with the components $p_\alpha(\varrho, \mathtt{X}) = \langle\varrho, \mathtt{P}_\alpha(\mathtt{X})\rangle$ given by $\mathtt{P}_\alpha(\mathtt{X}) = \mathtt{B}_\alpha + \mathrm{i}[\mathtt{Q}_\alpha, \mathtt{X}]$.

5. Optimal feedback control of purification

Note that the set $\operatorname{Arg\,max}\{\boldsymbol{p}\cdot\boldsymbol{u} - c(\boldsymbol{u})\}$ of the extremal points $\boldsymbol{u}^{\mathrm{o}}(\boldsymbol{p})$ is not empty, and any extremal point is good for each $\boldsymbol{p}$ as realizing the only possible global maximal value of the concave function $\boldsymbol{p}\cdot\boldsymbol{u} - c(\boldsymbol{u})$. It determines the diffusive HJB equation in the form

$$-\frac{\partial \mathsf{S}}{\partial t} + \mathsf{E}(\boldsymbol{p}(\varrho, \nabla_\varrho \mathsf{S})) + \langle \upsilon_m(\varrho), \nabla_\varrho \mathsf{S}\rangle = \frac{1}{2}\langle \varkappa(\varrho), \nabla_\varrho^{\otimes 2}\mathsf{S}\rangle.$$

In particular, in the case of constraint $\mathcal{B}_1 = \{\boldsymbol{u} : |\boldsymbol{u}| \leq 1\}$

$$-\frac{\partial \mathsf{S}}{\partial t} + |\boldsymbol{p}(\varrho, \nabla_\varrho \mathsf{S})| + \langle \upsilon_m(\varrho), \nabla_\varrho \mathsf{S}\rangle = \frac{1}{2}\langle \varkappa(\varrho), \nabla^{\otimes 2}\mathsf{S}\rangle.$$

This "zero rest mass" HJB equation can be obtained as the limit at $\mu \searrow 0$ of the "relativistic" HJB equation with $\mathsf{E}(\boldsymbol{p}) = \sqrt{\boldsymbol{p}^2 + \mu^2}$ corresponding to the cost gain $\mu\sqrt{1 - \boldsymbol{u}^2}$ under the constraint $|\boldsymbol{u}| \leq 1$, since such $\mathsf{E}(\boldsymbol{p})$ is the Legendre transform of

$$c(\boldsymbol{u}) = O_{\mathcal{B}_1}^{+}(\boldsymbol{u}) - \mu\sqrt{1 - \boldsymbol{u}^2}.$$

5.1. *Bequest function for purification*

Similarly, one can take $g(\boldsymbol{u}) = O_{\mathcal{B}_1}^{+}(\boldsymbol{u})$, or $O_{\mathcal{B}_1}^{+}(\boldsymbol{u}) - \nu\sqrt{1 - \boldsymbol{u}^2}$ with $\mathcal{B}_1 = \{\boldsymbol{u} : |\boldsymbol{u}| \leq 1\}$, and obtain the concave bequest functions

$$\mathsf{S}[\varrho] = -|\langle\varrho, \mathbf{F}\rangle|, \text{ or } \mathsf{S}[\varrho] = -\sqrt{\langle\varrho, \mathbf{F}\rangle^2 + \nu^2},$$

where $\mathbf{F} = (\mathrm{F}_\alpha)$. The first function in the form $\mathsf{S}[\varrho] = -|\vec{q}|$ corresponding to $\mathrm{F}_\alpha = \sigma_\alpha$ was used by Wiseman to optimize the quantum bit purification, and this can be done even nicer with $\mathsf{S}[\varrho] = -\sqrt{\vec{q}^2 + \nu^2}$.

Equally, one can use the Jacobs bequest function $\mathsf{S}[\varrho] = \sqrt{1 - \vec{q}^2}$ as the Legendre transform

$$\mathsf{S}[\varrho] = \inf_{\boldsymbol{u}}\left\{\sqrt{1 + \boldsymbol{u}^2} - \langle\varrho, \mathrm{F}_\alpha\rangle u^\alpha\right\} = \sqrt{1 - \langle\varrho, \mathbf{F}\rangle^2}$$

of the affine cost function given by $G(\boldsymbol{u}) = \vec{q}\cdot\vec{u}$.

Another possible choice is the entropy, say von Neumann entropy corresponding to $\mu = 2\mathrm{I}$:

$$\mathsf{S}[\varrho] = q \ln \sqrt{\frac{1-q}{1+q}} - \ln \sqrt{\frac{1-q^2}{4}}, \quad q = |\vec{q}| \ .$$

5.2. *Optimal feedback qubit control*

Let L_0 be $\mathsf{L} = \frac{\lambda}{2}\sigma_z = \mathsf{L}^\dagger$ $(\lambda = \overline{\lambda})$, and let us ignore the effect of environment by taking $R_j = 0$ for $j \neq 0$. We may also take cost $c(\vec{u}) = O^+_{\mathcal{B}_1}(\vec{u})$ of constraint $\mathcal{B}_1 = \{\vec{u} : |\vec{u}| \leq 1\}$ and any of the above bequest functions $\mathsf{S}[\varrho]$.

Explicitly we have $\langle v(u,\varrho), \mathrm{p}\rangle = \vec{v}(\vec{u},\vec{q}) \cdot \vec{p}$ linear in $\vec{u}$:

$$\vec{v}(\vec{u},\vec{q}) = \vec{v}_m + \vec{u} \times \vec{q}, \quad \vec{v}_m = \vec{m} - (\vec{m}\cdot\vec{q})\,\vec{q} - \frac{\lambda^2}{2}\vec{q}_z^{\perp}$$

which is maximized by the unit $\vec{u}^{\mathrm{o}} = \vec{q} \times \vec{p}/|\vec{q}\times\vec{p}|$ under the constraint $|\vec{u}| \leq 1$. This leads to the Hamiltonian function

$$\mathsf{H}_v(\vec{p},\vec{q}) = |\vec{q}\times\vec{p}| + \vec{v}_m \cdot \vec{p}.$$

Meanwhile, $\theta(\varrho) \equiv \frac{\lambda}{2}(\varrho\sigma_z + \sigma_z\varrho) - \langle \varrho, \lambda\sigma_z\rangle\,\varrho$ and so

$$\langle \theta(\varrho), \mathrm{p}\rangle = \lambda\left(1-z^2\right)p_z - \lambda z\,(xp_x + yp_y)\,.$$

5.3. *The qubit HJB equation*

With the customary abuse of notation we write $\mathsf{S}(t,\varrho) = \mathsf{S}(t,\vec{r})$ for $\vec{r} = -\vec{q}$ and $\theta(\varrho) = \lambda\left(-zx, -zy, 1-z^2\right)$ in terms of $\vec{r} = (x,y,z)$. The Itô term $\frac{1}{2}\left\langle \theta(\varrho)^{\otimes 2}, \nabla_\varrho^{\otimes 2}\mathsf{S}\right\rangle$ in HJB equation is then given by (with $\mathsf{S}_{xy} = \frac{\partial^2 \mathsf{S}}{\partial x \partial y}$, etc.)

$$\frac{\lambda^2}{2}\begin{pmatrix} -zx \\ -zy \\ 1-z^2 \end{pmatrix}^{\mathsf{T}} \begin{pmatrix} \mathsf{S}_{xx} & \mathsf{S}_{xy} & \mathsf{S}_{xz} \\ \mathsf{S}_{yx} & \mathsf{S}_{yy} & \mathsf{S}_{yz} \\ \mathsf{S}_{zx} & \mathsf{S}_{zy} & \mathsf{S}_{zz} \end{pmatrix} \begin{pmatrix} -zx \\ -zy \\ 1-z^2 \end{pmatrix}.$$

Putting everything together, we find that the HJB equation for optimal qubit control under the constraint is

$$\begin{aligned} &-\frac{\partial \mathsf{S}}{\partial t} + \left|\vec{p} \times \vec{\nabla}\mathsf{S}\right| + \left(\vec{m} - (\vec{m}\cdot\vec{r})\,\vec{r} + \frac{\lambda^2}{2}\vec{r}_z^{\perp}\right)\vec{\nabla}\mathsf{S} \\ &= \frac{\lambda^2}{2}\left(x^2z^2\frac{\partial^2\mathsf{S}}{\partial x^2} + y^2z^2\frac{\partial^2\mathsf{S}}{\partial y^2} + \left(1-z^2\right)^2\frac{\partial^2\mathsf{S}}{\partial z^2}\right. \\ &\left. + \, xyz^2\frac{\partial^2\mathsf{S}}{\partial x\partial y} - xz\left(1-z^2\right)\frac{\partial^2\mathsf{S}}{\partial x\partial z} - yz\left(1-z^2\right)\frac{\partial^2\mathsf{S}}{\partial y\partial z}\right). \end{aligned}$$

6. Discussion

Our analysis is based on the fact that quantum state is a sufficient coordinate not only for closed, but also for open unstable quantum systems under the Markov approximation, and this remains true even if the open system is under a continuous demolition observation. However we have to deal with differential equation of high or infinite dimensionality if $\dim \mathfrak{h} = \infty$.

Nevertheless, the Bellman principle can then be applied in much the same spirit as for classical states and we are able to derive the corresponding HJB theory for a wider class of cost functionals than traditionally considered in the literature.

When restricted to the Bloch sphere for the qubit with the cost being a quantum expectation, we recover the class of Bellman equations encountered so far in quantum feedback purification.

References

1. V. P. Belavkin: Theory of the Control of Observable Quantum Systems. Automatica and Remote Control 44 (2) 178–188 (1983). quant-ph/0408003,
2. V. P. Belavkin and S. Edwards: "Quantum Filtering and Optimal Control" . In: Quantum Stochastics, Filtering and Control, 143-206, World Scientific, 2008. Ed: V. Belavkin and M. Guta. Preprint in: quant-ph/0506018.
3. A. N. Sherstnev: Methods of Bilinear Forms in Noncommutative Theory of Measure and Integral. Moscow, Fizmatlit, 2008, 264p.
4. V. P. Belavkin: Non-demolition measurements, nonlinear filtering and dynamic programming of quantum stochastic processes, in *Proc of Bellman Continuum Workshop 'Modelling and Control of Systems', Sophia–Antipolis 1988*, ser. Lecture notes in Control and Inform Sciences, A.Blaquiere, Ed., vol. 121. Springer–Verlag, Berlin 1988, pp. 245–265.
5. V. P. Belavkin: A New Wave Equation for a Continuous Non-Demolition Measurement. Phys Letters A 140(3) 355–358 (1989). quant-ph/0512136.
6. V. P. Belavkin: Stochastic equations of quantum filtering, in *Prob. Theory and Math. Statistics*, B. et al., Ed: VSP/Mokslas, 1990, pp. 91–109.

QUANTUM ENTANGLEMENT AND CIRCULANT STATES

DARIUSZ CHRUŚCIŃSKI

Institute of Physics, Nicolaus Copernicus University,
Grudziądzka 5/7, 87–100 Toruń, Poland

We provide a large class of quantum $d \otimes d$ states which are positive under partial transposition (so called PPT states). The construction is based on certain direct sum decomposition of the total Hilbert space which is governed by a cyclic permutation from the symmetric group S_{d-1}. It turns out that partial transposition maps any such decomposition into another one corresponding to 'complementary' permutation. This class contains many well known examples of PPT states from the literature and gives rise to a huge family of completely new states. The class of circulat states gives rise to a nice example for the linear lifting and transition expectation.

1. Introduction

Quantum Entanglement is one of the key features which distinguish quantum mechanics from the classical one. Recent development of Quantum Information Theory [1] shows that quantum entanglement does have important practical applications and it serves as a basic resource for quantum cryptography, quantum teleportation, dense coding and quantum computing.

A fundamental problem in Quantum Information Theory is to test whether a given state of a composite quantum system is entangled or separable. Let us recall that a state represented by a density operator ρ living in the Hilbert space $\mathcal{H}_A \otimes \mathcal{H}_B$ is separable iff it can be represented as the following convex combination

$$\rho = \sum_k p_k \, \rho_k^{(A)} \otimes \rho_k^{(B)} \ , \tag{1}$$

where $\rho_k^{(A)}$ and $\rho_k^{(B)}$ are density operators living in $\mathcal{H}_A$ and $\mathcal{H}_B$, respectively, and $\{p_k\}$ stands for a probability distribution, that is, $p_k \geq 0$ and $\sum_k p_k = 1$. States which are not separable are called entangled. Surprisingly, this so called separability problem has no simple solution. Several operational criteria have been proposed to identify entangled states. Each of these criterion is only necessary and in general one needs to perform an infinite number of tests to be sure that a given state is separable (see e.g. [2] for the recent review). The most famous Peres-Horodecki criterion [3,4] is based on the operation of partial transposition: for any operator X living in $\mathcal{H}_A \otimes \mathcal{H}_B$ one defines its partial transposition $X^\tau = (\mathbb{1} \otimes \tau)X$ by

$$X^\tau_{i\alpha,j\beta} := X_{i\beta,j\alpha} \ , \tag{2}$$

where the corresponding matrix elements are computed with respect to fixed bases $\{e_i\}$ in $\mathcal{H}_A$ and $\{f_\alpha\}$ in $\mathcal{H}_B$. It is clear that this operation depends upon chosen bases $\{e_i\}$ and $\{f_\alpha\}$. However, as was observed by Peres [3], the positivity of X^τ is a universal property which is basis-independent. One calls a state ρ to be PPT (Positive Partial Transpose) if $X^\tau \geq 0$.

Now, if a state ρ is separable then its partial transposition $(\mathbb{1} \otimes \tau)\rho$ is evidently positive but the converse is not true. It was shown by Horodecki et al. [5] that PPT condition is both necessary and sufficient for separability for $2 \otimes 2$ and $2 \otimes 3$ systems (actually, this observation is based on the old mathematical result due to Woronowicz [6]).

Since all separable states belong to a set of PPT states, the structure of this set is of primary importance in Quantum Information Theory. Unfortunately, this structure is still unknown, that is, one may check whether a given state is PPT but we do not know how to construct a general quantum state with PPT property. There are several examples of PPT entangled states [4], [7–16] and its mathematical structure was studied in [17,18,19].

In the present paper, following [20] (see also [15]), we analyze a very simple construction of PPT states which is based on certain decomposition of the total Hilbert space $\mathbb{C}^d \otimes \mathbb{C}^d$ into direct sum of d-dimensional subspaces. This decomposition is controlled by some cyclic property, that is, knowing one subspace, say Σ_0, the remaining subspaces $\Sigma_1, \ldots, \Sigma_{d-1}$ are uniquely determined by applying a cyclic shift to elements from Σ_0. We call a density matrix ρ a *circulant state* if ρ is a convex combination of density matrices supported on Σ_α. The crucial observation is that a partial transposition of the circulant state has again a circular structure corresponding to another direct sum decomposition $\widetilde{\Sigma}_0 \oplus \ldots \oplus \widetilde{\Sigma}_{d-1}$.

Circulant states may be used to define a linear lifting [21], i.e. an operation from the space of states of one system living in $\mathbb{C}^d$ into the space of states of a composite system living in $\mathbb{C}^d \otimes \mathbb{C}^d$. We propose to call such particular type of lifting a *circulant lifting*. It lifts an arbitrary state of a single system to the circulant state of the composed system. Due to the duality between linear liftings and transition expectations circulant states provides also a nice example for the latter.

2. PPT states – simple example

The construction of PPT entangled states (PPTES) is a highly nontrivial task. An important example of PPTES in $3 \otimes 3$ was provided by Horodecki [4]. Here we start with another example which may be easily generalized. It is well know that the following unnormalized matrix

$$\rho = \left(\begin{array}{ccc|ccc|ccc} 1 & \cdot & \cdot & \cdot & 1 & \cdot & \cdot & \cdot & 1 \\ \cdot & 1/s & \cdot & \cdot & \cdot & \cdot & \cdot & \cdot & \cdot \\ \cdot & \cdot & s & \cdot & \cdot & \cdot & \cdot & \cdot & \cdot \\ \hline \cdot & \cdot & \cdot & s & \cdot & \cdot & \cdot & \cdot & \cdot \\ 1 & \cdot & \cdot & \cdot & 1 & \cdot & \cdot & \cdot & 1 \\ \cdot & \cdot & \cdot & \cdot & \cdot & 1/s & \cdot & \cdot & \cdot \\ \hline \cdot & \cdot & \cdot & \cdot & \cdot & \cdot & 1/s & \cdot & \cdot \\ \cdot & \cdot & \cdot & \cdot & \cdot & \cdot & \cdot & s & \cdot \\ 1 & \cdot & \cdot & \cdot & 1 & \cdot & \cdot & \cdot & 1 \end{array}\right) , \tag{3}$$

is PPTES for $s > 0$ (in order to have more transparent pictures we replaced all vanishing matrix elements by dots and we use this convention through out this paper). This example may be generalized as follows [15]: define

$$\rho = \rho_0 + \rho_{\rm D} , \tag{4}$$

where

$$\rho_0 = \sum_{i,j=0}^{d-1} a_{ij}\, e_{ij} \otimes e_{ij} , \tag{5}$$

and the diagonal $\rho_{\rm D}$ is given by

$$\rho_{\rm D} = \sum_{i\neq j=0}^{d-1} d_{ij} e_{ii} \otimes e_{jj} . \tag{6}$$

Throughout the paper we use the following convenient notation: e_i ($i = 0, \ldots, d-1$) denotes an orthonormal basis in $\mathbb{C}^d$ and $e_{ij} := |i\rangle < j|$. The positivity of ρ is guarantied by 1) the positivity of $d \otimes d$ matrix $[a_{ij}]$ and 2) $d_{ij} \geq 0$ for $i \neq j$. Since ρ_0 and $\rho_{\rm D}$ are supported on the orthogonal subspaces one has

$$[\rho_0, \rho_{\rm D}] = 0 . \tag{7}$$

For $d = 3$ the above construction gives the following picture

$$\rho \; = \; \left(\begin{array}{ccc|ccc|ccc} a_{00} & \cdot & \cdot & \cdot & a_{01} & \cdot & \cdot & \cdot & a_{02} \\ \cdot & d_{01} & \cdot & \cdot & \cdot & \cdot & \cdot & \cdot & \cdot \\ \cdot & \cdot & d_{02} & \cdot & \cdot & \cdot & \cdot & \cdot & \cdot \\ \hline \cdot & \cdot & \cdot & d_{10} & \cdot & \cdot & \cdot & \cdot & \cdot \\ a_{10} & \cdot & \cdot & \cdot & a_{11} & \cdot & \cdot & \cdot & a_{12} \\ \cdot & \cdot & \cdot & \cdot & \cdot & d_{12} & \cdot & \cdot & \cdot \\ \hline \cdot & \cdot & \cdot & \cdot & \cdot & \cdot & d_{20} & \cdot & \cdot \\ \cdot & \cdot & \cdot & \cdot & \cdot & \cdot & \cdot & d_{21} & \cdot \\ a_{20} & \cdot & \cdot & \cdot & a_{21} & \cdot & \cdot & \cdot & a_{22} \end{array} \right) , \tag{8}$$

and it is clear the (3) is a special example of the (8) with $a_{ij} \equiv 1$, $d_{01} = d_{12} = d_{20} = 1/s$ and $d_{02} = d_{10} = d_{21} = s$. Now, the crucial observation of [15] is that ρ defined in (4) is PPT iff the following condition is satisfied

$$|a_{ij}|^2 \leq d_{ij} d_{ji} \; , \qquad i \neq j \; . \tag{9}$$

Note, that this condition is trivially satisfied for (3). One has $d_{ij}d_{ji} = 1 = |a_{ij}|^2$. In the next section we provide important generalization of the above construction.

3. Circulant states

3.1. *Two qubits*

For pedagogical reason we start to present our construction for $d = 2$. Consider an operator living in $\mathbb{C}^2 \otimes \mathbb{C}^2$ of the following form

$$\rho = \rho_0 + \rho_1 \; , \tag{10}$$

where ρ_0 and ρ_1 are supported on two orthogonal subspaces

$$\begin{aligned} \Sigma_0 &= \mathrm{span}\, \{ e_0 \otimes e_0 \, , e_1 \otimes e_1 \} \; , \\ \Sigma_1 &= \mathrm{span}\, \{ e_0 \otimes e_1 \, , e_1 \otimes e_0 \} \; , \end{aligned} \tag{11}$$

where $\{e_0, e_1\}$ is a computational base in $\mathbb{C}^2$. It is clear that $\{\Sigma_0, \Sigma_1\}$ defines the direct sum decomposition of $\mathbb{C}^2 \otimes \mathbb{C}^2$, that is

$$\Sigma_0 \oplus \Sigma_1 \; = \; \mathbb{C}^2 \otimes \mathbb{C}^2 \; .$$

We call it a circulant decomposition because its structure is determined by the cyclic shift $S : \mathbb{C}^2 \to \mathbb{C}^2$ defined by

$$S \, e_i = e_{i+1} \; , \qquad (\mathrm{mod}\ 2) \; . \tag{12}$$

and hence

$$\Sigma_1 = (\mathbb{1} \otimes S)\, \Sigma_0 \ . \tag{13}$$

One has therefore

$$\rho_0 = \sum_{i,j=0}^{1} a_{ij}\, e_{ij} \otimes e_{ij} \ , \qquad \rho_1 = \sum_{i,j=0}^{1} b_{ij}\, e_{ij} \otimes S\, e_{ij} S^* \ ,$$

where one adds modulo 2. Now, since ρ_0 and ρ_1 are supported on two orthogonal subspaces Σ_0 and Σ_1 one has an obvious

Proposition 3.1. *ρ defined in (10) is a density matrix iff*

- *$a = [a_{ij}]$ and $b = [b_{ij}]$ are 2×2 semi-positive matrices, and*
- *$\mathrm{Tr}(a + b) = 1$.*

Now, the crucial observation is that both the original density matrix ρ and the partially transposed matrix $\rho^\tau = (\mathbb{1} \otimes \tau)\rho$ have exactly the same structure:

$$\rho = \left(\begin{array}{cc|cc} a_{00} & \cdot & \cdot & a_{01} \\ \cdot & b_{00} & b_{01} & \cdot \\ \hline \cdot & b_{10} & b_{11} & \cdot \\ a_{10} & \cdot & \cdot & a_{11} \end{array} \right) , \qquad \rho^\tau = \left(\begin{array}{cc|cc} \widetilde{a}_{00} & \cdot & \cdot & \widetilde{a}_{01} \\ \cdot & \widetilde{b}_{00} & \widetilde{b}_{01} & \cdot \\ \hline \cdot & \widetilde{b}_{10} & \widetilde{b}_{11} & \cdot \\ \widetilde{a}_{10} & \cdot & \cdot & \widetilde{a}_{11} \end{array} \right) , \tag{14}$$

where the matrices $\widetilde{a} = [\widetilde{a}_{ij}]$ and $\widetilde{b} = [\widetilde{b}_{ij}]$ read as follows

$$\widetilde{a} = \begin{pmatrix} a_{00} & b_{01} \\ b_{10} & a_{11} \end{pmatrix} , \qquad \widetilde{b} = \begin{pmatrix} b_{00} & a_{01} \\ a_{10} & b_{11} \end{pmatrix} , \tag{15}$$

that is, both ρ and ρ^τ are *circulant* bipartite operators. Therefore, one arrives at

Theorem 3.1. *A circulant state represented by (10) is PPT iff $\widetilde{a} = [\widetilde{a}_{ij}]$ and $\widetilde{b} = [\widetilde{b}_{ij}]$ are 2×2 semi-positive definite matrices.*

This simple construction reproduces many well known examples of 2-qubit states: for example Bell states, Werner [22] and isotropic states [23], Bell diagonal states, $O(2)$-invariant states [24], and many others.

3.2. *Two qutrits*

The construction of circulant states for two qubits may be easily generalized for two qutrits. It is based on the following direct sum decomposition

$$\mathbb{C}^3 \otimes \mathbb{C}^3 = \Sigma_0 \oplus \Sigma_1 \oplus \Sigma_2 \ , \tag{16}$$

where

$$\Sigma_0 = \mathrm{span}\,\{e_0 \otimes e_0\,, e_1 \otimes e_1\,, e_2 \otimes e_2\}\ , \tag{17}$$

and

$$\Sigma_1 = (\mathbb{1} \otimes S)\Sigma_0\ , \qquad \Sigma_2 = (\mathbb{1} \otimes S^2)\Sigma_0\ , \tag{18}$$

with $S : \mathbb{C}^3 \longrightarrow \mathbb{C}^3$ being a shift operator defined by

$$S\,e_i = e_{i+1}\ , \qquad (\mathrm{mod}\ 3)\ . \tag{19}$$

One easily finds

$$\begin{aligned} \Sigma_1 &= \mathrm{span}\,\{e_0 \otimes e_1\,, e_1 \otimes e_2\,, e_2 \otimes e_0\}\ , \\ \Sigma_2 &= \mathrm{span}\,\{e_0 \otimes e_2\,, e_1 \otimes e_0\,, e_2 \otimes e_1\}\ . \end{aligned} \tag{20}$$

Consider now an operator living in $\mathbb{C}^3 \otimes \mathbb{C}^3$ of the following form

$$\rho = \rho_0 + \rho_1 + \rho_2\ , \tag{21}$$

where

$$\rho_0 = \sum_{i,j=0}^{2} a_{ij}\, e_{ij} \otimes e_{ij}\ , \quad \rho_1 = \sum_{i,j=0}^{2} b_{ij}\, e_{ij} \otimes S\, e_{ij} S^*\ , \quad \rho_2 = \sum_{i,j=0}^{2} c_{ij}\, e_{ij} \otimes S^2\, e_{ij} S^{2*}\ .$$

Proposition 3.2. *ρ defined in (21) is a density matrix iff*

- *$a = [a_{ij}]$, $b = [b_{ij}]$ and $c = [c_{ij}]$ are 3×3 semi-positive matrices, and*
- $\mathrm{Tr}(a + b + c) = 1$.

The circulant $3 \otimes 3$ state ρ has therefore the following form

$$\rho = \left(\begin{array}{ccc|ccc|ccc} a_{00} & \cdot & \cdot & \cdot & a_{01} & \cdot & \cdot & \cdot & a_{02} \\ \cdot & b_{00} & \cdot & \cdot & \cdot & b_{01} & b_{02} & \cdot & \cdot \\ \cdot & \cdot & c_{00} & c_{01} & \cdot & \cdot & \cdot & c_{02} & \cdot \\ \hline \cdot & \cdot & c_{10} & c_{11} & \cdot & \cdot & \cdot & c_{12} & \cdot \\ a_{10} & \cdot & \cdot & \cdot & a_{11} & \cdot & \cdot & \cdot & a_{12} \\ \cdot & b_{10} & \cdot & \cdot & \cdot & b_{11} & b_{12} & \cdot & \cdot \\ \hline \cdot & b_{20} & \cdot & \cdot & \cdot & b_{21} & b_{22} & \cdot & \cdot \\ \cdot & \cdot & c_{20} & c_{21} & \cdot & \cdot & \cdot & c_{22} & \cdot \\ a_{20} & \cdot & \cdot & \cdot & a_{21} & \cdot & \cdot & \cdot & a_{22} \end{array} \right)\ , \tag{22}$$

and hence its partially transposed counterpart ρ^τ is given by

$$\rho^\tau = \left(\begin{array}{ccc|ccc|ccc} \widetilde{a}_{00} & \cdot & \cdot & \cdot & \cdot & \widetilde{a}_{01} & \cdot & \widetilde{a}_{02} & \cdot \\ \cdot & \widetilde{b}_{00} & \cdot & \widetilde{b}_{01} & \cdot & \cdot & \cdot & \cdot & \widetilde{b}_{02} \\ \cdot & \cdot & \widetilde{c}_{00} & \cdot & \widetilde{c}_{01} & \cdot & \widetilde{c}_{02} & \cdot & \cdot \\ \hline \cdot & \widetilde{b}_{10} & \cdot & \widetilde{b}_{11} & \cdot & \cdot & \cdot & \cdot & \widetilde{b}_{12} \\ \cdot & \cdot & \widetilde{c}_{10} & \cdot & \widetilde{c}_{11} & \cdot & \widetilde{c}_{12} & \cdot & \cdot \\ \widetilde{a}_{10} & \cdot & \cdot & \cdot & \cdot & \widetilde{a}_{11} & \cdot & \widetilde{a}_{12} & \cdot \\ \hline \cdot & \cdot & \widetilde{c}_{20} & \cdot & \widetilde{c}_{21} & \cdot & \widetilde{c}_{22} & \cdot & \cdot \\ \widetilde{a}_{20} & \cdot & \cdot & \cdot & \cdot & \widetilde{a}_{21} & \cdot & \widetilde{a}_{22} & \cdot \\ \cdot & \widetilde{b}_{20} & \cdot & \widetilde{b}_{21} & \cdot & \cdot & \cdot & \cdot & \widetilde{b}_{22} \end{array}\right) , \tag{23}$$

where the matrices $\widetilde{a} = [\widetilde{a}_{ij}]$, $\widetilde{b} = [\widetilde{b}_{ij}]$ and $\widetilde{c} = [\widetilde{c}_{ij}]$ read as follows

$$\widetilde{a} = \begin{pmatrix} a_{00} & c_{01} & b_{02} \\ c_{10} & b_{11} & a_{12} \\ b_{20} & a_{21} & c_{22} \end{pmatrix} , \quad \widetilde{b} = \begin{pmatrix} b_{00} & a_{01} & c_{02} \\ a_{10} & c_{11} & b_{12} \\ c_{20} & b_{21} & a_{22} \end{pmatrix} , \quad \widetilde{c} = \begin{pmatrix} c_{00} & b_{01} & a_{02} \\ b_{10} & a_{11} & c_{12} \\ a_{20} & c_{21} & b_{22} \end{pmatrix} . \tag{24}$$

Note, that

$$\rho^\tau = \widetilde{\rho}_0 + \widetilde{\rho}_1 + \widetilde{\rho}_2 , \tag{25}$$

where $\widetilde{\rho}_k$ are supported on three orthogonal subspaces of $\mathbb{C}^3 \otimes \mathbb{C}^3$:

$$\begin{aligned} \widetilde{\Sigma}_0 &= \text{span}\,\{e_0 \otimes e_0\,, e_1 \otimes e_2\,, e_2 \otimes e_1\} \ , \\ \widetilde{\Sigma}_1 &= \text{span}\,\{e_0 \otimes e_1\,, e_1 \otimes e_0\,, e_2 \otimes e_2\} \ , \\ \widetilde{\Sigma}_2 &= \text{span}\,\{e_0 \otimes e_2\,, e_1 \otimes e_1\,, e_2 \otimes e_0\} \ . \end{aligned} \tag{26}$$

It is clear that $\widetilde{\Sigma}_\alpha$ defines again circulant direct sum decomposition

$$\mathbb{C}^3 \otimes \mathbb{C}^3 = \widetilde{\Sigma}_0 \oplus \widetilde{\Sigma}_1 \oplus \widetilde{\Sigma}_2 , \tag{27}$$

and

$$\widetilde{\Sigma}_\alpha = (\mathbb{1} \otimes S^\alpha)\widetilde{\Sigma}_0 . \tag{28}$$

It shows that partial transposition $(\mathbb{1} \otimes \tau)$ maps any circulant density matrix $\rho = \rho_0 + \rho_1 + \rho_2$ into another circulant matrix $\rho^\tau = \widetilde{\rho}_0 + \widetilde{\rho}_1 + \widetilde{\rho}_2$. Now, contrary to the $d = 2$ case, both circulant structures are different. One has therefore

Theorem 3.2. *A circulant $3 \otimes 3$ state ρ is PPT iff the matrices $\widetilde{a}$, $\widetilde{b}$ and $\widetilde{c}$ are semi-positive definite.*

4. General case – two qudits

Now, we are ready to construct circular states in $d \otimes d$. The basic idea is to decompose the total Hilbert space $\mathbb{C}^d \otimes \mathbb{C}^d$ into a direct sum of d orthogonal d-dimensional subspaces related by a certain cyclic property. It turns out that there are $(d-1)!$ different cyclic decompositions and it is therefore clear that they may be labeled by permutations from the symmetric group S_{d-1}. Take any permutation π from the symmetric group S_d and fix its action on '0' by $\pi(0) = 0$. This way it is effectively a permutation from S_{d-1}. Let us define Σ_0^π which is spanned by

$$e_0 \otimes e_{\pi(0)}\, , e_1 \otimes e_{\pi(1)}\, , \ldots\, , e_{d-1} \otimes e_{\pi(d-1)}\ . \tag{29}$$

Moreover, let us define $d-1$ orthogonal subspaces by

$$\Sigma_\alpha^\pi = (\mathbb{1} \otimes S^\alpha)\Sigma_0^\pi\ ,$$

where S is a circulant matrix corresponding to shift in $\mathbb{C}^d$:

$$S\, e_i = e_{i+1}\ , \qquad (\mathrm{mod}\ d)\ , \tag{30}$$

and $\alpha = 0, 1, \ldots, d-1$. It is clear that

$$\Sigma_0^\pi \oplus \Sigma_1^\pi \oplus \ldots \oplus \Sigma_{d-1}^\pi = \mathbb{C}^d \otimes \mathbb{C}^d\ . \tag{31}$$

To construct a circulant state corresponding to this decomposition let us introduce d positive $d \times d$ matrices $a^{(\alpha)} = [a_{ij}^{(\alpha)}]$; $\alpha = 0, 1, \ldots, d-1$. Now, define d positive operators ρ_α^π supported on Σ_α^π by

$$\begin{aligned}
\rho_\alpha^\pi &= \sum_{i,j=0}^{d-1} a_{ij}^{(\alpha)}\, e_{ij} \otimes S^\alpha\, e_{\pi(i),\pi(j)}\, S^{*\alpha} \\
&= \sum_{i,j=0}^{d-1} a_{ij}^{(\alpha)}\, e_{ij} \otimes e_{\pi(i)+\alpha,\pi(j)+\alpha}\ .
\end{aligned} \tag{32}$$

Finally, one observes that the following operator

$$\rho_\pi = \rho_0^\pi + \rho_1^\pi + \ldots + \rho_{d-1}^\pi\ , \tag{33}$$

defines the circulant state (corresponding to π). Normalization of ρ_π is equivalent to the following condition for matrices $a^{(\alpha)}$

$$\mathrm{Tr}\left(a^{(0)} + a^{(1)} + \ldots + a^{(d-1)}\right) = 1\ .$$

Theorem 4.1. *If ρ_π is a circulant state corresponding to permutation π with $\pi(0) = 0$, then its partial transposition ρ_π^τ is also circulant with respect to another decomposition corresponding to permutation $\widetilde{\pi}$ such that*

$$\pi(i) + \widetilde{\pi}(i) = d\ , \tag{34}$$

for $i = 1, 2, \ldots, d-1$, and $\widetilde{\pi}(0) = 0$, that is

$$(\mathbb{1} \otimes \tau)\rho_\pi = \sum_{\alpha=0}^{d-1} \sum_{i,j=0}^{d-1} \widetilde{a}_{ij}^{(\alpha)} e_{ij} \otimes S^\alpha e_{\widetilde{\pi}(i),\widetilde{\pi}(j)} S^{\alpha *} . \tag{35}$$

Moreover

$$\widetilde{a}_{ij}^{(\alpha)} = a_{ij}^{(\alpha + [\widetilde{\pi}(j) - \widetilde{\pi}(i)])} , \tag{36}$$

where we add modulo d.

From (34) it is clear that $\widetilde{\pi}$ is the inverse of π, i.e. $\widetilde{\pi} \circ \pi = \mathrm{id}$. This way partial transposition relates circulant structures related by mutually inverse permutations.

5. Circulant liftings

Circulant states provide interesting example of a linear lifting. Let $\mathcal{A}_1$ and $\mathcal{A}_2$ be $\mathbb{C}^*$-algebras. Recall, that a lifting [21] from $\mathcal{A}_1$ to $\mathcal{A}_2 \otimes \mathcal{A}_1$ is a map

$$\mathcal{E}^\sharp \ \mathcal{S}(\mathcal{A}_1) \longrightarrow \mathcal{S}(\mathcal{A}_2 \otimes \mathcal{A}_1) . \tag{37}$$

If the map $\mathcal{E}^\sharp$ is affine and its dual $\mathcal{E} : \mathcal{A}_1 \otimes \mathcal{A}_2 \longrightarrow \mathcal{A}_1$ is completely positive, then one calls $\mathcal{E}^\sharp$ a linear lifting. If $\mathcal{E}^\sharp$ maps pure states into pure states, one calls it a pure lifting. A linear completely positive map $\mathcal{E} : \mathcal{A}_2 \otimes \mathcal{A}_1 \longrightarrow \mathcal{A}_1$ is called a transition expectation if in addition it satisfies

$$\mathcal{E}(\mathbb{1}_2 \otimes \mathbb{1}_1) = \mathbb{1}_1 , \tag{38}$$

where $\mathbb{1}_k$ denotes an identity element in $\mathcal{A}_k$.

Denote by M_d a $\mathbb{C}^*$-algebra of $d \times d$ complex matrices and consider the following lifting

$$\mathcal{E}^\sharp \ S(M_d) \longrightarrow S(M_d \otimes M_d) , \tag{39}$$

defined by

$$\mathcal{E}^\sharp(\rho) = \sum_{\alpha=0}^{d-1} c_{ij}^{(\alpha)} \, e_{ij} \otimes S^\alpha e_{i\alpha} \, \rho \, e_{\alpha j} S^{\alpha *} , \tag{40}$$

where $[c_{ij}^{(\alpha)}]$ are $d \times d$ positive matrices for $\alpha = 0, 1, \ldots, d-1$ such that $\mathrm{Tr} c^{(\alpha)} = 1$. Note that

$$\mathcal{E}^\sharp(\rho) = \sum_{\alpha=0}^{d-1} a_{ij}^{(\alpha)} \, e_{ij} \otimes S^\alpha e_{ij} S^{\alpha *} , \tag{41}$$

where

$$a_{ij}^{(\alpha)} = p_\alpha \, c_{ij}^{(\alpha)} \ , \tag{42}$$

and

$$p_\alpha = \rho_{\alpha\alpha} \ . \tag{43}$$

It shows that for any ρ its lifting $\mathcal{E}^\sharp(\rho)$ defines a circulant state. Now, if $c^{(\alpha)}$ are rank-1 projectors, i.e. $c_{mn}^{(\alpha)} = c_m^{(\alpha)} \overline{c}_n^{(\alpha)}$ are the Grahm matrices for the d complex d-vectors $c^{(\alpha)}$, the above formula simplifies to

$$\mathcal{E}^\sharp(\rho) \ := \ V \, \mathcal{D}(\rho) \, V^* \ , \tag{44}$$

where

$$\mathcal{D}(\rho) := \sum_{i=0}^{d-1} e_{ii} \, \rho \, e_{ii} \ , \tag{45}$$

is the projection onto the diagonal part of ρ, and

$$V \ \mathbb{C}^d \ \longrightarrow \ \mathbb{C}^d \otimes \mathbb{C}^d \tag{46}$$

is defined by

$$V e_\alpha \ = \ \sum_{j=0}^{d-1} c_j^{(\alpha)} \, e_j \otimes e_{j+\alpha} \tag{47}$$

Note, that due the trace condition $\mathrm{Tr} c^{(\alpha)} = 1$ the linear operator V defines an isometry

$$V^* V = \mathbb{I} \ . \tag{48}$$

It should be stressed that the above circulant lifting is never pure. An example of a pure lifting is given by

$$\mathcal{E}^\sharp_{\text{pure}}(\rho) \ := \ V \rho V^* \ . \tag{49}$$

Note, however, that for an arbitrary ρ the corresponding lifting $\mathcal{E}^\sharp_{\text{pure}}(\rho)$ is in general not a circulant state.

6. Conclusions

We have constructed a large class of PPT states in $d \otimes d$ which correspond to circular decompositions of $\mathbb{C}^d \otimes \mathbb{C}^d$ into direct sums of d-dimensional subspaces. This class significantly enlarges the previous class defined in [15] constructed by a completely different method. It contains several known examples from the literature. The simplest one is a set of generalized Bell states in $d \otimes d$: for any $\mu, \nu \in \{0, 1, \ldots, d-1\}$ let us define

$$\psi_{\mu\nu} = (\Omega^{\mu} \otimes S^{\nu})\psi_d^+ , \tag{50}$$

where ψ_d^+ is a canonical maximally entangled state in $\mathbb{C}^d \otimes \mathbb{C}^d$, that is

$$\psi_d^+ = \frac{1}{\sqrt{d}} \sum_{k=0}^{d-1} e_k \otimes e_k , \tag{51}$$

and Ω is defined by

$$\Omega e_k = \omega^k e_k , \tag{52}$$

with $\omega = e^{2\pi i/d}$. It is easy to see that each 1-dimensional projector

$$P_{\mu\nu} = |\psi_{\mu\nu}\rangle < \psi_{\mu\nu}| , \tag{53}$$

defines a circulant state. It is therefore clear that any mixture

$$\rho = \sum_{\mu,\nu=0}^{d-1} c_{\mu\nu} P_{\mu\nu} , \tag{54}$$

with $c_{\mu\nu} \geq 0$ and $\sum_{\mu,\nu=0}^{d-1} c_{\mu\nu} = 1$ is again circulant.

Other examples are provided by Werner and isotropic states, entangled PPT states in $4 \otimes 4$ constructed by Ha [17] and Fei et. al. [13] (for detailed discussion we refer to our recent paper [20]). It is clear that the presented class of circulant states produces a highly nontrivial family of new states (we stress however that the seminal Horodecki $3 \otimes 3$ entangled PPT state [4] does not belong to our class).

There are still many open problems connected with this class: the basic question is how to detect entanglement within this class of PPT states. One may expect that there is special class of entanglement witnesses which are sensitive to entanglement encoded into circular decompositions, that is, one may look for an entanglement witness which has again a circulant structure but now not all matrices $a^{(\alpha)}$ are positive. The condition that entanglement witness should be positive on separable states has to replace positivity of all $a^{(\alpha)}$ by some weaker conditions (for recent discussion of entanglement witnesses see [25] and [26]).

It is interesting to explore the possibility of other characteristic decompositions leading to new classes of PPT states. Finally, it would be very interesting to look for multipartite generalization, that is, to construct circulant states in $\mathbb{C}^d \otimes \ldots \otimes \mathbb{C}^d$ for an arbitrary number of copies.

Actually, a subclass of circulant states were also investigated from a different perspective by Baumgartner, Hiesmayr and Narnhofer [27,28,29].

Acknowledgments

This work was partially supported by the Polish Ministry of Science and Higher Education Grant No 3004/B/H03/2007/33.

References

1. M. A. Nielsen and I. L. Chuang, *Quantum computation and quantum information*, Cambridge University Press, Cambridge, 2000.
2. R. Horodecki, P. Horodecki, M. Horodecki and K. Horodecki, *Quantum entanglement*, arXiv: quant-ph/0702225.
3. A. Peres, Phys. Rev. Lett. **77**, 1413 (1996).
4. P. Horodecki, Phys. Lett. A **232**, 333 (1997).
5. M. Horodecki, P. Horodecki and R. Horodecki, Phys. Lett. A **223**, 1 (1996);
6. S. L. Woronowicz, Rep. Math. Phys. **10**, 165 (1976).
7. P. Horodecki, M. Horodecki, and R. Horodecki, Phys. Rev. Lett. **82**, 1056 (1999).
8. A. C. Doherty, P.A. Parrilo, and F. M. Spedalieri, Phys. Rev. Lett. **88**, 187904 (2002).
9. D. P. DiVincenzo, P. W. Shor, J. A. Smolin, B. M. Terhal and A. V. Thapliyal, Phys. Rev. A **61**, 062312 (2000).
10. P. Horodecki and M. Lewenstein, Phys. Rev. Lett. **85**, 2657 (2000).
11. D. Bruss and A. Peres, Phys. Rev. A **61**, 030301(R) (2000).
12. S. Yu and N. Liu, Phys. Rev. Lett, **95**, 150504 (2005).
13. S.-M. Fei, X. Li-Jost, B.-Z. Sun, Phys. Lett. A **352**, 321 (2006).
14. M. Piani and C. Mora, Phys. Rev. A **75**, 02305 (2007).
15. D. Chruściński and A. Kossakowski, Phys. Rev. A **74**, 022308 (2006).
16. L. Clarisse, Phys. Lett. A **359** (2006) 603.
17. K.-C. Ha, Publ. RIMS, Kyoto Univ. **34**, 591 (1998)
18. K. Ha, S.-H Kye, and Y. Park, Phys. Lett. A **313**, 163 (2003).
19. K. Ha, S.-H. Kye, Phys. Lett. A **325** 315 (2004); J. Phys. A: Math. Gen. **38**, 9039 (2005).
20. D. Chruściński and A. Kossakowski, Phys. Rev. A **76**, 032308 (2007).
21. L. Accardi and M. Ohya, Appl. Math. Opt. **39**, 33 (1999).
22. R.F. Werner, Phys. Rev. A **40**, 4277 (1989).
23. M. Horodecki and P. Horodecki, Phys. Rev. A **59**, 4206 (1999).
24. K.G.H. Vollbrecht and R.F. Werner, Phys. Rev. A **64**, 062307 (2001).
25. D. Chruściński and A. Kossakowski, Open Systems and Inf. Dynamics, **14**, 275 (2007).
26. D. Chruściński and A. Kossakowski, J. Phys. A: Math. Theor. **41**, 145301 (2008).

27. B. Baumgartner, B. C. Hiesmayr and H. Narnhofer, Phys. Rev. A **74**, 032327 (2006).
28. B. Baumgartner, B. C. Hiesmayr and H. Narnhofer, J. Phys. A: Math. Theor. **40**, 7919 (2007).
29. B. Baumgartner, B. C. Hiesmayr and H. Narnhofer, Phys. Lett. A **372**, 2190 (2008).

THE COMPOUND FOCK SPACE AND ITS APPLICATION IN BRAIN MODELS

KARL-HEINZ FICHTNER
Friedrich Schiller University Jena,
School of Mathematics and Computer Science, Institute of Applied Mathematics, Ernst-Abbe-Platz 2, 07743 Jena, Germany,
E-mail: fichtner@mathematik.uni-jena.de

WOLFGANG FREUDENBERG
Brandenburg Technical University Cottbus,
Department of Mathematics, PO box 101344, 03013 Cottbus, Germany,
E-Mail: freudenberg@math.tu-cottbus.de

We introduce the compound Fock space as the Fock space over the Fock space. Though this investigation might be of some interest by itself the main purpose to deal with the compound Fock space was to model the space of the memory in a quantum statistical model of the recognition process.

1. Introduction

In 1932 in the pioneering paper [10] the physicist V. A. Fock suggested a way of passing from quantum states of single objects to states of collections of these objects. If the basic objects are from the Hilbert space $\mathcal{H}$ by this procedure - called second quantization - the objects are lifted to a new Hilbert space $\Gamma(\mathcal{H})$, the so-called Fock space over $\mathcal{H}$ representing sets of objects from $\mathcal{H}$. The second quantization $\Gamma(\Gamma(\mathcal{H}))$ of the Hilbert space $\Gamma(\mathcal{H})$ we will call compound Fock space. Since the Fock space $\Gamma(\mathcal{H})$ is a Hilbert space the compound Fock space again is a usual Fock space, namely the Fock space over the Hilbert space $\widetilde{\mathcal{H}} := \Gamma(\mathcal{H})$. However, because of the special structure of $\widetilde{\mathcal{H}}$ one gets a lot of special properties of the compound Fock spaces, and the present paper is devoted to a deeper analysis of this "Fock space over the Fock space".

Though this investigation might be of some interest by itself the main purpose to deal with the compound Fock space was to model the space of the memory in a quantum statistical model of the recognition process.

The papers [2, 3] represents the first attempts to explain the process of recognition in terms of quantum statistics whereby the model is compared with models widely accepted in psychology and neurology (cf. [14, 15]). Now, in a (still incomplete) series of papers (cf. [5, 4, 6]) where the present paper is one part of it we present the mathematical prerequisites for the models.

Starting point will be a metric space G equipped with a finite diffuse measure. The set G represents the space where the process of recognition and processing of the signals takes place, and this is not necessarily the location of the neurons. For the mathematical model it is irrelevant what is the concrete structure of G. To start with a general space G has the advantage that it can be used as a model for very different aspects of the recognition processes in the brain. The elements of the Hilbert space $\mathcal{H} = L^2(G)$ can be interpreted as functions of the excited neurons. The bosonic (symmetric) Fock space $\Gamma(\mathcal{H})$ represents the space of signals. The memory consists of sets of signals. That's why we model the states of the memory by states on the symmetric Fock space over $\Gamma(\mathcal{H})$ which we call the compound Fock space over $\mathcal{H} = L^2(G)$. For an interpretation of these spaces and of the model of the recognition process which is based on a comparison of incoming signals with signals chosen from memory we refer to the above mentioned papers.

2. The Space of Signals

2.1. *The Boson Fock Space*

Let G denote the space where the process of recognition and processing of the signals takes place. For the mathematical model it is irrelevant what is the concrete structure of G. To start with a general space has the advantage that it can be used as a model for very different recognition processes in the brain. So let G be an arbitrary complete separable metric space and $\mathfrak{G}$ its σ-algebra of Borel sets. Further, let μ be a fixed finite diffuse measure on $[G, \mathfrak{G}]$, i.e. $\mu(G) < \infty$ and $\mu(\{x\}) = 0$ for all singletons $x \in G$. Especially, we are concerned with the case where G is a compact subset of $\mathbb{R}^d$ and μ is the $d-$dimensional Lebesgue measure restricted to G.

We denote by δ_x the Dirac measure in the point $x \in G$, i. e.

$$\delta_x(B) = \begin{cases} 1 \text{ if } x \in B \\ \\ 0 \text{ if } x \notin B \end{cases} \qquad (B \in \mathfrak{G}).$$

Observe that a point x may be equivalently described by the Dirac measure δ_x. Analagously, a *symmetric* configuration $(x_1, \ldots, x_n)$ of n points from G is fully characterized by the *counting measure* $\varphi = \delta_{x_1} + \cdots + \delta_{x_n}$.

For each $n \in \mathbb{N}$ ($\mathbb{N}$ is the set of natural numbers) we denote by $M_n(G)$ the set of all symmetric n-particle configurations from G, i. e.

$$M_n(G) = \{\varphi = \delta_{x_1} + \cdots + \delta_{x_n} \;:\; x_1, \ldots, x_n \in G\}.$$

The empty configuration $\mathfrak{o}$ is described by the null meaure on G, i. e. we have $\mathfrak{o}(G) = 0$, and we set

$$M(G) = \bigcup_{n=0}^{\infty} M_n(G) \tag{1}$$

with $M_0(G) = \{\mathfrak{o}\}$. The elements of $M(G)$ can be interpreted as finite (symmetric) configurations in G. We equip $M(G)$ with its canonical σ-algebra $\mathfrak{M}(G)$ — the smallest σ-algebra containing all sets of the form $\{\varphi \in M(G) : \varphi(K) = n\}$, $K \in \mathfrak{G}$, $n \in \mathbb{N}$. Observe that $\varphi(K) = n$ means that the configuration φ has exactly n points in the subset K of the phase space G. On $[M(G), \mathfrak{M}(G)]$ we introduce a measure F_μ by setting

$$F_\mu(Y) := \mathbb{1}_Y(\mathfrak{o}) + \sum_{n\geq 1} \frac{1}{n!} \int_{G^n} \mathbb{1}_Y\Big(\sum_{j=1}^{n} \delta_{x_j}\Big)\mu^n(d[x_1, \ldots, x_n]) \qquad (Y \in \mathfrak{M}(G)). \tag{2}$$

Hereby, $\mathbb{1}_Y$ denotes the indicator function of a set Y, and $\mu^n = \otimes_{i=1}^n \mu$ denotes the n-fold product measure of μ. Observe that F_μ restricted to n-particle configurations is (up to symmetrization) just the product measure μ^n. Further, F_μ is a finite measure. Indeed,

$$F_\mu(M(G)) = 1 + \sum_{n\geq 1} \frac{\mu^n(G^n)}{n!} = \sum_{n\geq 0} \frac{(\mu(G))^n}{n!} = e^{\mu(G)} < \infty.$$

Let us still mention that F_μ is concentrated on the set

$$M(G)_s := \{\varphi \in M(G) : \varphi(\{x\}) \leq 1 \;\text{ for all }\; x \in G\} \tag{3}$$

of so-called simple counting measures (i.e. without multiple points). We denote by $L^2(M(G)) = L^2(M(G), \mathfrak{M}(G), F_\mu)$ the space of square integrable complex-valued functions on G, i. e.

$$L^2(M(G)) = \left\{\Psi : M(G) \longrightarrow \mathbb{C} : \int_{M(G)} |\Psi(\varphi)|^2 F_\mu(d\varphi) < \infty\right\} \tag{4}$$

equipped with the usual scalar product

$$\langle \Psi, \Phi \rangle := \int_{M(G)} \overline{\Psi(\varphi)} \cdot \Phi(\varphi) F_\mu(d\varphi)$$

where $\overline{z}$ denotes the complex conjugate of $z \in \mathbb{C}$.

Definition 2.1. The space $L^2(M(G))$ is called the Boson (or symmetric) Fock space over G.

Usually one defines the symmetric Fock space $\Gamma(\mathcal{H})$ over a Hilbert space $\mathcal{H}$ as the direct sum of the symmetrized tensor products $\mathcal{H}^{\otimes n}_{\text{symm}}$ of the underlying Hilbert space $\mathcal{H}$, i.e.

$$\Gamma(\mathcal{H}) = \bigoplus_{n=0}^{\infty} \frac{1}{n!} \mathcal{H}^{\otimes n}_{\text{symm}}.$$

In our case the basic Hilbert space $\mathcal{H}$ (the space of a single quantum particle) will be the space $\mathcal{H} = L^2(G) = L^2(G, \mathfrak{G}, \mu)$.

It was shown in [8] that $\Gamma(L^2(G))$ and $L^2(M(G))$ are isomorphic. So in the sequel we will identify both spaces. The above formula shows that the Fock space $\Gamma(L^2(G))$ is again a Hilbert space. For details we refer to [8] or to [12] where a similar definition is given. Dealing with Fock spaces over subspaces B of G one just has to replace in the definitions above $[G, \mathfrak{G}, \mu]$ by $[B, \mathfrak{B}, \mu_{|B}]$ where $\mu_{|B}$ denotes the restriction of μ to $\mathfrak{B} = \mathfrak{G} \cap B$.

2.2. *Exponential Vectors – Coherent States*

The Boson Fock space $L^2(M(G))$ describes random finite *systems of particles* localized in G. Pure states are given by wave functions $\Psi \in L^2(M(G))$ with $\|\Psi\| = 1$. More precisely, a state is a linear functional of the form $< \Psi, \cdot \Psi >$ on the space $\mathcal{L}(L^2(M(G)))$ of linear bounded operators on the Fock space $L^2(M(G))$. An especially important class of functions in $L^2(M(G))$ are the so-called *exponential vectors*. The notion of a signal will be defined with the aid of such exponential vectors. Roughly speaking, exponential vectors describe states of systems of quantum particles where each particle is in the same one-particle state.

For a measurable function $g : G \longrightarrow \mathbb{C}$ we define $\exp\{g\} : M(G) \longrightarrow \mathbb{C}$ by setting

$$\exp\{g\}(\varphi) = \begin{cases} \prod_{x \in \varphi} g(x) \text{ if } \varphi \neq \mathfrak{o},\ \varphi \in M(G), \\ 1 \qquad\qquad \text{if } \varphi = \mathfrak{o}. \end{cases} \tag{5}$$

The function $\exp\{g\}$ is called *exponential vector* corresponding to g. Hereby, $\mathfrak{o}$ again denotes the empty configuration ($\mathfrak{o}(G) = 0$), and $x \in \varphi$ stands for $\varphi(\{x\}) > 0$. Observe that $= e^{-\|g\|^2/2} \cdot \exp\{g\}$ is the normalization of $\exp\{g\}$. So $e^{-\|g\|^2/2} \cdot \exp\{g\}$ defines a pure state of a quantum system.

Definition 2.2. For $g \in L^2(G)$ we denote by ω_g the pure state given by the wave function $e^{-\|g\|^2/2} \cdot \exp\{g\}$, i. e. for $A \in \mathcal{L}(L^2(M(G)))$

$$\omega_g(A) = e^{-\|g\|^2} \cdot \langle \exp\{g\},\, A \exp\{g\}\rangle.$$

The state ω_g is called the *coherent state* corresponding to g. The state ω_0 is called the *vacuum state* in $L^2(M(G))$

A coherent state ω_g describes a system of identical independent particles being all in the *"one-particle state"* $f = g/\|g\|$.

2.3. *The Space of Elementary Signals*

In application to brain models we have to consider finite disjoint partitions $G_1, \ldots, G_n$ of G into areas being responsible for different tasks, and we have to consider the tensor product space $L^2(M(G_1)) \otimes \ldots \otimes L^2(M(G_n))$. However, in the present paper we restrict ourselves to the case $n = 1$. We interpret the Boson Fock space $L^2(M(G))$ as the *space of signals.* A pure state of such a signal is given by a wave function in this space, i. e. by a function $\Psi \in L^2(M(G))$ with $\|\Psi\| = 1$. So *Signals* are states on $L^2(M(G))$. The *vacuum state* ω_0 represents the *empty signal.* The Fock space $L^2(M(G))$ is an infinite dimensional Hilbert space. It seems to be natural to restrict all considerations to finite dimensional subspaces.

Observe that for arbitrary $f,\ g \in L^2(G_r)$ we have $\langle \exp\{f\},\ \exp\{g\}\rangle = e^{<f,g>} \neq 0$. Consequently, there do not exist exponential vectors being orthogonal. Now, let us be given a finite orthonormal system $f_1, \ldots, f_N$, $N \geq 1$ of functions from $L^2(G)$. For $i \neq j$ it holds $\langle \exp\{f_i\},\ \exp\{f_j\}\rangle = 1$. To obtain orthogonal elements in $L^2(M(G))$ we have to restrict the exponential vectors to non-void configurations, i. e. we have to take out the vacuum part from $\exp\{f_j\}$. Given the above fixed orthonormal system we set

$$|j\,\rangle := \frac{\exp\{f_j\} - \exp\{0\}}{\sqrt{e-1}} \quad (j \in \{1, \ldots, N\}) \tag{6}$$

$$|0\rangle := \exp\{0\}. \tag{7}$$

The sequence $(|j\rangle)_{j=0}^{N}$ is an *orthonormal system* of functions from $L^2(M(G))$. The elements $|j\rangle$are interpreted as *elementary signals*, the (finite-dimensional) Hilbert space $\mathcal{H}_{\text{sig}} \subset L^2(M(G))$with orthonormal basis $(|j\rangle)_{j=0}^{N}$is called the *signal space.*

3. The Memory

3.1. *The Compound Fock space*

In the above section we passed from the *one-particle space* $L^2(G)$to the *configuration space* $\Gamma(L^2(G))$which occurred to be again a L^2-space, namely $L^2(M(G))$. We passed from single points $x \in G$to the set $M(G)$of finite configurations $\varphi = \delta_{x_1} + \ldots + \delta_{x_n}$of points $x_1, \ldots, x_n$from G. Signals were defined to be elements from $L^2(M(G))$, i. e. functions on the configuration space. The information stored in the memory is more complex. We have to consider sets of configurations. First we replace single points $x$from $G$by configurations $\varphi$from $\tilde{M}(G) := M(G) \setminus \{\mathfrak{o}\}$. We do not allow $\varphi$to be the empty configuration since it does not make sense to store empty signals in the memory (though the memory itself or parts of it may be empty) . These configurations φwill be our starting point and, for each $n \in \mathbb{N}$we pass from the single configurations $\varphi_1, \ldots, \varphi_n$from $\tilde{M}(G)$to the set

We will define now step by step in analogy to the considerations in Section 2.1 this basic space for states in the memory. We equip $\tilde{M}(G)$with its canonical σ-algebra $\tilde{\mathfrak{M}}(G) = \mathfrak{M}(G) \cap \tilde{M}(G)$where $\mathfrak{M}(G)$was defined in Section 2.1. The restriction of the measure F_μ(defined by (2)) from $M(G)$to $\tilde{M}(G)$we will denote again by F_μ.

$$F_\mu(Y) = \sum_{n\geq 1} \frac{1}{n!} \int_{G^n} \mathbb{I}_Y(\sum_{j=1}^{n} \delta_{x_j})\mu^n(d[x_1, \ldots, x_n]) \qquad (Y \in \tilde{\mathfrak{M}}(G)). \tag{8}$$

Definition 3.1. The Hilbert space given by

$$L^2(\tilde{M}(G), F_\mu) := \left\{ \Psi : \tilde{M}(G) \longrightarrow \mathbb{C} : \int_{\tilde{M}(G)} |\Psi(\varphi)|^2 F_\mu(d\varphi) < \infty \right\} \tag{9}$$

we call the *truncated* bosonic Fock space over G.

The space $\widetilde{\mathcal{H}}_{\text{sig}} := \mathcal{H}_{\text{sig}} \cap L^2(\tilde{M}(G_r), F_\mu)$ represents the corresponding spaces of non-void regular signals. The only difference to Definition 4 is the replacement of $M(G)$ by $\tilde{M}(G) := M(G) \setminus \{\mathfrak{o}\}$. The vacuum vector $\exp\{0\}$ restricted to $\tilde{M}(G)$ is identically zero, and exponential vectors

$\exp\{f\}$, $\exp\{g\}$ are orthogonal in the truncated Fock space $L^2(\tilde{M}(G), F_\mu)$ if and only if f and g are orthogonal in $L^2(G)$.

We equip $\mathcal{M}(G)$ defined by (??) with its canonical σ-algebra $\mathfrak{M}(\mathcal{M})$ – the smallest σ-algebra containing all sets of the form $\{\Phi \in \mathcal{M}(G) : \Phi(K) = n\}$, $K \in \tilde{\mathfrak{M}}(G)$, $n \in \mathbb{N}$. Hereby, $\Phi(K) = n$ means that the configuration system Φ has exactly n configurations from the subset K of the phase space $\tilde{M}(G)$.

On $[\mathcal{M}(G), \mathfrak{M}(\mathcal{M})]$ we introduce a measure $\mathbb{F}$ by setting for $\mathbf{Y} \in \mathfrak{M}(\mathcal{M})$

$$\mathbb{F}(\mathbf{Y}) := \mathbb{I}_{\mathbf{Y}}(\mathfrak{o}) + \sum_{n \geq 1} \frac{1}{n!} \int_{\tilde{M}(G)^n} \mathbb{I}_{\mathbf{Y}}(\sum_{i=1}^{n} \delta_{\varphi_i}) F_\mu^n([\varphi_1, \ldots, \varphi_n]). \tag{10}$$

Hereby, $\mathbb{I}_{\mathbf{Y}}$ denotes again the indicator function of a set $\mathbf{Y}$, and $F_\mu^n = \otimes_{i=1}^n F_\mu$ denotes the n-fold product measure of F_μ. Observe that $\mathbb{F}$ restricted to n-particle configurations is (up to symmetrization) just the product measure F_μ^n. Using the notation from Section 2.1 we could write $\mathbb{F} = \mathbb{F}_{F_\mu}$. Further, $\mathbb{F}$ is a finite measure. Indeed, we already observed that F_μ is a finite measure since μ was assumed to be finite. We get

$$\begin{aligned}\mathbb{F}(\mathcal{M}(\mathbb{G})) &= 1 + \sum_{n \geq 1} \frac{F_\mu^n\left((\tilde{M}(G))^n\right)}{n!} = \sum_{n \geq 0} \frac{\left(F_\mu(\tilde{M}(G))\right)^n}{n!} \\ &= \exp\{F_\mu(\tilde{M}(G)\} < \infty.\end{aligned}$$

Let us still mention that $\mathbb{F}$ is concentrated on the set

$$\mathcal{M}(G)_s := \left\{\Phi \in \mathcal{M}(G) : \Phi(\{\varphi\}) \leq 1 \text{ for all } \varphi \in \tilde{M}(G)\right\} \tag{11}$$

of simple configuration systems (i.e. without multiple configurations). We denote by $L^2(\mathcal{M}(G)) := L^2(\mathcal{M}(G), \mathfrak{M}(\mathcal{M}), \mathbb{F})$ the space of square integrable complex-valued functions on $\mathcal{M}(G)$. The scalar product in $L^2(\mathcal{M}(G))$ is given by

$$\langle W, V \rangle := \int_{\mathcal{M}(G)} \overline{W(\Phi)} \cdot V(\Phi) \mathbb{F}(d\Phi)$$

where again $\overline{z}$ denotes the complex conjugate of z.

Definition 3.2. The space $L^2(\mathcal{M}(G))$ we will call the *compound (symmetric) Fock space over* G or also the (symmetric) Fock space over the (symmetric) Fock space over G.

As remarked already in Section 2.1 usually one defines the symmetric Fock space $\Gamma(\mathcal{H})$ over a Hilbert space $\mathcal{H}$ as the direct sum of the symmetrized tensor products $\mathcal{H}^{\otimes n}_{\text{symm}}$ of the underlying Hilbert space $\mathcal{H}$. In our case the basic Hilbert space $\mathcal{H}$ (the space of a single configuration of particles) will be the space $L^2(\tilde{M}(G), F_\mu)$ defined by (9). It was shown in [8] that for arbitrary complete separable metric spaces G (equipped with a locally finite measure μ) the spaces $\Gamma(L^2(G))$ and $L^2(M(G))$ are isomorphic. Replacing the metric space G by $\tilde{M}(G) = M(G) \setminus \{\mathfrak{o}\}$ which can be shown to be again a complete metric separable space (cf. [13]) we see that $\Gamma(L^2(\tilde{M}(G)))$ and $L^2(\mathcal{M}(G)))$ are isomorphic. So we may identify both spaces:

$$\Gamma(L^2(\tilde{M}(G)) = L^2(\mathcal{M}(G)). \tag{12}$$

In application to brain models we have to take into account arbitrary finite partitions of G into disjoint areas $G_1, \ldots, G_n$. It is easy to see that one can make the identification

$$L^2(M(G)) = L^2(M(G_1)) \otimes \ldots \otimes L^2(M(G_n)). \tag{13}$$

However, the identification (13) is not true for the truncated spaces $\tilde{M}(G_r) = M(G_r) \setminus \{\mathfrak{o}\}$:

$$L^2(\mathcal{M}(G)) \neq L^2(\mathcal{M}(G_1)) \otimes \ldots \otimes L^2(\mathcal{M}(G_n)). \tag{14}$$

While the domains of functions from $L^2(\mathcal{M}(G))$ consist of systems of configurations being non-void in G functions from $L^2(\mathcal{M}(G_1)) \otimes \ldots \otimes L^2(\mathcal{M}(G_n))$ are restricted to domains consisting of systems of configurations being non-void in each region G_r.

A very important property of the symmetric Fock space is that multiple integrals can be reduced to single ones.

Lemma 3.1. . *Let* $f : \tilde{M}(G) \times \mathcal{M}(G) \longrightarrow \mathbb{C}$ *be integrable with respect to* $F_\mu \otimes \mathbb{F}$. *Then*

$$\int\limits_{\tilde{M}(G)\times\mathcal{M}(G)} F_\mu \otimes \mathbb{F}([\varphi, \lneqq]) \mho(\varphi, \gneqq) = \int\limits_{\mathcal{M}(\mathbb{G})} \mathbb{F}(\gneqq) \sum_{\varphi \in \lneqq} \mho(\varphi, \lneqq - \delta_\varphi). \tag{15}$$

Hereby, $\varphi \in \Phi$ *means that* $\Phi(\{\varphi\}) > 0$, *and we set* $\sum_{\varphi \in \mathfrak{D}} = 0$.

The above lemma was shown for general Fock spaces e.g. in [7, Lemma 3.2], in [9, Lemma 12.1] and also in [12]. The above lemma is called $\not{\sum}$-lemma. A function W from $L^2(\mathcal{M}(G))$ is thus a complex-valued function defined on the space of finite collections Φ of configurations (being themselves finite point systems from G). A normalized function W from $L^2(\mathcal{M}(G))$ represents the (pure) state of a system of signals in the memory. By

$$P_W(\mathbf{Y}) := \int\limits_{\mathbf{Y}} |W(\Phi)|^2 \, \mathbb{F}(\lneqq) \qquad (\mathbf{Y} \in \mathfrak{M}(\mathcal{M}))$$

there is defined a probability measure on $[\mathcal{M}(G), \mathfrak{M}(\mathcal{M})]$ that can be interpreted as the *position distribution* of the state W ($\|W\| = 1$). It is the distribution of a random system of point configurations.

In Section 2.2 we introduced for functions $g \in L^2(G)$ the exponential vectors $\exp\{g\}$ in the Fock space $L^2(M(G))$. Analogously, we define them in the compound Fock space $L^2(\mathcal{M}(G))$. We start with some (not necessarily normalized) function $R \in L^2(\tilde{M}(G))$ and define a function $\mathrm{Exp}\{R\} \in L^2(\mathcal{M}(G))$ by setting

$$\mathrm{Exp}\{R\}(\Phi) := \begin{cases} \prod\limits_{\varphi \in \Phi} R(\varphi) \text{ if } \mathfrak{O} \neq \Phi \in \mathcal{M}(G), \\ \\ 1 \qquad\qquad \text{if } \Phi = \mathfrak{O}. \end{cases} \tag{16}$$

This means that for $\Phi = \varphi_1 + \ldots + \varphi_n$ one obtains

$$\mathrm{Exp}\{R\}(\Phi) = R(\varphi_1) \cdot \ldots \cdot R(\varphi_n).$$

Without proof we will state some well-known properties of exponential vectors:

Lemma 3.2. *Let R and S be functions from $L^2(\tilde{M}(G))$ and Φ, Φ_1, Φ_2 be elements from $\mathcal{M}(G)$. Then we have*

$$\exp\{R\}(\Phi_1 + \Phi_2) = \exp\{R\}(\Phi_1) \cdot \exp\{R\}(\Phi_2), \tag{17}$$

$$\exp\{R + S\}(\Phi) = \sum_{\hat{\Phi} \subseteq \Phi} \exp\{R\}(\hat{\Phi}) \cdot \exp\{S\}(\Phi - \hat{\Phi}), \tag{18}$$

$$\exp\{R \cdot S\}(\Phi) = \exp\{R\}(\Phi) \cdot \exp\{S\}(\Phi), \tag{19}$$

$$||\exp\{R\}||^2_{L^2(\mathcal{M}(G))} = e^{||g||^2_{L^2(M(G))}} \qquad (R \in L^2(M(G))), \tag{20}$$

$$\langle \exp\{R\}, \exp\{S\} \rangle_{L^2(M(G))} = e^{\langle R,\ S \rangle_{L^2(M(G))}} \quad (R,\ S \in L^2(M(G))). \tag{21}$$

The symbol $\hat{\Phi} \subseteq \Phi$ means that $\hat{\Phi}$ is a subconfiguration of Φ, i. e. $\Phi - \hat{\Phi} \in \mathcal{M}(G)$. Thus the summation in (18) runs over all splittings of the System Φ of configurations into two parts $\hat{\Phi}$ and $\Phi - \hat{\Phi}$. Observe that $\exp\{R\} \in L^2(\mathcal{M}(G))$ if and only if $R \in L^2(M(G))$. Moreover, exponential vectors are total in the compound Fock space $L^2(\mathcal{M}(G))$. Thus each function from $L^2(\mathcal{M}(G))$ can be approximated by linear combinations of exponential vectors, i. e. by functions of the type

$$\alpha_1 \exp\{R_1\} + \ldots + \alpha_n \exp\{R_n\}$$

with $R_1, \ldots, R_n \in L^2(M(G))$, $\alpha_1, \ldots, \alpha_n \in \mathbb{C}$, $n \in \mathbb{N}$.

3.2. *States of the Memory*

The compound Fock space introduced above will serve as the basic space for states in the memory. Each state on the Hilbert space $L^2(\mathcal{M}(G))$ one could call a state of the memory. However, $L^2(\mathcal{M}(G))$ is an infinite dimensional space, and as in the case of the space of signals we will restrict our considerations to a finite dimensional subspace connected with the defined space of signals. In Section 2.3 we considered the (finite dimensional) subspace $\mathcal{H}_{\text{sig}} \subset L^2(M(G))$ of signals. If $R_1,\ R_2 \in \mathcal{H}_{\text{sig}}$ are orthogonal then $\text{Exp}\{R_1\} - \text{Exp}\{0\}$ and $\text{Exp}\{R_2\} - \text{Exp}\{0\}$ again will be orthogonal. Observe that the elementary signals $|j\,\rangle$ and except $\exp\{0\} = |0\,\rangle$all basis elements of $\mathcal{H}_{\text{sig}} \subseteq L^2(M(G))$even belong to the smaller space $L^2(\tilde{M}(G))$.

Definition 3.3. The space of memory is the Hilbert space given by

$$\mathcal{H}_{\text{mem}} := \text{Lin}\left\{\text{Exp}\{R\} - \text{Exp}\{0\} :\ R \in \mathcal{H}_{\text{sig}} \cap L^2(\tilde{M}(G))\right\}. \tag{22}$$

where $\text{Lin}(A)$ denotes (the closure of) the linear span of the elements of A. A state on $\mathcal{H}_{\text{mem}}$ we will call a state of the memory.

A state of the memory can be considered as a state on $L^2(\mathcal{M}(G))$ concentrated on $\mathcal{H}_{\text{mem}} = \Gamma(\mathcal{H}_{\text{sig}} \cap L^2(\tilde{M}(G)))$.

Example 3.1. Let $|j, r\rangle$ be given by (6), and set

$$R := \frac{1}{\sqrt{N}} \sum_{j=1}^{N} |j\,\rangle\,. \tag{23}$$

Then the pure state given by the normalisation of $\text{Exp}\{R\} - \text{Exp}\{0\}$ could be interpreted as the state of the memory with no specific concrete knowledge but being able to "learn everything".

3.3. *Some Basic Operators*

The recognition process is based on a comparison of signals: one signal will be the input signal obtained by accumulation of elementary signals, the other one is taken arbitrarily from the memory. If both signals coincide (completely or partially) there will be a *collapse* indicating the recognition of the signal. A detailed discussion and interpretation of this mechanism one can find in [2, 3, 5, 4, 6]. In the present paper we describe only one operator being important in this process of selecting a signal from the memory.

Definition 3.4. Let $\mathcal{D} : L^2(\mathcal{M}(G)) \longrightarrow L^2(\tilde{M}(G) \times \mathcal{M}(G))$ be defined for all $W \in L^2(\mathcal{M}(G))$ by

$$\mathcal{D}W(\varphi, \Phi) := \frac{1}{\sqrt{|\Phi| + 1}} \cdot W(\Phi + \delta_\varphi) \qquad (\varphi \in \tilde{M}(G),\ \Phi \in \mathcal{M}(G)). \quad (24)$$

In the above definition we have put (as in the notation $|\varphi|$ describing the number of points from G in the configuration φ)

$$|\Phi| := \Phi(\tilde{M}(G)) \qquad (\Phi \in \mathcal{M}(G)), \quad (25)$$

i. e. $|\Phi|$ denotes the number of point configurations in the system Φ.

Remark 3.1. The difference between the operator $\mathcal{D}$ and the Malliavin derivative on a general Fock space (which is an unbounded operator) is the normalizing factor $(|\Phi| + 1)^{-1/2}$.

Proposition 3.1. *The operator $\mathcal{D}$ defined above is an isometry. Its adjoint $\mathcal{D}^* : L^2(\tilde{M}(G) \times \mathcal{M}(G)) \longrightarrow L^2(\mathcal{M}(G))$ is given for $Z \in L^2(\tilde{M}(G) \times \mathcal{M}(G))$ by*

$$\mathcal{D}^* Z(\Phi) = \frac{1}{\sqrt{|\Phi|}} \cdot \sum_{\varphi \in \Phi} Z(\varphi, \Phi - \delta_\varphi) \qquad (\Phi \in \mathcal{M}(G) \setminus \mathfrak{O}), \quad (26)$$

$$\mathcal{D}^* Z(\mathfrak{O}) = 0.$$

We omit the proof of Lemma 3.1

Now, let $W \in \mathcal{H}_{\text{mem}}$ be a (pure) state of the memory. Then because of Proposition 3.1 the image $\mathcal{D}W$ describes a (pure) state on $L^2(\tilde{M}(G) \times \mathcal{M}(G)) = L^2(\tilde{M}(G)) \otimes L^2(\mathcal{M}(G))$. But even more is true. If $W \in \mathcal{H}_{\text{mem}}$ then $\mathcal{D}W \in \mathcal{H}_{\text{sig}} \otimes \mathcal{H}_{\text{mem}}$. This selection of a signal from the memory will be the basis for the comparison procedure between this signal produced by our senses and the incoming signal.

4. Processing – Recognition of Signals

Now, let ρ_1 be a state on $\mathcal{H}_{\text{sig}} \subseteq L^2(M(G))$. The state ρ_1 describes an input signal obtained by an accumulation of elementary signals. Further, let ρ_2 be a state of the memory, i. e. ρ_2 is a state on $\mathcal{H}_{\text{mem}} \subseteq L^2(\mathcal{M}(G))$. Using the operator $\mathcal{D}$ arbitrarily a signal is separated from the memory. This yields to a state $\rho_1 \otimes \mathcal{D}\rho_2\mathcal{D}^*$ on $\mathcal{H} := \mathcal{H}_{\text{sig}} \otimes \mathcal{H}_{\text{sig}} \otimes \mathcal{H}_{\text{mem}} \subseteq L^2(M(G)) \otimes L^2(M(G)) \otimes L^2(\mathcal{M}(G))$. States on $\mathcal{H}$ will be the starting point for the processing mechanism. Choosing one of these signals with the aid of the operator $\mathcal{D}$ may be interpreted as generating a hypothesis concerning an "expected view of the world". If there is another signal coming from our senses then the brain compares this signal with the signal chosen from the memory. That procedure done with a splitting operator on $\mathcal{H}_{\text{sig}} \otimes \mathcal{H}_{\text{sig}}$ changes the state of both signals. If both signals (the input and the signal created from memory) coincide (completely or partially) there will be a *collapse* indicating the recognition of the signal. Finally, the changed signal will be transformed back to the memory by the operator $\mathcal{D}^*$. For more details we refer to the papers mentioned above. This procedure can be described schematically in the following way (I denotes the identical operator):

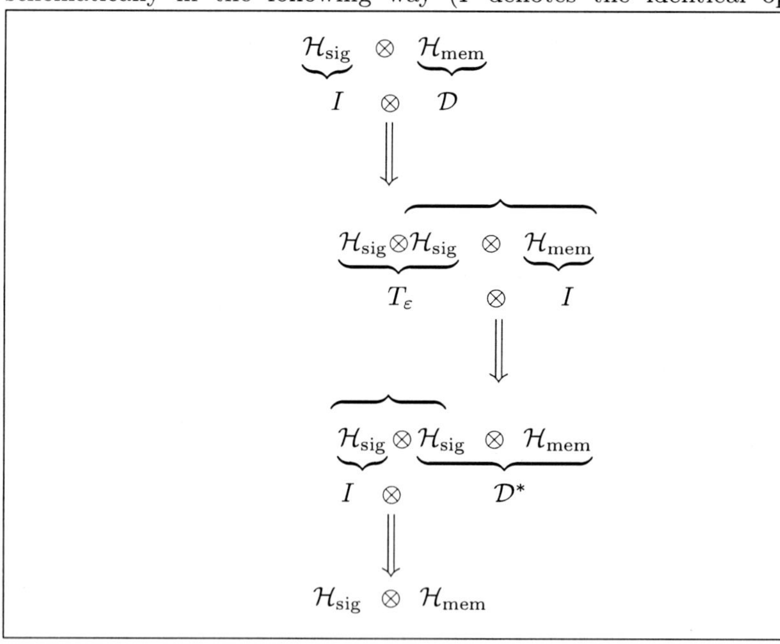

References

1. K.-H. Fichtner. On the position distribution of the ideal Bose gas. *Math. Nachr.*, 151:59 – 67, 1991.
2. K.-H. Fichtner and L. Fichtner. Bosons and a quantum model of the brain. Jenaer Schriften zur Mathematik und Informatik Math/Inf/08/05, FSU Jena, Faculty of Mathematics and Informatics, 07737 Jena, Germany, sep 2005.
3. K.-H. Fichtner and L. Fichtner. Quantum markov chains and the process of recognition. Jenaer Schriften zur Mathematik und Informatik Math/Inf/02/07, FSU Jena, Faculty of Mathematics and Informatics, 07737 Jena, Germany, 2007.
4. K.-H. Fichtner, L. Fichtner, W. Freudenberg, and M. Ohya. On a quantum model of the recognition process. In L. Accardi, W. Freudenberg, and M.Ohya, editors, *Quantum Bio-Informatics*, volume XXI of *QP–PQ: Quantum Probability and White Noise Analysis*, pages 64 – 84, New Jersey London Singapore. World Scientifc.
5. K.-H. Fichtner, L. Fichtner, W. Freudenberg, and M. Ohya. On a mathematical model of brain activities. In *Quantum Theory, Reconsideration of Foundations - 4*, volume 962 of *AIP Conference Proceedings*, pages 85 – 90, Melville, New York, 2007. American Institute of Physics.
6. K.-H. Fichtner, L. Fichtner, W. Freudenberg, and M. Ohya. Quantum models of the recognition process - a convergence theorem. Technical report, 2007. 32 pages.
7. K.-H. Fichtner and W. Freudenberg. Point processes and normal states of boson systems. *Naturwiss.-Technisches Zentrum NTZ*, Leipzig, 1986. 56 p.
8. K.-H. Fichtner and W. Freudenberg. Point processes and the position distribution of infinite boson systems. *J. Stat. Phys.*, 47:959—978, 1987.
9. K.-H. Fichtner and W. Freudenberg. Characterization of states of infinite boson systems I. On the construction of states of boson systems. *Commun. Math. Phys.*, 137:315—357, 1991.
10. V. A. Fock. Konfigurationsraum und zweite Quantelung. *Zeitschrift für Physik A*, Bd. 75, N 9-10:622–647, 1932 (Erratum: Zeitschrift für Physik Phys. 1932, Bd. 76, N 11-12, S. 852).
11. A. Liemant, K. Matthes, and A. Wakolbinger. *Equilibrium Distributions of Branching Processes.* Akademie-Verlag, Berlin, 1988.
12. H. Maassen. Quantum Markov processes on Fock space described by integral kernels. In L. Accardi and W. von Waldenfels, editors, *Quantum Probability and Applications II*, volume 1136 of *Lecture Notes in Mathematics*, pages 361 – 374, Berlin, Heidelberg, New York, 1985. Springer.
13. K. Matthes, J. Kerstan, and J. Mecke. *Infinitely Divisible Point Processes.* Wiley, Chichester, 1978.
14. Wolf Singer. Consciousness and the structure of neuronal representations. *Phil. Trans. R. Soc. Lond.*, B 353:1829–1840, 1998.
15. Wolf Singer. *Der Beobachter im Gehirn. Essays zur Hirnforschung.* Suhrkamp Verlag, Frankfurt a.M., 2002.

CHARACTERISATION OF BEAM SPLITTERS

LARS FICHTNER
Friedrich Schiller University Jena,
School of Social and Behavioral Sciences,
nstitute of Psychology, Am Steiger 3,
07743 Jena, Germany,
E-mail: lars.fichtner@uni-jena.de

MARKUS GÄBLER
Brandenburg Technical University Cottbus,
Department of Mathematics, PO box 101344,
03013 Cottbus, Germany,
E-mail: gaebler@math.tu-cottbus.de

For an arbitrary finite measurable decomposition of a finite measure space G into disjoint $G_1, \ldots, G_n$ the corresponding Fock space $\Gamma\big(L^2(G)\big)$ factorises as $\Gamma\big(L^2(G)\big) = \bigotimes_{k=1}^{n} \Gamma\big(L^2(G_k)\big)$. We study unitary operators V on $\Gamma\big(L^2(G)\big) \otimes \Gamma\big(L^2(G)\big)$ that also factorise into $V = \bigotimes_{k=1}^{n} V_{G_k}$, where V_{G_k} are unitary, vacuum-preserving operators on $\Gamma\big(L^2(G_k)\big) \otimes \Gamma\big(L^2(G_k)\big)$. It is shown, that the only operators having this property are the unitary beam splitters well-known in quantum optics. The result is related to a quantum model of recognition and brain activities introduced in [6].

1. Introduction and Main Result

Consider the bosonic (symmetric) Fock space $\Gamma\big(L^2(G,\mu)\big)$, where G is assumed to be a complete separable metric space equipped with a finite, diffuse (non-atomic) measure μ, i.e. $\mu(G) < \infty$ and $\mu(\{x\}) = 0$ for all singletons $\{x\} \subseteq G$. For example, take G to be a bounded subset of R^d and μ Lebesque measure.

For an arbitrary finite measurable decomposition of G into disjoint

$G_1, \ldots, G_n$ the corresponding Fock space factorises as

$$\Gamma\big(L^2(G)\big) = \bigotimes_{k=1}^{n} \Gamma\big(L^2(G_k)\big). \tag{1.1}$$

Using (1.1) we find the following identification

$$\Gamma\big(L^2(G)\big) \otimes \Gamma\big(L^2(G)\big) = \bigotimes_{k=1}^{n} \Big(\Gamma\big(L^2(G_k)\big) \otimes \Gamma\big(L^2(G_k)\big)\Big). \tag{1.2}$$

We will study unitary operators V that also factorise in the like manner, i.e. for any finite measurable decomposition $G_1, \ldots, G_n$ of G there exist unitary operators V_{G_k} on $\Gamma\big(L^2(G_k)\big) \otimes \Gamma\big(L^2(G_k)\big)$ such that in the sense of identification (1.2) we have

$$V = \bigotimes_{k=1}^{n} V_{G_k}. \tag{1.3}$$

There is a particular family of unitary operators known as beam splitters in quantum optics having property (1.3). In recalling their definition we make use of the fact that the set of exponential vectors or coherent functions $\{\exp(f) : f \in L^2(G)\}$ is linearly independent and its linear span is dense in $\Gamma\big(L^2(G)\big)$. Hence bounded operators on $\Gamma\big(L^2(G)\big) \otimes \Gamma\big(L^2(G)\big)$ may be well-defined by their action on $\{\exp(f) \otimes \exp(g) : f, g \in L^2(G)\}$.

Consider measurable functions $a_1, a_2, b_1, b_2 : G \to C$ with the property that

$$|a_1|^2 + |a_2|^2 \equiv 1 \equiv |b_1|^2 + |b_2|^2 \quad \text{and} \quad a_1\overline{b_1} + a_2\overline{b_2} \equiv 0, \tag{1.4}$$

where $\overline{z}$ denotes the complex conjugate of $z \in C$.

Definition 1. A unitary operator V on $\Gamma\big(L^2(G)\big) \otimes \Gamma\big(L^2(G)\big)$ is called beam splitter, if there exist functions $a_1, a_2, b_1, b_2 : G \to C$ satisfying (1.4) and

$$V \exp(f) \otimes \exp(g) = \exp(a_1 f + a_2 g) \otimes \exp(b_1 f + b_2 g) \qquad \big(f, g \in L^2(G)\big). \tag{1.5}$$

The purpose of this article is to show that beam splitters are fully characterised by property (1.3) and vacuum preservation, i.e.

$$V_{G_k} \exp\big(0_{|G_k}\big) \otimes \exp\big(0_{|G_k}\big) = \exp\big(0_{|G_k}\big) \otimes \exp\big(0_{|G_k}\big) \qquad (k = 1, \ldots, n), \tag{1.6}$$

where $f_{|A}$ denotes the restriction of $f \in L^2(G)$ to measurable $A \subseteq G$.

We have:

Theorem 1. *Let G be a complete separable metric space equipped with its Borel σ-algebra $\mathfrak{G}$ and a finite diffuse measure μ. V a unitary operator on $\Gamma(L^2(G)) \otimes \Gamma(L^2(G))$. Then V is a beam splitter according to Definition 1 if and only if for an arbitrary finite measurable decomposition $G_1, \ldots, G_n$ of G there exist unitary operators V_{G_k} on $\Gamma(L^2(G_k)) \otimes \Gamma(L^2(G_k)), k = 1, \ldots, n$, with the following two properties*

(V0) $V_{G_k} \exp(0_{|G_k}) \otimes \exp(0_{|G_k}) = \exp(0_{|G_k}) \otimes \exp(0_{|G_k}) \qquad (k = 1, \ldots, n)$

(V1) $V = \bigotimes_{k=1}^{n} V_{G_k}$. ■

That (V0) and (V1) are necessary is immediate from Definition 1 by setting

$$V_A \exp(f) \otimes \exp(g) := \exp(a_{1|A} f + a_{2|A} g) \otimes \exp(b_{1|A} f + b_{2|A} g) \qquad (f, g \in L^2(A), A \in \mathfrak{G}). \quad (1.7)$$

This paper mainly deals with proving that they are also sufficient.

Having the modelling of brain activities as a particular application for Theorem 1 in mind, we also present an equivalent condition for (V0). Denoting with I_A the natural embedding of $\Gamma(L^2(A)) \otimes \Gamma(L^2(A))$ into $\Gamma(L^2(G)) \otimes \Gamma(L^2(G))$, given by

$$I_A \psi := \psi \otimes \Big(\exp(0_{|A^c}) \otimes \exp(0_{|A^c})\Big) \qquad \Big(\psi \in \Gamma(L^2(A)) \otimes \Gamma(L^2(A)), A \in \mathfrak{G}\Big), \quad (1.8)$$

we have

Corollary 1. *V is a beam splitter if and only if (V1) holds for an arbitrary finite measurable decomposition $G_1, \ldots, G_n$ of G, where*

$$V_A := I_A^{-1} V I_A \qquad (A \in \mathfrak{G}) \quad (1.9)$$

is a well-defined operator on $\Gamma(L^2(A)) \otimes \Gamma(L^2(A))$, independent of the particular decomposition of G to which A might belong. ■

Remark 1. Studying operators of the above kind was motivated by a quantum model of recognition described in [2, 3, 1, 4, 6, 8]. In this model, a comparison of two signals, one arising from the senses and the other generated from the memory, is modelled by a unitary operator V on $\Gamma(L^2(G)) \otimes \Gamma(L^2(G))$, G representing the physical space where recognition takes place, i.e. the brain. For example, take G to be a bounded subset of R^3 and μ Lebesque measure.

Neuroscience suggests that the operator V factorises according to (V1), which corresponds to a parallel processing of partial signals belonging to different regions of the brain. Hereby, requiring (V1) to hold for any finite measurable decomposition $G_1, \ldots, G_n$ of G, and not just a fixed one, accounts for the fact that decomposition of G into different regions not only depends on the individual but also may change over time. Furthermore, processing in the different regions is governed by the same principles, "a unified algorithm". Mathematically, this leads to (1.9). Corollary 1 therefore shows, that beam splitters are the only candidates for this kind of comparison operator. By reasons beyond the scope of this paper (for example symmetry and homogenity conditions), the operator used in the model is the so-called symmetric beam splitter given by $a_1 \equiv a_2 \equiv b_1 \equiv -b_2 \equiv \frac{\sqrt{2}}{2}$.

An immediate consequence of (V0) is, that V preserves the vacuum, i.e.

$$V \exp(0) \otimes \exp(0) = \exp(0) \otimes \exp(0)\,, \tag{1.10}$$

which has an interesting interpretation: If both memory and senses produce the empty signal, the organism is dead. Hence there is no more change in this state.

Section 2 introduces the bosonic Fock space and its properties, followed by the proofs of Corollary 1 and Theorem 1 in sections 3 and 4, respectively. The proofs of the two main auxillary results used in the proof of Theorem 1, namely Propositions 1 and 4 will be given in a forthcoming paper.

2. The Bosonic Fock Space

For a separable Hilbert space H let

$$\Gamma(H) := C \oplus \bigoplus_{n=1}^{\infty} \frac{1}{\sqrt{n!}} H_{sym}^{\otimes n}, \tag{2.1}$$

where $H_{sym}^{\otimes n}$ denotes the n-fold symmetric tensor product of H [14, 15]. $\Gamma(H)$ is again a separable Hilbert space.

Definition 2. $\Gamma(H)$ is called the Boson (or symmetric) Fock space over H.

2.1. *Representation of* $\mathbf{\Gamma(L^2(G))}$ *as an* $\mathbf{L^2}$*-Space*

We are interested in the case $H := L^2(G)$, where $(G, \mathfrak{G}, \mu)$ is a measure space consisting of the complete separable metric space G, the corresponding σ-algebra of Borel sets $\mathfrak{G}$ and a finite diffuse (non-atomic) measure μ, i.e. $\mu(G) < \infty$ and $\mu(\{x\}) = 0$ for all $x \in G$. In this case, $\Gamma\big(L^2(G)\big)$ itself may be represented as an L^2-Space [10, 9, 12].

Denote by $M(G)$ the set of finite counting measures on $(G, \mathfrak{G})$, i.e.

$$M(G) := \left\{ \varphi : \mathfrak{G} \to N_0 : \varphi = \sum_{k=1}^{n} \delta_{x_k} \text{ for some } n \in N_0 \text{ and } x_k \in G \right\}, \tag{2.2}$$

where δ_x denotes the Dirac measure in x and $\sum_{k=1}^{0} \delta_{x_k} := \mathfrak{o}$ the zero measure, i.e. $\mathfrak{o}(G) = 0$. The elements of $M(G)$ are interpreted as finite symmetric point configurations in G, $\mathfrak{o}$ representing the empty configuration. Define

$$M_n(G) := \{\varphi \in M(G) : \varphi(G) = n\} \qquad (n \in N_0) \tag{2.3}$$

to be the subset of configurations having exactly n points. By giving $M_{\geq n}(G)$, $M_{<n}(G), \ldots$ the obvious meaning we have

$$M(G) = M_{\geq 0}(G) = \bigcup_{n=0}^{\infty} M_n(G). \tag{2.4}$$

We equip $M(G)$ with its canonical σ-algebra $\mathfrak{M}(G)$, that is, the smallest σ-algebra containing all sets of the form $\{\varphi \in M(G) : \varphi(B) = k\}$ for $B \in \mathfrak{G}$ and $k \in N_0$, and the measure

$$F_\mu = \delta_{\mathfrak{o}} + \sum_{n=1}^{\infty} \frac{\mu^n \circ T_n^{-1}}{n!}, \tag{2.5}$$

where $T_n : G^n \to M(G), T_n(x_1, \ldots, x_n) := \sum_{k=1}^{n} \delta_{x_i}$ and μ^n is the n-fold product measure of μ on G^n. Since μ is finite so is F_μ and we have

$$F_\mu\big(M(G)\big) = \sum_{n=0}^{\infty} F_\mu\big(M_n(G)\big) = \sum_{n=0}^{\infty} \frac{\big(\mu(G)\big)^n}{n!} = e^{\mu(G)}.$$

By $\mathcal{M}(G)$ we denote the space of square integrable complex-valued functions on $M(G)$, i. e.

$$\begin{aligned}\mathcal{M}(G) &:= L^2\big(M(G)\big) = L^2\big(M(G), \mathfrak{M}(G), F_\mu\big) \\ &= \left\{\Psi : M(G) \to C : \int_{M(G)} F_\mu(d\varphi)\, |\Psi(\varphi)|^2 < \infty\right\} \qquad (2.6)\end{aligned}$$

with scalar product

$$\langle \Psi, \Phi \rangle := \int_{M(G)} F_\mu(d\varphi)\, \overline{\Psi(\varphi)} \cdot \Phi(\varphi) \qquad \big(\Psi, \Phi \in \mathcal{M}(G)\big), \qquad (2.7)$$

where $\overline{z}$ denotes the complex conjugate of $z \in C$.

It was shown in [9] that $\mathcal{M}(G)$ and $\Gamma\big(L^2(G)\big)$ are isomorphic. Hence we will call $\mathcal{M}(G)$ the Boson (or symmetric) Fock space over $L^2(G)$.

For $f : G \to C$ denote with

$$\exp(f)\,(\varphi) := \begin{cases} 1 & \text{if } \varphi = \mathbf{o} \\ \prod\limits_{k=1}^{n} f(x_k) & \text{if } \varphi = \sum\limits_{k=1}^{n} \delta_{x_k} \end{cases} \qquad \big(\varphi \in M(G)\big) \qquad (2.8)$$

the coherent function generated by f. It is well known that $\exp(f) \in \mathcal{M}(G)$ if and only if $f \in L^2(G)$. In this case $\exp(f)$ is called exponential vector.

Immediately from definition we have the following property:

Lemma 2.1. *The map* $\exp(\cdot) : L^2(G) \to \mathcal{M}(G)$ *is one-to-one.* ■

The importance of exponential vectors mainly rests on the fact that the set $\{\exp(f) : f \in L^2(G)\}$ is linearly independent and total (i.e. its linear span is dense) in $\mathcal{M}(G)$ and therefore may be used to define bounded linear operators on $\mathcal{M}(G)$.

For example let T be an isometry on $L^2(G)$. The unique bounded linear operator on $\mathcal{M}(G)$ defined through

$$\Gamma(T) \exp(f) := \exp(Tf) \qquad \big(f \in L^2(G)\big) \qquad (2.9)$$

is called second quantisation of T.

Lemma 2.2. $\Gamma(T)$ *is unitary if and only if* T *is.* ■

2.2. $\mathcal{M}(A)$ and its Factorisation

If, for $A \in \mathfrak{G}$, $(G, \mathfrak{G}, \mu)$ is replaced with $(A, \mathfrak{A}, \mu_{|A})$ in the above definitions, where $\mu_{|A}$ denotes the restriction of μ to $\mathfrak{A} := A \cap \mathfrak{G}$, respective definitions for $M(A), \mathfrak{M}(A), \mathcal{M}(A), \ldots$ are obtained. Hence we find the following identification, which will always be assumed hereafter

$$\mathcal{M}(A) := L^2\big(M(A)\big) = \Gamma\big(L^2(A)\big) \qquad (A \in \mathfrak{G}). \tag{2.10}$$

Denote with

$$\varphi_{|A} := \varphi(\cdot \cap A) \qquad \big(\varphi \in M(G), A \in \mathfrak{G}\big) \tag{2.11}$$

the subconfiguration of φ obtained by ignoring all points outside of A and identify $M(A)$ with the set of point configurations concentrated on A, i.e.

$$M(A) \cong \{\varphi \in M(G) : \varphi = \varphi_{|A}\} \qquad (A \in \mathfrak{G}). \tag{2.12}$$

Then we have a unique representation of

$$\varphi = \varphi_{|A_1} + \cdots + \varphi_{|A_n} \qquad \big(\varphi \in M(A), (A_1, \ldots, A_n) \in \Sigma(A), A \in \mathfrak{G}\big), \tag{2.13}$$

where

$$\Sigma(A) := \{(A_1, \ldots, A_n) : A_1 \cup \cdots \cup A_n = A, A_1, \ldots, A_n \in \mathfrak{G}, \text{disjoint}\} \tag{2.14}$$

denotes the set of all finite measurable decompositions of $A \in \mathfrak{G}$ into disjoint $A_1, \ldots, A_n \in \mathfrak{G}$.

Moreover (2.8) and (2.13) imply

$$\exp(f)\,(\varphi) = \bigotimes_{k=1}^{n} \exp\big(f_{|A_k}\big)\,\big(\varphi_{|A_1}, \ldots, \varphi_{|A_n}\big)$$
$$\big(f \in L^2(A), \varphi \in M(A), (A_1, \ldots, A_n) \in \Sigma(A), A \in \mathfrak{G}\big), \tag{2.15}$$

which, together with the fact that $\{\exp(f) : f \in L^2(A)\}$ and $\left\{\bigotimes_{k=1}^{n} \exp\big(f_{|A_k}\big) : f \in L^2(A)\right\}$ are total in $\mathcal{M}(A)$ and $\bigotimes_{k=1}^{n} \mathcal{M}(A_k)$, respectively, leads to a natural identification of

$$\mathcal{M}(A) = \bigotimes_{k=1}^{n} \mathcal{M}(A_k) \qquad \big((A_1, \ldots, A_n) \in \Sigma(A), A \in \mathfrak{G}\big) \tag{2.16}$$

and

$$\exp(f) = \bigotimes_{k=1}^{n} \exp\big(f_{|A_k}\big) \qquad \big(f \in L^2(A), (A_1, \ldots, A_n) \in \Sigma(A), A \in \mathfrak{G}\big). \tag{2.17}$$

2.3. *Tensor Products of* $\mathcal{M}(A)$

Now define

$$\mathcal{M}^d(A) := \bigotimes_{i=1}^{d} \mathcal{M}(A) \qquad (A \in \mathfrak{G}, d \in N), \tag{2.18}$$

leading to

$$\mathcal{M}^d(A) = \bigotimes_{k=1}^{n} \mathcal{M}^d(A_k) \qquad \big((A_1, \ldots, A_n) \in \Sigma(A), A \in \mathfrak{G}, d \in N\big). \tag{2.19}$$

For $A \in \mathfrak{G}$ and $d \in N$ set

$$(\tilde{G}, \tilde{\mathfrak{G}}, \tilde{\mu}) := \Big(A \times \{1, \ldots, d\}, \mathfrak{A} \otimes \mathcal{P}\big(\{1, \ldots, d\}\big), \mu_{|A} \otimes \textstyle\sum_{i=1}^{d} \delta_i\Big), \tag{2.20}$$

where $\mathcal{P}(\cdot)$ stands for the power set operation.

Since $\big(A \times \{1\}, \ldots, A \times \{d\}\big) \in \Sigma(\tilde{G})$ and using (2.13), we may identify

$$\varphi = (\varphi_1, \ldots, \varphi_d) \qquad (\varphi \in M(\tilde{G})), \tag{2.21}$$

where

$$\varphi_i := \varphi_{|A \times \{i\}} \in M\big(A \times \{i\}\big) = M(A) \qquad (1 \leq i \leq d). \tag{2.22}$$

Similarly,

$$f = (f_1, \ldots, f_d) \qquad (f \in L^2(\tilde{G})), \tag{2.23}$$

where

$$f_i := f_{|A \times \{i\}} \in L^2\big(A \times \{i\}\big) = L^2(A) \qquad (1 \leq i \leq d), \tag{2.24}$$

and therefore

$$\exp(f) = \bigotimes_{i=1}^{d} \exp(f_i) \qquad \big(f \in L^2(\tilde{G})\big). \tag{2.25}$$

Thus,

$$[M(A)]^d = M\big(A \times \{1, \ldots, d\}\big) \tag{2.26}$$

and

$$\mathcal{M}^d(A) = \mathcal{M}\big(A \times \{1, \ldots, d\}\big) = \Gamma\Big(L^2\big(A, C^d\big)\Big), \tag{2.27}$$

i.e. the tensor product of d identical Fock spaces is again a Fock space.

2.4. *Embedding $\mathcal{M}^2(A)$ into $\mathcal{M}^2(G)$*

We will make use of the natural embedding of $\mathcal{M}^2(A)$ into $\mathcal{M}^2(G)$ given by

$$I_A\psi := \psi \otimes \Big(\exp(0_{|A^c}) \otimes \exp(0_{|A^c})\Big) \qquad (\psi \in \mathcal{M}^2(A)), \tag{2.28}$$

i.e.

$$\mathcal{M}^2(A) \cong I_A(\mathcal{M}^2(A)) =: \mathcal{M}^2_G(A). \tag{2.29}$$

Denote with $I_A^{-1} : \mathcal{M}^2_G(A) \to \mathcal{M}^2(A)$ the inverse of I_A, i.e.

$$I_A^{-1} I_A = \mathrm{Id}_{\mathcal{M}^2(A)} \quad \text{and} \quad I_A I_A^{-1} = \mathrm{Id}_{\mathcal{M}^2_G(A)}. \tag{2.30}$$

3. Proof of Corollary 1

Assume V to be a unitary operator for which (V1) holds. We will show equivalence of (V0) and (1.9).

Suppose (V0). Then (V1), (2.28) and (V0) imply

$$V I_{G_k}\psi_k = (V_{G_k}\psi_k) \otimes \bigotimes_{k\neq l=1}^{n} \Big(V_{G_l}\exp(0_{|G_l}) \otimes \exp(0_{|G_l})\Big) = I_{G_k} V_{G_k}\psi_k$$
$$\big((G_1,\ldots,G_n) \in \Sigma(G), \psi_k \in \mathcal{M}^2(G_k), k = 1,\ldots,n\big), \tag{3.1}$$

Applying $I_{G_k}^{-1}$ to both sides of (3.1) shows

$$V_{G_k} = I_{G_k}^{-1} V I_{G_k} \qquad \big((G_1,\ldots,G_n) \in \Sigma(G), k = 1,\ldots,n\big). \tag{3.2}$$

Since the right-hand side of (3.2) only depends on G_k, but not the decomposition as a whole, it makes sense to consider the operator

$$V_A = I_A^{-1} V I_A \qquad (A \in \mathfrak{G}), \tag{3.3}$$

on $\mathcal{M}^2(A)$ independent of any particular decomposition of G to which A might belong. This shows (1.9).

Conversely, assume (1.9) or, equivalently,

$$I_A V_A = V I_A \qquad (A \in \mathfrak{G}). \tag{3.4}$$

We obviously have unitarity of V_A. We are to show

$$V_A \exp(0_{|A}) \otimes \exp(0_{|A}) = \exp(0_{|A}) \otimes \exp(0_{|A}) \qquad (A \in \mathfrak{G}). \tag{3.5}$$

But this is immediate, because by (2.28), (3.4) and (V1),

$$(V_{A^c}\psi)\otimes\Big(\exp(0_{|A})\otimes\exp(0_{|A})\Big)=I_{A^c}V_{A^c}\psi=VI_{A^c}\psi$$
$$=(V_{A^c}\psi)\otimes\Big(V_A\exp(0_{|A})\otimes\exp(0_{|A})\Big)\qquad\big(\psi\in\mathcal{M}^2(A^c),A\in\mathfrak{G}\big),\quad(3.6)$$

thus completing the proof.

4. Proof of Theorem 1

Throughout this section V will always denote a unitary operator on $\mathcal{M}^2(G)$ for which (V0) and (V1) hold.

Denote with χ_Y, where

$$\chi_Y(y):=\begin{cases}1 & y\in Y\\ 0 & y\notin Y\end{cases}$$

the indicator function of any measurable set Y and with O_f the bounded operator of multiplication on a Hilbert space $H=L^2(\Omega,\mathcal{F},\nu)$ with bounded $f\in H$, i.e.

$$O_f g(\omega):=f(\omega)\cdot g(\omega)\qquad\big(\omega\in\Omega,f,g\in H,f\text{ bounded}\big).$$

Proposition 1. *Let V be a unitary operator on $\mathcal{M}^2(G)$ fullfilling (V0) and (V1). Then for arbitrary $f,g\in L^2(G)$ there exist $\tilde{f},\tilde{g}\in L^2(G)$ such that*

$$V\exp(f\chi_A)\otimes\exp(g\chi_A)=\exp\Big(\tilde{f}\chi_A\Big)\otimes\exp(\tilde{g}\chi_A)\qquad\big(A\in\mathfrak{G}\big).\quad(4.1)$$

Proof. In a forthcoming paper. ∎

We identify

$$\mathcal{M}^2(G)=\mathcal{M}\big(G\times\{1,2\}\big).$$

With this identification,

$$\exp(f)\otimes\exp(g)=\exp(f,g)\qquad\big(f,g\in L^2(G)\big).\quad(4.2)$$

Hence we have

Proposition 2. *For arbitrary $f, g \in L^2(G)$ there exist $\tilde{f}, \tilde{g} \in L^2(G)$ such that*

$$V \exp(f\chi_A, g\chi_A) = \exp\left(\tilde{f}\chi_A, \tilde{g}\chi_A\right) \qquad (A \in \mathfrak{G}). \tag{4.3}$$

■

Lemma 2.1 shows that $\tilde{f}, \tilde{g} \in L^2(G)$ from (4.3) are in fact unique and hence may be used to define a function $B : L^2\big(G \times \{1,2\}\big) \to L^2\big(G \times \{1,2\}\big)$ by

$$B(f, g) := \big(\tilde{f}, \tilde{g}\big) \qquad \big(f, g \in L^2(G)\big). \tag{4.4}$$

Applying (4.4) to (f, g) we get from (4.3)

$$V \exp(f\chi_A, g\chi_A) = \exp\big(O_{(\chi_A, \chi_A)} B(f, g)\big) \qquad \big(f, g \in L^2(G), A \in \mathfrak{G}\big). \tag{4.5}$$

On the other hand, (4.4) applied to $(f\chi_A, g\chi_A)$ and (4.3) imply

$$V \exp(f\chi_A, g\chi_A) = \exp(B(f\chi_A, g\chi_A)) \qquad \big(f, g \in L^2(G), A \in \mathfrak{G}\big). \tag{4.6}$$

Using (4.4), (4.5), (4.6) and Lemma 2.1 we have therefore shown

Proposition 3. *There exists an everywhere defined function $B : L^2\big(G \times \{1,2\}\big) \to L^2\big(G \times \{1,2\}\big)$ such that*

$$V \exp(f, g) = \exp(B(f, g)) \qquad \big(f, g \in L^2(G)\big) \tag{4.7}$$

and

$$B(f\chi_A, g\chi_A) = O_{(\chi_A, \chi_A)} B(f, g) \qquad \big(f, g \in L^2(G), A \in \mathfrak{G}\big). \tag{4.8}$$

■

Proposition 4. *B is unitary.*

Proof. In a forthcoming paper. See also [11], Proposition B.4.2. ■

Now linearity of B and (4.8) imply

$$B(\chi_{A_1}, \chi_{A_2}) = O_{(\chi_{A_1}, \chi_{A_1})} B(1, 0) + O_{(\chi_{A_2}, \chi_{A_2})} B(0, 1) \qquad (A_1, A_2 \in \mathfrak{G}). \tag{4.9}$$

Since μ is finite, constant functions belong to $L^2(G, \mu)$. Defining

$$(a_1, b_1) := B(1, 0) \qquad \text{and} \qquad (a_2, b_2) := B(0, 1),$$

we therefore have from (4.9)

$$B(\chi_{A_1}, \chi_{A_2}) = (a_1\chi_{A_1} + a_2\chi_{A_2}, b_1\chi_{A_1} + b_2\chi_{A_2}) \qquad (A_1, A_2 \in \mathfrak{G}). \quad (4.10)$$

But together with unitarity of B and the fact, that $\{(\chi_{A_1}, \chi_{A_2}) : A_1, A_2 \in \mathfrak{G}\}$ is total in $L^2(G \times \{1,2\})$, this is equivalent to

$$B(f, g) = (a_1 f + a_2 g, b_1 f + b_2 g) \qquad (f, g \in L^2(G)), \quad (4.11)$$

and, using (4.7) and (4.2),

$$V \exp(f) \otimes \exp(g) = \exp(a_1 f + a_2 g) \otimes \exp(b_1 f + b_2 g) \qquad (f, g \in L^2(G)). \quad (4.12)$$

Comparing (4.12) with the operator of independent exchange from [5] we find that $V = U_B$ and hence B is unitary if and only if a_1, a_2, b_1, b_2 fulfill (1.4). The proof is complete.

References

1. K.-H. Fichtner, L. Fichtner, W. Freudenberg, and M. Ohya. On a mathematical model of brain activities. In *Quantum Theory, Reconsideration of Foundations - 4*, volume 962 of *AIP Conference Proceedings*, pages 85 – 90, Melville, New York, 2007. American Institute of Physics.
2. K.-H. Fichtner, L. Fichtner, W. Freudenberg, and M. Ohya. Quantum models of the recognition process - a convergence theorem. Technical report, 2007. 32 pages.
3. K.-H. Fichtner, L. Fichtner, W. Freudenberg, and M. Ohya. Quantum models of the recognition process - mathematical prerequisites. Technical report, 2007. 44 pages.
4. K.-H. Fichtner, L. Fichtner, W. Freudenberg, and M. Ohya. On a quantum model of the recognition process. In L. Accardi, W. Freudenberg, and M.Ohya, editors, *Quantum Bio-Informatics*, volume XXI of *QP–PQ: Quantum Probability and White Noise Analysis*, pages 64 – 84, New Jersey London Singapore, 2008. World Scientifc.
5. K.-H. Fichtner, W. Freudenberg, and V. Liebscher. On exchange mechanisms for bosons. *Random Operators and Stochastic Equations*, 12:no. 4, 331–348, 2004.
6. Karl-Heinz Fichtner and Lars Fichtner. Bosons and a quantum model of the brain. *Jenaer Schriften zur Mathematik und Informatik*, 08 2005.
7. Karl-Heinz Fichtner and Uwe Schreiter. Locally independent Boson systems. Quantum probability and applications V, Proc. 4th Workshop, Heidelberg/FRG 1988, Lect. Notes Math. 1442, 145-161 (1990)., 1990.
8. Lars Fichtner. Ein Quantenmodell der Signalerkennung im Hirn. Dissertation thesis FSU Jena, in preparation.

9. W. Freudenberg. Punktprozesse und Zustände unendlicher Bosonensysteme. *Dissertation B (Habilitationsschrift)*, 1986. Universität Jena.
10. Alain Guichardet. *Symmetric Hilbert spaces and related topics. Infinitely divisible positive definite functions, continuous products and tensor products, Gaussian and Poissonian stochastic processes.* Lecture Notes in Mathematics. 261. Berlin-Heidelberg-New York: Springer-Verlag. V, 197 p., 1972.
11. Volkmar Liebscher. Beam splittings, coherent states and quantum stochastic differential equations. Habilitation thesis FSU Jena, 1998.
12. Hans Maassen. Quantum Markov processes on Fock space described by integral kernels. In *Quantum probability and applications, II (Heidelberg, 1984)*, volume 1136 of *Lecture Notes in Math.*, pages 361–374. Springer, Berlin, 1985.
13. Klaus Matthes, Johannes Kerstan, and Joseph Mecke. *Infinitely divisible point processes. English translation: B. Simon.* Chichester etc.: John Wiley; Sons. XII, 532 p., 1978.
14. Paul-André Meyer. *Quantum probability for probabilists*, volume 1538 of *Lecture Notes in Mathematics*. Springer-Verlag, Berlin, 1993.
15. K. R. Parthasarathy. *An introduction to quantum stochastic calculus*, volume 85 of *Monographs in Mathematics*. Birkhäuser Verlag, Basel, 1992.
16. U. M. Titulaer and R. J. Glauber. Density operators for coherent fields, May 1966.

APPLICATION OF ENTROPIC CHAOS DEGREE TO A COMBINED QUANTUM BAKER'S MAP

KEI INOUE*, MASANORI OHYA† AND IGOR V. VOLOVICH‡

* *Department of Electronics and Computer Science,*
Tokyo University of Science, Yamaguchi,
Sanyo-onoda, Yamaguchi 756-0884, JAPAN
inoue@ed.yama.tus.ac.jp

† *Department of Information Sciences,*
Tokyo Univerisity of Science,
Noda, Chiba 278-8510, JAPAN
ohya@is.noda.tus.ac.jp

‡ *Steklov Mathematical Institute,*
Russian Academy of Sciences, Gubkin St. 8,
117966 Moscow, Russia
volovich@mi.ras.ru

Quantum baker's map is a theoretical model that produces chaos in a quantum system. We have been studied quantum dynamics for the quantum baker's map. We used the Schack and Caves symbolic description of the quantum baker's map. We showed an exact expression for the expectation value of the time dependent position operator in the previous paper. In this paper, we introduce a combined quantum baker's map. Chaos of the combined dynamics for quantum baker's map is studied by the entropic chaos degree.

Keywords :*quantum baker's map, entropic chaos degree, quantum symbolic dynamics, combination of quantum baker's maps*

1. Introduction

Chaos shows a complicated and difficult behavior to predict since the chaos has a property of exponential sensibility to initial conditions. The property means that the divergence of infinitesimally nearby initial conditions grows exponentially. Dynamics showing such a chaotic behavior is called a chaotic classical dynamics. A quantum-classical correspondence for a chaotic dynamics has been studied in the field of study of quantum chaos for many years [1-9], [12-13], [15-21], [23-31]. The quantum baker's map is a theoretical model in the quantum chaos [5], and has been generalized by Schack and Caves [27].

A chaotic classical dynamics is often studied by computing the Lyapunov exponents of the dynamics. A Chaos degree has been introduced to measure degree of chaos in dynamics in the framework of Information Dynamics [10] by one of the authors [20]. In particular an entropic chaos degree, which was applied to some dynamical models, has several advantages over the Lyapunov exponent since the chaos degree is easy to compute for any time series.

In this paper, an expectation value for a combined quantum baker's map is introduced. Semiclassical properties for the combined quantum baker's map are considered by the entropic chaos degree. Further details will be discussed in [14].

2. Classical Baker's Transformation

The classical baker's transformation maps the unit square $[0,1]\times[0,1]$ onto itself according to

$$(q,p)\rightarrow\begin{cases}(2q,p/2), & (0\leq q<1/2)\\ (2q-1,(p+1)/2), & (1/2\leq q\leq 1)\end{cases}.$$

This corresponds to compressing the unit square in the p direction and stretching it in the q direction, while preserving the area, then cutting it vertically and stacking the right part on top of the left part.

The classical baker's transformation has a simple description in terms of its symbolic dynamics [3]. Each point (q,p) is represented by a symbolic string with a dot

$$\xi=\cdots\xi_{-2}\xi_{-1}\xi_0\cdot\xi_1\xi_2\cdots$$

where $\xi_k\in\{0,1\}$ and

$$q=\sum_{k=1}^{\infty}\xi_k 2^{-k},\quad p=\sum_{k=0}^{\infty}\xi_{-k}2^{-k-1}.$$

The action of the classical baker's transformation on a symbolic string ξ is given by the shift map (Bernoulli shift) U defined by

$$U\xi=\acute{\xi}=\cdots\xi_{-2}\xi_{-1}\xi_0\xi_1\cdot\xi_2\cdots.$$

This means that, at each time step, the dot is shifted one place to the right while the entire string remains fixed. After m steps, the q coordinate becomes

$$q_m=\sum_{k=1}^{\infty}\xi_{m+k}2^{-k}. \tag{1}$$

This relation defines the classical trajectory with the initial data

$$q = q_0 = \sum_{k=1}^{\infty} \xi_k 2^{-k}.$$

3. Quantum Baker's Map

Quantum baker's maps have been defined on the D-dimensional Hilbert space of the quantized unit square [5].

To quantize the unit square the Weyl unitary displacement operators $\hat{U}$ and $\hat{V}$, which produce displacements in the position and momentum directions, respectively, have been defined on the D-dimensional Hilbert space. The operators $\hat{U}$ and $\hat{V}$ satisfy the commutation relation

$$\hat{U}\hat{V} = \varepsilon \hat{V}\hat{U}$$

where $\varepsilon = e^{(2\pi i/D)}$. The operators $\hat{U}$ and $\hat{V}$ can be written

$$\hat{U} = e^{2\pi i \hat{q}}, \; \hat{V} = e^{2\pi i \hat{p}}.$$

For consistency with the commutation relation, the quantum scale $\hbar$ has been set to satisfy $2\pi\hbar = 1/D$.

Corresponding to antiperiodic boundary conditions, eigenvalues of the position and momentum operators $\hat{q}$ and $\hat{p}$ have been chosen to $q_j = (j + \frac{1}{2})/D$, $j = 0, \ldots, D-1$, $p_k = (k+\frac{1}{2})/D$, $k = 0, \ldots, D-1$, respectively. The dimension D of space has been assumed $D = 2^N$, which is the dimension of the Hilbert space of the N-qubits.

The basis in the $D = 2^N$ dimensional Hilbert space is

$$|q_j\rangle = |\xi_1\rangle \otimes |\xi_2\rangle \otimes \cdots \otimes |\xi_N\rangle$$

where $j = \sum_{l=1}^{N} \xi_l 2^{N-l}$, $\xi_l \in \{0, 1\}$. Each qubit ξ_k has basis

$$|0\rangle = \begin{pmatrix} 1 \\ 0 \end{pmatrix}, |1\rangle = \begin{pmatrix} 0 \\ 1 \end{pmatrix}$$

and q_j can be written $q_j = 0.\xi_1\xi_2 \ldots \xi_N 1$ as binary numbers. If we define

$$|.\xi_1\xi_2 \cdots \xi_N\rangle = e^{i\pi/2}|q_j\rangle$$

where $e^{i\pi/2}$ is the phase factor derived from the antiperiodic boundary conditions [27], the eigenvectors of the position and momentum operators $\hat{q}$ and $\hat{p}$ are related through the quantum Fourier transformation [23]: $F|q_k\rangle = |p_k\rangle$.

Applying the quantum Fourier transfer operator to the n rightmost bits of the position eigenstate $|.\xi_{n+1}\cdots\xi_N\xi_n\cdots\xi_1\rangle$, produces the following family of states [27].

$$
\begin{aligned}
|\xi_1\cdots\xi_n.\xi_{n+1}\cdots\xi_N\rangle \equiv e^{i\pi(0.\xi_n\cdots\xi_1 1)}&|\xi_{n+1}\rangle\otimes\cdots\otimes|\xi_N\rangle\\
&\otimes\sqrt{1/2}\left(|0\rangle+e^{2\pi i(0.\xi_1 1)}|1\rangle\right)\\
&\otimes\sqrt{1/2}\left(|0\rangle+e^{2\pi i(0.\xi_2\xi_1 1)}|1\rangle\right)\\
&\otimes\cdots\cdots\\
&\otimes\sqrt{1/2}\left(|0\rangle+e^{2\pi i(0.\xi_n\cdots\xi_1 1)}|1\rangle\right)
\end{aligned}
\tag{2}
$$

where $1 \le n \le N-1$. The state (2) is localized in both position and momentum.

For each n satisfied with $0 \le n \le N-1$, a quantum baker's map can be defined by [27]

$$
B_n|\xi_1\cdots\xi_n.\xi_{n+1}\cdots\xi_N\rangle = |\xi_1\cdots\xi_{n+1}.\xi_{n+2}\cdots\xi_N\rangle
$$

where the dot is shifted one place to the right. In particular, the map B_{N-1} for $n = N-1$ is the original quantum baker's map [5].

In the sequel the map B_0 for $n = 0$ is only considered. The map B_0 is represented by the unitary operator T with the following matrix elements [28]:

$$
\langle\xi|T|\eta\rangle = \frac{1-i}{2}\exp\left(\frac{\pi}{2}i|\xi_1-\eta_N|\right)\prod_{k=2}^{N}\delta(\xi_k-\eta_{k-1}). \tag{3}
$$

where $|\xi\rangle = |\xi_1\xi_2\cdots\xi_N\rangle$, $|\eta\rangle = |\eta_1\eta_2\cdots\eta_N\rangle$, δ is Kronecker delta: $\delta(0)=1$, $\delta(x)=0$ for $x \ne 0$.

4. Dynamics of Quantum Baker's Map

In this section, an expectation value of the position operator $\hat{q}$ unitarily evolved by the quantum baker's map T is considered. To study the classical limit $\hbar \to 0$ of the quantum baker's map T, the following expectation value of the position operator $\hat{q}$ for time $m = 0, 1, \ldots$ with respect to a vector $|\xi\rangle$ has been introduced.

$$
r_m^{(N)} = \langle\xi|T^m\hat{q}T^{-m}|\xi\rangle
$$

where $|\xi\rangle = |\xi_1\xi_2\cdots\xi_N\rangle$.

From Eq. (3) one gets for $m = 0, 1, \ldots, N-1$,

$$\langle\xi|T^m|\eta\rangle = \left(\frac{1-i}{2}\right)^m \left(\prod_{k=1}^{N-m} \delta(\xi_{m+k} - \eta_k)\right) \left(\prod_{l=1}^{m} \exp\left(\frac{\pi}{2} i \left|\xi_l - \eta_{N-m+l}\right|\right)\right) \tag{4}$$

and for $m = N$

$$\langle\xi|\, T^N \,|\eta\rangle = \left(\frac{1-i}{2}\right)^N \left(\prod_{l=1}^{N} \exp\left(\frac{\pi}{2} i |\xi_l - \eta_l|\right)\right). \tag{5}$$

Using this formula the following theorem has been proved [13].

Theorem 4.1. *For each* $m = kN + p$ $(p = 0, 2, \ldots, N-1,\ k \in \mathbf{N} \cup \{0\})$, *we have*

$$r_m^{(N)} = \begin{cases} \sum_{j=1}^{N-p} \xi_{p+j} 2^{-j} + \frac{2^p}{2^{N+1}} & (k = 0 \pmod 4) \\ \sum_{j=N-p+1}^{N} \eta_{j-(N-p)} 2^{-j} + \frac{2^N - 2^p + 1}{2^{N+1}} & (k = 1 \pmod 4) \\ \sum_{j=1}^{N-p} \eta_{p+j} 2^{-j} + \frac{2^p}{2^{N+1}} & (k = 2 \pmod 4) \\ \sum_{j=N-p+1}^{N} \xi_{j-(N-p)} 2^{-j} + \frac{2^N - 2^p + 1}{2^{N+1}} & (k = 3 \pmod 4) \end{cases} \tag{6}$$

where $\eta_j = \xi_j + 1 \pmod 2$.

5. Entropic Chaos Degree

In usual quantum system, $\mathcal{A}$ is the set $\mathbf{B}(\mathcal{H})$ of all bounded operators on a Hilbert space $\mathcal{H}$ and $\mathfrak{S}$ is the set $\mathfrak{S}(\mathcal{H})$ of all density operators on the $\mathcal{H}$ where an extreme decomposition of $\rho \in \mathfrak{S}(\mathcal{H})$ is a Schatten decomposition $\rho = \sum_k p_k E_k$ (i.e., $\{E_k\}$ are one dimensional orthogonal projections such that $\sum E_k = I$). Then the entropic chaos degree of ρ and the channel Λ^* is defined by

$$D(\rho; \Lambda^*) \equiv \inf\left\{\sum_k p_k S(\Lambda^* E_k); \{E_k\}\right\}$$

where the infimum is taken over all possible Schatten decompositions of ρ and S is von Neumann entropy [10].

In classical discrete system, the Schatten decomposition of ρ is unique $\rho = \sum_k p_k \delta_k$ with delta measure $\delta_k(j) \equiv \begin{cases} 1 & (k=j) \\ 0 & (k \neq j) \end{cases}$, so that the entropic chaos degree is given as

$$D(\varphi; \Lambda^*) = \sum_k p_k S(\Lambda^* \delta_k)$$

where ρ is the probability distribution of the orbit obtained from a dynamics of a system and the channel Λ^* is generated from the dynamics [20].

Let $\{x_m\}$ be an orbit, F be a map from x_m to x_{m+1}. A dynamics F of the orbit produces the above channel Λ^*. Now we consider a difference equation system (i.e. $x_{m+1} = F(x_m)$) represented by a map F on $I \equiv [a,b]^{\mathbf{N}} \subset \mathbf{R}^{\mathbf{N}}$. Take a finite partition of I such as

$$I \equiv \bigcup_k B_k \ (B_i \cap B_j = \emptyset \ (i \neq j)).$$

For an initial value x and a given finite partition $B = \{B_i\}$, the state at time m of the orbit determined by the difference equation is defined by the probability distribution $p_B^{(m)} \equiv \left(p_{i,B}^{(m)}\right)$ of the orbit $\{F^n(x); n = 0, 1, \cdots\}$;

$$p_{i,B}^{(m)} \equiv \frac{1}{M+1} \sum_{k=m}^{M+m} 1_{B_i}\left(F^k x\right)$$

where 1_A is a characteristic function. When the initial value x is distributed due to a measure ν on I, the above $p_i^{(m)}$ is given as

$$p_i^{(m)} \equiv \frac{1}{M+1} \int_I \sum_{k=m}^{M+m} 1_{B_i}\left(F^k x\right) d\nu.$$

In the case that F is a classical baker's map, if the orbit isn't stable and periodic, then it is shown that the $M \to \infty$ limit of $p_i^{(m)}$ exists and equals to a natural invariant measure for a fixed $m \in \mathbf{N}$ [22].

The joint probability distribution $\left(p_{ij}^{(m,m+1)}\right)$ between the time m and $m+1$ is defined by

$$p_{ij,B}^{(m,m+1)} \equiv \frac{1}{M+1} \sum_{k=m}^{M+m} 1_{B_i}\left(F^k x\right) 1_{B_j}\left(F^{k+1} x\right) \qquad (7)$$

or

$$p_{ij}^{(m,m+1)} \equiv \frac{1}{M+1} \int_I \sum_{k=m}^{M+m} 1_{B_i}\left(F^k x\right) 1_{B_j}\left(F^{k+1} x\right) d\nu.$$

Then the channel Λ_m^* at the time m is determined by

$$\Lambda_{m,B}^* \equiv \left(\frac{p_{ij,B}^{(m,m+1)}}{p_{i,B}^{(m)}} \right) \Longrightarrow p_B^{(m+1)} = \Lambda_{m,B}^* p_B^{(m)}$$

and the entropic chaos degree D at the time m is given as [20]

$$D_B\left(x;F\right) \equiv D\left(p_B^{(m)};\Lambda_{m,B}^*\right) = \sum_i p_{i,B}^{(m)} S(\Lambda_{m,B}^*\delta_i) = \sum_{i,j} p_{ij,B}^{(m,m+1)} \log \frac{p_{i,B}^{(m)}}{p_{ij,B}^{(m,m+1)}}$$

We can judge if the dynamics causes chaos or not by the value of D as [20].

$$D > 0 \Longleftrightarrow \textit{chaotic},$$
$$D = 0 \Longleftrightarrow \textit{stable}.$$

The entropic chaos degree has been applied to some dynamical maps such as a logistic map, a baker's transformation and a Tinkerbell map, it could explain their chaotic characters [20,11]. Our entropic chaos degree is easily applicable to a dynamics which is given as only a time series, compared with usual measures such as the Lyapunov exponent.

6. A Combined Quantum Baker's Map and Its Entropic Chaos Degree

We consider an expectation value of a combined position operator,

$$\begin{aligned}(c_r)_m^{(N)} &= \left\langle \xi \left| (1-a)T^m \hat{q} T^{-m} + \frac{a}{2}\left(T^{(m-1)}\hat{q}T^{-(m-1)} + T^{(m+1)}\hat{q}T^{-(m+1)}\right) \right| \xi \right\rangle \\ &= (1-a)r_m^{(N)} + \frac{a}{2}\left(r_{m-1}^{(N)} + r_{m+1}^{(N)}\right), \quad a \in [0,1]\end{aligned}$$

as an orbit value of a combined quantum baker's map. The expectation value $(c_r)_m^{(N)}$ gets the classical value $(c_q)_m$ at the classical limit: $N \to \infty$.

Then the probability distribution $\left(p_i^{(m)}\right)$ at the time m and the joint probability distribution $\left(p_{ij}^{(m,m+1)}\right)$ between the time m and $m+1$ are given as

$$p_i^{(m)} \equiv \frac{1}{M+1} \sum_{k=m}^{M+m} 1_{B_i}\left((c_r)_k^{(N)}\right),$$

$$p_{ij}^{(m,m+1)} \equiv \frac{1}{M+1} \sum_{k=m}^{M+m} 1_{B_i}\left((c_r)_k^{(N)}\right) 1_{B_j}\left((c_r)_{k+1}^{(N)}\right)$$

for an initial value $r_0^{(N)} \in [0,1]$, respectively. Then the entropic chaos degree of the combined quantum baker's map is computed as

$$D_q\left(p^{(m)};\Lambda_m^*\right) = \sum_{i,j} p_{ij}^{(m,m+1)} \log \frac{p_i^{(m)}}{p_{ij}^{(m,m+1)}}. \tag{8}$$

In the sequel, a quantum and classical initial values $r_0^{(N)}$ and q_0 of the baker's map are set as

$$r_0^{(N)} = \sum_{l=1}^{N} \xi_l 2^{-l} + 1/2^{N+1} = 0.\xi_1\xi_2\cdots\xi_N 1$$

$$q_0 = \sum_{l=1}^{\infty} \xi_l 2^{-l} = 0.\xi_1\xi_2\cdots\cdots$$

where ξ_i is a pseudo-number 0 or 1, respectively. Then the quantum and classical initial values $(c_r)_1^{(N)}$ and $(c_q)_1$ of the combined baker's map become

$$(c_r)_1^{(N)} = (1-a)r_1^{(N)} + \frac{a}{2}\left(r_0^{(N)} + r_2^{(N)}\right),$$

$$(c_q)_1 = (1-a)q_1 + \frac{a}{2}\left(q_0 + q_2\right).$$

According to the classical setting, we divide the interval $[0,1]$ into 100 equi-partition $\{B_i\}$, and we set that $m = 1$ and the number M of orbit points $(c_r)_k$ is 10000. Then we can write the entropic chaos degree D_q as $D_q(N,a)$ since D_q depends on two parameters N and a.

The dependence of the entropic chaos degree on a and N in this section will be discussed in [14] for further details.

7. Dependence of the Entropic Chaos Degree on the Combination Parameter *a*

We study the dependence of the entropic chaos degree on the combination parameter a for some fixed values of N. Then the entropic chaos degree is denoted by a one-parameter function $D_q(a)$ of a. Figure 1 shows the entropic chaos degree $D_q(a)$ versus a on $[0,1]$ for some fixed values of N. The entropic chaos degree for the combined quantum baker's map takes smaller value than that of the entropic chaos degree for the combined classical baker's transformation.

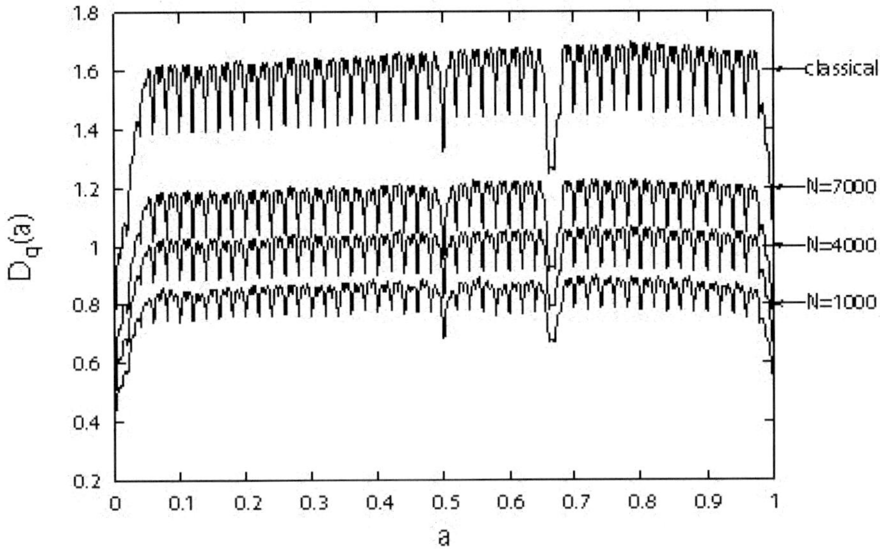

Fig.1. $D_q(a)$ versus a on $[0, 1]$

8. Dependence of the Entropic Chaos Degree on N

From theorem 4.1, there is some correspondence between the quantum expectation value r_k and the classical value q_k for baker's map for $0 \leq k \leq N$, while there is no correspondence between their values for $N+1 \leq k \leq M$ where M is the number of all orbit points. Therefore if we assume N such that $N = M$, then the entropic chaos degree for the combined quantum baker's map probably takes the same value as that of the chaos degree for the combined classical baker's transformation.

In the sequel, we study the dependence of the chaos degree on N for $a = 0.1$. Then the chaos degree is denoted by a one-parameter function $D_q(N)$ of N.

Figure 2 shows that the entropic chaos degree $D_q(N)$ oscillates and increases for $1 \leq N < 5000$ and monotonously increase for $5000 \leq N \leq M$, converging to the entropic chaos degree for the classical combined baker's transformation.

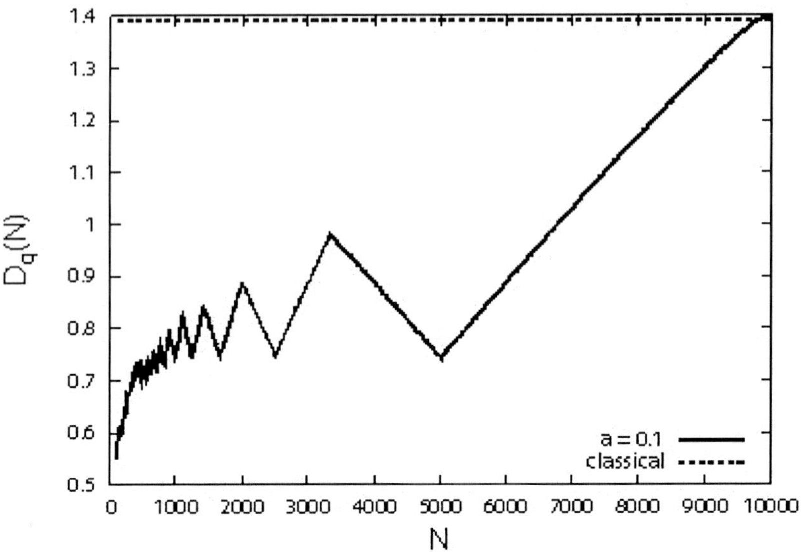

Fig.2. $D_q(N)$ versus N for $1 \leq N \leq M$

References

1. D.V.Anosov and V.I.Arnold (eds.), Dynamical Systems, INITI, Moscow, 1996.
2. I.Y.Arefeva, P.B.Medvedev, O.A.Rytchkov and I.V.Volovich, Chaos in m(atrix) theory, Chaos, Solitons & Fractals, **10**, 213–223, 1999.
3. V.M.Alekseev and M.N.Yakobson, *Symbolic dynamics and hyperbolic dynamic systems*, Phys. Reports, **75**, 287-325, 1981.
4. M.V.Berry, Some Quantum to Classical Asymptotics, In Giannoni, M.J. Voros, A. and Justi, Zinn, editors, Les Houches Summer School "chaos and quantum physics", North-Holland, Amsterdam, 1991.
5. N.L.Balazs and A.Voros, The quantized baker's transformation, Ann. Phys., **190**, 1–31, 1989.
6. G.Casati and B.V.Chirikov (eds.), Quantum Chaos: between Order and Disorder, Cambridge Univ. Press, Cambridge, 1995.
7. F.M.Dittes, E.Doron and U.Smilansky, Long-time behavior of the semiclassical baker's map, Phys. Rev. E, **49**, R963–R966,1994.
8. J.Emerson and L.E.Ballentine, Quantum-Classical Correspondence for the Equilibrium Distributions of Two Interacting Spins, Phys. Rev. E **64**, 026217–026227, 2001.
9. S.Habib, K.Shizume and W.H.Zurek, Decoherence, chaos, and the correspondence principle, Phys. Rev. Lett. **80**, 4361–4365 , 1998.
10. R.S.Ingarden, A.Kossakowski and M.Ohya, *Information Dynamics and Open Systems*, Kluwer Academic Publishers, 1997.

11. K.Inoue, M.Ohya and K.Sato, Application of chaos degree to some dynamicalsystems, Chaos, Soliton and Fractals, **11**, 1377–1385, 2000.
12. K.Inoue, M.Ohya and I.V.Volovich, On quantum classical correspondence for baker's map, quant-ph/0108107.
13. K.Inoue, M.Ohya and I.V.Volovich, Semiclassical properties and chaos degree for the quantum baker's map, Journal of Mathematical Physics, **43**, No.2, 734–755, 2002.
14. K.Inoue, M.Ohya and I.V.Volovich, On a combined quantum baker's map and its characterization by entropic chaos degree, SUT preprint.
15. L.Kaplan and E.J.Heller, Overcoming the wall in the semiclassical baker's map, Phys. Rev. Lett., **76**, 1453–1456, 1996.
16. Z.P.Karkuszewski, J.Zakrzewski and W.H.Zurek, Breakdown of correspondence in chaotic systems: Ehrenfest versus localization times, Phys. Rev. A **65**, 042113–042117, 2002.
17. M.G.E.da Luz and A.M.Ozorio de Almeida, Path integral for the quantum baker's map, Nonlinearity, **8**, 43–64, 1995.
18. A.Mariano, P.Facchi and S.Pascazio, Decoherence and fluctuations in quantum interference experiments, Fortschr. Phys. **49** , 1033-1039, 2001.
19. D.Monteoliva and J.P.Paz, Decoherence in a classically chaotic quantum system: entropy production and quantum-classical correspondence, Phys. Rev. E 64, 056238–56250, 2001.
20. M.Ohya, Complexities and their applications to characterization of chaos, International Journal of Theoretical Physics, **37**, No.1, 495–505, 1998.
21. A.M.Ozorio de Almeida and M.Saraceno, Periodic orbit theory for the quantized baker's map, Ann. Phys., **210**, 1–15, 1991.
22. E. Ott, Chaos in Dynamical Systems, Cambridge Univ.Press, 1993.
23. M.Saraceno, Classical structures in the quantized baker transformation, Ann. Phys., **199**, 37–60, 1990.
24. M.Saraceno and A.Voros, Towards a semiclassical theory of the quantum baker's map, Physica D, **79**, 206–268, 1994.
25. Ya.G.Sinai, Introduction to Ergodic Theory, Fasis, Moscow, 1996.
26. R. Schack and C.M.Caves, Hypersensitivity to perturbations in the quantum baker's map, Phys. Rev. Lett., **71**, 525–528, 1993.
27. R.Schack and C.M.Caves, Shifts on a finite qubit string: A class of quantum baker's maps, Applicable Algebra in Engineering, Communication and Computing, **AAECC 10**, 305–310, 2000.
28. A.N.Soklakov and R.Schack, Classical limit in terms of symbolic dynamics for the quantum baker's map, Phys. Rev. E, **61**, 5108–5114, 2000.
29. A.N.Soklakov and R.Schack, Decoherence and linear entropy increase in the quantum baker's map, Phys. Rev. E **66**, 036212–036222 ,2002.
30. G.M.Zaslavskii, Stochasticity of Dynamical Systems, Nauka, Moscow, 1984.
31. W.H.Zurek, Decoherence, einselection, and the quantum origins of the classical, Rev. Mod. Phys. **75**, 715–775, 2003.

ON QUANTUM ALGORITHM FOR MULTIPLE ALIGNMENT OF AMINO ACID SEQUENCES

SATOSHI IRIYAMA AND MASANORI OHYA
Department of Information Sciences, Tokyo University of Science 2641, Yamazaki, Noda City, Chiba, Japan

The alignment of genome sequences or amino acid sequences is one of fundamental operations for the study of life. Usual computational complexity for the multiple alignment of N sequences with common length L by dynamic programming is $O(L^N)$. This alignment is considered as one of the NP problems, so that it is desirable to find a nice algorithm of the multiple alignment. Thus in this paper we propose the quantum algorithm for the multiple alignment based on the works [12,1,2] in which the NP complete problem was shown to be the P problem by means of quantum algorithm and chaos information dynamics.

1. Introduction

When we analyze life in gene level, we examine the homology of genome or amino acid sequences to compare these sequences, for which we have to align the sequences. The alignment of two sequences is called the pairwise alignment, and that for sequences more than three is called the multiple alignment. To align the sequences, we insert gap (*) into the position of a sequence where an base or amino acid is considered to deviate. Such alignment should be first done to analyze genome sequences or amino acid sequences, so that it is one of the fundamental operations for study of life.

At the present stage, the algorithms of the pairwise alignment are done by applying dynamic programming [9,10,11]. However, it is rather difficult to use the similar algorithm for the multiple alignment because the computational complexity of the N sequences with their length L by dynamic programming becomes $O(L^N)$, whose alignment will be very difficult as N increases. Therefore, the various methods have been considered to reduce the computational complexity. Among those, the Simulated Annealing has been used in [12,13,14]. The simulated annealing is one of the methods solving some combinatorics optimization problems such as travelling salesman problem. Even if the simulated annealing effectively works, it is difficult to demonstrate the multiple alignment in polynomial time of N, so that the multiple alignment is considered as one of the NP problems. Recently Ohya and Volovich [1,2] could prove that the NP complete problem becomes P problem by quantum algorithm and chaos information dynamics. In this paper, we discuss the quantum algorithm for the multiple alignment, that is, how we can construct quantum gate to make the sequences align. Though we are not able to implement the quantum algorithm on the present computer, our present work will be effective to study life once quantum computer is realized.

2. Outline of Alignment

Let

$$\begin{aligned} \mathcal{A} &: \mathbf{MNPWYSTWQYT} \\ \mathcal{B} &: \mathbf{MNPQYTVWPY} \\ \mathcal{C} &: \mathbf{MNWYSTQPYV} \end{aligned}$$

be the amino acid sequences of three organisms or identical proteins.

These sequences $\mathcal{A}$, $\mathcal{B}$ and $\mathcal{C}$ look not so close each other. It is considered that some amino acids are changed, deleted or inserted during the course of the biological evolution from a common origin of $\mathcal{A}$, $\mathcal{B}$ and $\mathcal{C}$. Therefore it is important to align the sequences $\mathcal{A}$, $\mathcal{B}$ and $\mathcal{C}$ to study similarity or difference of organisms properly. For example, after an operation of alignment they become

$$\begin{aligned} \mathcal{A} &: \mathbf{MNPWYST} * \mathbf{WQYT} \\ \mathcal{B} &: \mathbf{MNPQY} * \mathbf{TVWPY} * \\ \mathcal{C} &: \mathbf{MN} * \mathbf{WYST} * \mathbf{QPYV} \end{aligned}$$

For a set of amino acid sequences, we define a suitable function f_d indicating the similarity of sequences in the set. And we look for the alignment which makes the value of f_d minimum. However, we must consider a lot of variety of alignment because of length of sequence and the number of gaps. For given the length of amino acid sequence L and the number of gaps g, the number of possible alignments becomes

$$\begin{aligned} {}_{L+1}H_g &= {}_{L+g}C_g \\ &= \frac{(L+g)!}{g!L!} \end{aligned}$$

For instance, $L = 100$ and $g = 10$, we have

$$_{101}H_{10} = 46897636623981$$

variations. We explain a language expression of alignment problem below.

In this paper, we propose a quantum algorithm to solve this problem in polynomial time of the length L and the number of sequences N.

2.1. *Language classes of alignment*

A language l is a sequence of alphabets and we say that l is recognized by a Turing machine M if and only if there exists M that halts for $x \in l$ and does not halt $x \notin l$. In this subsection we explain language classes and in which language class the alignment problems belong to.

Here we review well known language classes P and NP.

Definition 2.1. We say that the language is in class P if it is recognized by a deterministic Turing machine in polynomial time of input size.

Definition 2.2. We say that the language is in class NP if there is a non-deterministic Turing machine in polynomial time of input size. Besides, if all languages in NP reduces to l_1 in polynomial time, a language l_1 is called NP-complete.

Let $\mathcal{A}_1\mathcal{A}_2\cdots\mathcal{A}_N$ be amino acid sequences, $\mathcal{A} \equiv \mathcal{A}_1\mathcal{A}_2\cdots\mathcal{A}_N$ a total sequence, ϕ an alignment for $\mathcal{A}$ and s positive real number, we define the language $l_{alignment}$ as

$$l_{alignment} = \{< \mathcal{A}, f_d, s > | \exists \phi \text{ s.t. } f_d\left(\phi\left(\mathcal{A}\right)\right) \leq s\}$$

where f_d is an objective function and $< x >$ means a representation of x.

These theorem are proven in [13,11,10].

Theorem 2.1. *$l_{alignment}$ belongs to P when $N \leq 2$.*

Theorem 2.2. *$l_{alignment}$ belongs to NP-complete when $card\left(\mathcal{A}\right) \geq 3$ and f_d is SP-score*

Theorem 2.3. *The problems to find the set of alignments that minimize $f_d\left(\phi\left(\mathcal{A}\right)\right)$ is called a Max-NP hard problem.*

3. Quantum Algorithm

In this section we explain mathematical foundations of a quantum algorithm following the paper[16].

For mathematical expression of the problem, we construct the quantum algorithm in the following steps:

(1) Define a Hilbert space for computation.
(2) Construct an initial state.
(3) Construct unitary operators to solve the problem.
(4) Apply them for the initial state and obtain the result.
(5) Measure an observable with the result state.

In the first step, we define the Hilbert space depending on the problem. Let $\mathbb{C}^2$ be a Hilbert space spanned by $|0\rangle = \begin{pmatrix} 1 \\ 0 \end{pmatrix}$ and $|1\rangle = \begin{pmatrix} 0 \\ 1 \end{pmatrix}$, a normalized vector $|\psi\rangle = \alpha|0\rangle + \beta|1\rangle$ $(\alpha, \beta \in \mathbb{C}, |\alpha|^2 + |\beta|^2 = 1)$on this space is called a qubit. Since we can use a superposition of $|0\rangle$ and $|1\rangle$ as an initial state vector, the quantum algorithm is more effective than classical one.

We represent two or more sequence of qubits as

$$|x, y\rangle \equiv |x\rangle \otimes |y\rangle,$$

$$|x^n\rangle = \underbrace{|x\rangle \otimes \cdots \otimes |x\rangle}_{n}$$

Here, we consider a basis of Hilbert space $(\mathbb{C}^2)^{\otimes n}$ such as

$$\begin{aligned} |e_0\rangle &= |0, 0, \cdots, 0\rangle \\ |e_1\rangle &= |1, 0, \cdots, 0\rangle \\ |e_2\rangle &= |0, 1, \cdots, 0\rangle \\ &\vdots \\ |e_{2^n-1}\rangle &= |1, 1, \cdots, 1\rangle \end{aligned}$$

We can treat 2^n integer numbers by using this basis.

One can apply Hadamard transformation

$$H = \frac{1}{\sqrt{2}} \begin{pmatrix} 1 & 1 \\ 1 & -1 \end{pmatrix}$$

to create a superposition of one qubit. For $|0\rangle$ and $|1\rangle$, it works as

$$H|0\rangle = \frac{1}{\sqrt{2}}|0\rangle + \frac{1}{\sqrt{2}}|1\rangle$$
$$H|1\rangle = \frac{1}{\sqrt{2}}|0\rangle - \frac{1}{\sqrt{2}}|1\rangle .$$

Hadamard transformation has a very important role in quantum algorithm. For example, we apply $H^{\otimes n}$ to

$$|\psi\rangle = |0^n\rangle ,$$

then we obtain

$$\frac{1}{\sqrt{2^n}} \sum_{i=0}^{2^n-1} |e_i\rangle .$$

Here we introduce logical gates on the qubit space, which are NOT gate, C-NOT gate and CC-NOT gate. We call these gates fundamental gates. We can also construct AND and OR gate by considering the product of fundamental gates and some implementations. The NOT gate U_{NOT} is defined on one qubit Hilbert space $\mathbb{C}^2$ as

$$U_{NOT} = |1\rangle\langle 0| + |0\rangle\langle 1| .$$

C-NOT U_{CN} gate and CC-NOT U_{CCN} are given on two and three qubit Hilbert space as

$$U_{CN} = |0\rangle\langle 0| \otimes I + |1\rangle\langle 1| \otimes U_{NOT}$$
$$U_{CCN} = |0\rangle\langle 0| \otimes I \otimes I + |1\rangle\langle 1| \otimes |0\rangle\langle 0| \otimes I + |1\rangle\langle 1| \otimes |1\rangle\langle 1| \otimes U_{NOT},$$

respectively. The logical gates U_{OR} and U_{AND} are constructed by the combinations of these gates [15].

3.1. *Chaos Amplification*

After applying unitary operator to the initial state, we obtain the result state. We use chaos amplification proposed in [5,1,2] in order to obtain the result state effective. It is useful to quantum computing in which the result probability of unitary computation is very small.

Let

$$\rho = q^2 P_1 + \left(1 - q^2\right) P_0$$

be a result of the quantum algorithm where P_0 and P_1 are projectors to the state vectors $|0\rangle$ and $|1\rangle$. One has to notice that P_1 and P_0 generate an Abelian algebra which can be considered as a classical system. Consider the so called logistic map which is given by the equation

$$x_{n+1} = ax_n\left(1 - x_n\right) \equiv g_a\left(x\right), \quad x_n \in [0,1]$$

The properties of this map depend on the parameter a. Then the density matrix ρ above is interpreted as the initial data ρ_0, and we apply the logistic map as

$$\rho_m = \frac{(I + g_a^m(\rho_0)\sigma_3)}{2}$$

where I is the identity matrix and σ_3 is the z-component of Pauli matrix on $\mathbb{C}^2$. In fact,

$$\begin{aligned} g_a\left(\begin{pmatrix} 1-q^2 & 0 \\ 0 & q^2 \end{pmatrix}\right) &= a\begin{pmatrix} 1-q^2 & 0 \\ 0 & q^2 \end{pmatrix}\left(\begin{pmatrix} 1 & 0 \\ 0 & 1 \end{pmatrix} - \begin{pmatrix} 1-q^2 & 0 \\ 0 & q^2 \end{pmatrix}\right) \\ &= a\begin{pmatrix} 1-q^2 & 0 \\ 0 & q^2 \end{pmatrix}\begin{pmatrix} q^2 & 0 \\ 0 & 1-q^2 \end{pmatrix} \\ &= aq^2\left(1-q^2\right)I \end{aligned}$$

So, the state ρ_m can be written as

$$\rho_m = \frac{(I + g_a^m(q^2)\sigma_3)}{2}$$

To find a proper value m we finally measure the value of σ_3 in the state ρ_m such that

$$M_m \equiv tr\rho_m\sigma_3.$$

In OMV SAT algorithm, we apply the unitary operator U_{SAT} to the initial state ρ and measure the last qubit which store the value $f\left(C\right)$. After the measurement, we have the final state $\rho' = |v_{out}\rangle\langle v_{out}| = q^2 P_1 + \left(1-q^2\right)P_0$.

Theorem 3.1. *Let C be the set of clauses on $X = \{x_1, \ldots, x_n, \bar{x}_1, \ldots, \bar{x}_n\}$, ρ is an initial state of quantum algorithm of SAT and ρ_f is the state after computation satisfying $\Lambda_M^* \Lambda_U^*\left(\rho\right) = \rho_f$. For all m, it holds*

$$M_m \begin{cases} = 0 & \textit{iff } C \textit{ is not SAT} \\ > 0 & \textit{iff } C \textit{ is SAT} \end{cases}$$

Theorem 3.2. *Let* $x_0 = \frac{l}{2^n}$ *and we put* $|t(C)| = l$. *If and only if* C *is SAT, then there exists* m *satisfying the following inequation*

$$\left\lceil \frac{n-1-\log_2 l}{\log_2 3.71 - 1} \right\rceil \leq m \leq \left\lceil \frac{5}{4}(n-1) \right\rceil \tag{1}$$

When $l = 1$, m has the maximum $\frac{n-1-\log_2 l}{\log_2 3.71-1} < \frac{5}{4}(n-1)$. We apply this chaos amplification to the result of quantum algorithm of multiple alignment.

4. Quantum Algorithm of Multiple Alignment

We discuss the quantum algorithm of the multiple alignment by simulated annealing[13] in this section. Let us consider N amino acid sequences $\mathcal{A}_1, \mathcal{A}_2, \cdots, \mathcal{A}_N$. These are represented as follows.

$$\begin{aligned} \mathcal{A}_1 &: a_1^1 a_2^1 \cdots a_{m_1}^1 \\ \mathcal{A}_2 &: a_1^2 a_2^2 \cdots a_{m_2}^2 \\ &\cdots \\ \mathcal{A}_N &: a_1^N a_2^N \cdots a_{m_N}^N, \end{aligned}$$

where $m_i (i = 1, \cdots, N)$ are the numbers of amino acids for each sequences.

In order to make the quantum algorithm of the multiple alignment, we need to modify the input data for the alignment. Let us explain this process for the following three sequences:

GGIPG
GGQPIGA
GIPQIG

First, we add some gaps at the end of the amino acid sequences to make all sequences have the same length and to be useful for the alignment such as

GGIPG $****$
GGQPIGA $**$
GIPQIG $***$

Here, the maximum number of the gaps needed for the multiple alignment is due to the rules considered in [14].

Then let L be the length of the arranged amino acid sequences, so that all amino acid sequences can be written by

$$\begin{aligned} \mathcal{A}_1' &: a_1^1 a_2^1 \cdots a_L^1 \\ \mathcal{A}_2' &: a_1^2 a_2^2 \cdots a_L^2 \\ &\cdots \\ \mathcal{A}_N' &: a_1^N a_2^N \cdots a_L^N . \end{aligned}$$

Let $\pi : \Sigma^L \to \Sigma^L$ be a function, we call π an alignment if π_k keeps the order of amino acids. For example,

$$\pi\left(\mathbf{GGIPG} * * * **\right) = *\mathbf{G} * \mathbf{GIP} * \mathbf{G} * *$$

We take π_k for k-th amino acid sequence arbitrary and apply it for each sequences. We write this operation as

$$\phi\left(\mathcal{A}\right) = \pi_1\left(\mathcal{A}_1\right)\pi_2\left(\mathcal{A}_2\right)\cdots\pi_N\left(\mathcal{A}_N\right)$$

Then we obtain the alignmented sequences $\mathcal{A}_1'', \cdots \mathcal{A}_N''$.

Here we define the objective function $f_d(\mathcal{A})$ to apply the simulated annealing by

$$f_d\left(\mathcal{A}\right) = \sum_{k=1}^{L} \left\{ \frac{\sum_{i=1}^{N-1} \sum_{j=i+1}^{N} d\left(a_k^i, a_k^j\right)}{{}_N C_2} \right\},$$

where $d\left(a_k^i, a_k^j\right)$ is a distance defined as

$$d(a_k^i, a_k^j) = \left\{ \begin{array}{l} 0\left(a_k^i = a_k^j\right) \\ 1\left(a_k^i \neq a_k^j \text{ and } \left(a_k^i \neq * \text{ and } a_k^j \neq *\right)\right) \\ w\left(a_k^i \neq a_k^j \text{ and } \left(a_k^i = * \text{ or } a_k^j = *\right)\right) \end{array} \right\}.$$

In the above definition, w is called the weight having a value in $(0, 2]$, we take $w = 2$ in the sequel. This objective function f_d is an averaged difference among all sequences.

In order to compute the minimum value of $f_d(\mathcal{A})$ by the simulated annealing[12], we first replace each amino acids with 1 and gaps with 0, and keep the order of amino acids in the other working space. The resulting sequence of 0 and 1 is called a labelled sequence. After this operation, we

have

$$\begin{array}{lll} \mathcal{A}'_1 : \mathbf{GGIPG} * * & & \mathcal{B}_1 : 1111100 \\ \mathcal{A}'_2 : \mathbf{GGPQIG}* & \Longrightarrow & \mathcal{B}_2 : 1111110 \\ \mathcal{A}'_3 : \mathbf{GIPQIG}* & & \mathcal{B}_3 : 1111110 \end{array}$$

The perturbation in the simulated annealing means here to exchange some 0 and 1 in the labelled sequences as

$$\begin{array}{lllll} \mathcal{B}_1 & & \mathcal{B}'_1 : 1111001 & & \mathcal{A}''_1 : \mathbf{GGIP} * *\mathbf{G} \\ \mathcal{B}_2 & \Longrightarrow & \mathcal{B}'_2 : 1101111 & \Longrightarrow & \mathcal{A}''_2 : \mathbf{GG} * \mathbf{PQIG} \\ \mathcal{B}_3 & & \mathcal{B}'_3 : 1011111 & & \mathcal{A}''_3 : \mathbf{G} * \mathbf{IPQIG} \end{array}$$

We can see the position of bold style 0 is changed in the sequence $\mathcal{B}_2$ and $\mathcal{B}_3$. Note that the order of the amino acids in the sequences is not changed by this perturbation and the perturbated sequence is denoted by $\mathcal{A}'$.

When we apply the quantum algorithm to the multiple alignment, the symbols 0 and 1 are considered as the vectors $|0\rangle$ and $|1\rangle$ of qubits. Therefore the Hilbert space for the quantum algorithm is defined as

$$\mathcal{H} = \left(\mathbb{C}^2\right)^{\otimes NL+X+\log s+1}$$

where L is the length of amino acid sequence, N is a number of sequences, and s is a threshold of the objective function. An initial state vector for quantum algorithm is

$$\begin{aligned} |\psi_0\rangle = \left|1^{k_1}, 0^{g_1}\right\rangle \otimes \left|1^{k_2}, 0^{g_2}\right\rangle \otimes \cdots \\ \otimes \left|1^{k_N}, 0^{g_N}\right\rangle \otimes \left|0^X\right\rangle \otimes |s\rangle \otimes |0\rangle \end{aligned}$$

where k_i (g_i) is the number of amino acids (gaps) of i-th sequence respectively, and X is the number of qubit of work space.

Step 1

This step is composed of the following three steps **S(1.1)**, **S(1.2)** and **S(1.3)**.

S(1.1)

Construct the unitary operator U_P^i for the perturbation for i-th sequence. Make a pair of $|0\rangle$ and $|1\rangle$ by choosing arbitrary for each amino acid sequences. Then, repeat this operation until all $|0\rangle$ make pairs. Let m be the position of the $|1\rangle$ and n be the position of $|0\rangle$.

$$U_P^i = \prod_{\text{all pair}(m_i,n_i)} U_{m_i,n_i}^i$$

$$\begin{aligned}
U_{m_i,n_i}^i &= \otimes^{(i-1)L} I \\
&\otimes \left\{ \otimes^{n_i-1} I \otimes |0\rangle\langle 0| \otimes^{L-n_i} I \right. \\
&\left. + \otimes^{m_i-1} I \otimes (|0\rangle\langle 1| + |1\rangle\langle 0|) \otimes^{n_i-m_i-1} I \otimes |1\rangle\langle 1| \otimes^{L-n_i} I \right\} \\
&\otimes^{(N-i)L} I \otimes^{X+\log s+1} I
\end{aligned}$$

For example, let

$$\overbrace{|1\rangle \otimes |1\rangle \otimes |1\rangle \otimes |1\rangle \otimes |1\rangle}^{\text{amino acid}} \otimes \overbrace{|0\rangle \otimes |0\rangle \otimes |0\rangle}^{\text{gap}}.$$

be a amino acid sequence, we make pairs of $|0\rangle$ and $|1\rangle$ as

$$\begin{aligned}
\text{Pair1} &\equiv (2,7), \\
\text{Pair2} &\equiv (5,6), \\
\text{Pair3} &\equiv (3,8),
\end{aligned}$$

where the first number of the pair denotes the position of $|1\rangle$ and the second number denotes the position of $|0\rangle$.

Then U_P has the form as

$$U_P \equiv U_{2,7} \cdot U_{5,6} \cdot U_{3,8}.$$

S(1.2)

Apply Hadamard transformation H to $|0\rangle$ (the part of the gap). This gate is given by

$$U_H = \otimes_{i=1}^N \left(\otimes^{k_i} I \otimes^{g_i} H \right) \otimes^{X+\log s+1} I$$

In the example, we have

$$
\begin{aligned}
&(\otimes^5 I \otimes^3 H)\,(|1\rangle \otimes |1\rangle \otimes |1\rangle \otimes |1\rangle \otimes |1\rangle \otimes |0\rangle \otimes |0\rangle \otimes |0\rangle)\\
&= \frac{1}{\sqrt{2^3}}(|1\rangle \otimes |1\rangle \otimes |1\rangle \otimes |1\rangle \otimes |1\rangle \otimes |0\rangle \otimes |0\rangle \otimes |0\rangle\\
&\quad + |1\rangle \otimes |1\rangle \otimes |1\rangle \otimes |1\rangle \otimes |1\rangle \otimes |0\rangle \otimes |0\rangle \otimes |1\rangle\\
&\quad + |1\rangle \otimes |1\rangle \otimes |1\rangle \otimes |1\rangle \otimes |1\rangle \otimes |0\rangle \otimes |1\rangle \otimes |0\rangle\\
&\quad + |1\rangle \otimes |1\rangle \otimes |1\rangle \otimes |1\rangle \otimes |1\rangle \otimes |0\rangle \otimes |1\rangle \otimes |1\rangle\\
&\quad + |1\rangle \otimes |1\rangle \otimes |1\rangle \otimes |1\rangle \otimes |1\rangle \otimes |1\rangle \otimes |0\rangle \otimes |0\rangle\\
&\quad + |1\rangle \otimes |1\rangle \otimes |1\rangle \otimes |1\rangle \otimes |1\rangle \otimes |1\rangle \otimes |0\rangle \otimes |1\rangle\\
&\quad + |1\rangle \otimes |1\rangle \otimes |1\rangle \otimes |1\rangle \otimes |1\rangle \otimes |1\rangle \otimes |1\rangle \otimes |0\rangle\\
&\quad + |1\rangle \otimes |1\rangle \otimes |1\rangle \otimes |1\rangle \otimes |1\rangle \otimes |1\rangle \otimes |1\rangle \otimes |1\rangle).\\
&= |\varphi\rangle
\end{aligned}
$$

S(1.3)

The third step is to apply U_P to the above state $|\varphi\rangle$. The resulting state $U_P\,|\varphi\rangle$ is called the perturbated state of the sequence.

For the example, it holds that

$$
\begin{aligned}
U_P\,|\varphi\rangle = \frac{1}{\sqrt{2^3}}(&|1\rangle \otimes |1\rangle \otimes |1\rangle \otimes |1\rangle \otimes |1\rangle \otimes |0\rangle \otimes |0\rangle \otimes |0\rangle\\
&+ |1\rangle \otimes |1\rangle \otimes |\mathbf{0}\rangle \otimes |1\rangle \otimes |1\rangle \otimes |0\rangle \otimes |0\rangle \otimes |\mathbf{1}\rangle\\
&+ |1\rangle \otimes |\mathbf{0}\rangle \otimes |1\rangle \otimes |1\rangle \otimes |1\rangle \otimes |0\rangle \otimes |\mathbf{1}\rangle \otimes |0\rangle\\
&+ |1\rangle \otimes |\mathbf{0}\rangle \otimes |\mathbf{0}\rangle \otimes |1\rangle \otimes |1\rangle \otimes |0\rangle \otimes |\mathbf{1}\rangle \otimes |\mathbf{1}\rangle\\
&+ |1\rangle \otimes |\mathbf{0}\rangle \otimes |\mathbf{0}\rangle \otimes |1\rangle \otimes |1\rangle \otimes |0\rangle \otimes |\mathbf{1}\rangle \otimes |\mathbf{1}\rangle\\
&+ |1\rangle \otimes |1\rangle \otimes |\mathbf{0}\rangle \otimes |1\rangle \otimes |\mathbf{0}\rangle \otimes |\mathbf{1}\rangle \otimes |0\rangle \otimes |\mathbf{1}\rangle\\
&+ |1\rangle \otimes |\mathbf{0}\rangle \otimes |1\rangle \otimes |1\rangle \otimes |\mathbf{0}\rangle \otimes |\mathbf{1}\rangle \otimes |\mathbf{1}\rangle \otimes |0\rangle\\
&+ |1\rangle \otimes |\mathbf{0}\rangle \otimes |\mathbf{0}\rangle \otimes |1\rangle \otimes |\mathbf{0}\rangle \otimes |\mathbf{1}\rangle \otimes |\mathbf{1}\rangle \otimes |\mathbf{1}\rangle).
\end{aligned}
$$

Let us do the multiple alignment of N sequences with their common sequence length L.

Step2(computation)

S(2.1)

For the superposition state, compute $f_d\,(\phi\,(\mathcal{A}))$ by the unitary operator U_C by the following algorithm

(1) Initialize counters : $k = 1, i = 1$ and $j = 2$
(2) For $k = L$ do (step3,4,5,6,7)

(3) For $i = N - 1$ do (step4,5,6)
(4) For $j = N$ do (step5,6)
(5) Calculate d for $(i-1)\,L + k - 1$-th and $jL + k - 1$-th amino acid of $\mathcal{A}$
(6) Store the result at the workspace
(7) Take the sum of d for i and j, divide ${}_N C_2$ into it, and store the result at the workspace
(8) Compute $f_d\left(\phi\left(\mathcal{A}\right)\right)$ form these result

This algorithm is given by classical one, so there exists the unitary operator to achieve this.

S(2.2)

Define the unitary operator U_A as the following

$$
\begin{aligned}
&U_A \otimes_1^{LN} I \otimes \left|f_d\left(\phi\left(\mathcal{A}\right)\right)\right\rangle \otimes |s\rangle \otimes |0\rangle \\
&\equiv \otimes_1^{LN} I \otimes \left|f_d\left(\phi\left(\mathcal{A}\right)\right)\right\rangle \otimes |s\rangle \otimes \left\{ \begin{array}{l} |1\rangle \ \left(s > f_d\left(\phi\left(\mathcal{A}\right)\right)\right) \\ |0\rangle \ \left(s \leq f_d\left(\phi\left(\mathcal{A}\right)\right)\right) \end{array} \right\}.
\end{aligned}
$$

Apply U_A to the state $U_C U_P U_H \left|\psi_0\right\rangle$, the check bit of the sequences satisfying $s \geq f_d(\phi_i\left(\mathcal{A}\right))$, namely the sequences accepted, becomes $|1\rangle$. If the acceptance probability is very small (about less than $\frac{1}{2^{LN}}$), we use the chaotic dynamics to amplify the probability [1,2].

Step3(observation)

We observe the check vector whether it is $|1\rangle$ or not by a similar way given in [1,2]. Then we obtain the aligned sequences which is the result of the multiple alignment.

5. Computational Complexity of Multiple Alignment

Finally, we discuss the difference between quantum algorithm and classical algorithm for the multiple alignment. Table 1 shows the computational costs of the classical algorithm, the classical algorithm of the simulated annealing and the quantum algorithm, from which we can easily see the advance of the quantum algorithm.

	Classical algorithm of simulated annealing	Quantum algorithm
Perturbation	$2^{N \times L}$	$N \times L$
Calculation	$2^{N \times L} \times L \times_N C_2$	$L \times_N C_2$
Total	$(2 + L \times_N C_2) \times 2^{N \times L}$	$(2N +_N C_2) \times L$

It is shown that this quantum algorithm is finished in polynomial order of the size N of input sequences and the sequence length L.

References

1. M.Ohya and I.V.Volovich, Quantum computing and chaotic amplification, J. opt. B, 5,No.6 639-642, 2003.
2. M.Ohya and I.V.Volovich, New quantum algorithm for studying NP-complete problems, Rep.Math.Phys., 52, No.1,25-33 2003.
3. M.Ohya and N.Masuda, NP problem in Quantum Algorithm, Open Systems and Information Dynamics, 7 No.1 33-39, 2000.
4. M.Ohya, Complexities and Their Applications to Characterization of Chaos, Int. Journ. of Theoret. Physics, 37 495, 1998.
5. L.Accardi and M.Ohya, Compound channels, transition expectations, and liftings, Appl. Math. Optim., Vol.39, 33-59, 1999.
6. L.Accardi and M.Ohya, A Stochastic Limit Approach to the SAT Problem,Open Systems and Information dynamics, 11,1-16, 2004.
7. L.Accardi and R.Sabbadini, On the Ohya–Masuda quantum SAT Algorithm, Preprint Volterra, N. 432, 2000.
8. M.Ohya and I.V.Volovich, Mathematical Foundation of Quantum Information and Quantum Computation, to be published.
9. M.Ohya and Y.Uesaka : Amino Acid Sequences and DP Matching, International Journal of Information Science.63,139-151 1992
10. Sellers P.H.: On the Theory and Computation of Evolutionary Distances, SIAM J.Appl.Math, 1974
11. S.B.Needleman and C.D. Wunsch : A general method applicable to search for similarities in the amino acid sequence of two proteins., J.Mol.Biol.,88,pp.233-258, 1986
12. M.Ohya, S.Miyazaki and K.Ogata : On Multiple Alignment of Genome Sequences. I.E.I.C.E Trans.Commun, vol.E75-B, No.6, 453-457, 1992
13. E. Aarts and J. Korst : Simulated Annealing and Boltzmann Machines, John Wiley and Sons, 1989
14. M.Ohya and K,Ogata : Multiple Alingment for Amino Acid Sequences by Dynamic Program. Electronics and Communications in Japan, Part3, vol.81, No4, 1998
15. S.Iriyama and M.Ohya, Rigorous Estimation of the Computational Complexity for OMV SAT Algorithm, Open System & Information Dynamics, 15, 2, 173-187, 2008.
16. S.Iriyama and M.Ohya, Language Classes Defined by Generalized Quantum Turing Machine, to appear in OSID, 2008
17. S.Iriyama, M.Ohya and I.V.Volovich, Generalized Quantum Turing Machine and its Application to the SAT Chaos Algorithm, QP-PQ:Quantum Prob. White Noise Anal., Quantum Information and Computing, 19, World Sci. Publishing, 204-225, 2006.

QUANTUM-LIKE MODELS FOR DECISION MAKING IN PSYCHOLOGY AND COGNITIVE SCIENCE

ANDREI. KHRENNIKOV *

International Center for Mathematical Modelling
in Physics and Cognitive Sciences
University of Växjö, S-35195, Sweden
E-mail: Andrei.Khrennikov@vxu.se

We show that (in contrast to rather common opinion) the domain of applications of the mathematical formalism of quantum mechanics is not restricted to physics. This formalism can be applied to the description of various quantum-like (QL) information processing. In particular, the calculus of quantum (and more general QL) probabilities can be used to explain some paradoxical statistical data which was collected in psychology and cognitive science. The main lesson of our study is that one should sharply distinguish the mathematical apparatus of QM from QM as a physical theory. The domain of application of the mathematical apparatus is essentially wider than quantum physics. Quantum-like representation algorithm, formula of total probability, interference of probabilities, psychology, cognition, decision making

Keywords : *Quantum-like representation algorithm, formula of total probability, interference of probabilities, psychology, cognition, decision making*

1. Introduction

Already Bohr [1] pointed out to the possibility to apply the mathematical formalism of quantum mechanics outside of physics, in particular, in psychology, see also correspondence between Pauli and Jung [2]. We point out that the complementarity principle is a general philosophical principle. In applications to quantum physics it is quantatively exhibited through *interference phenomenon.*

In purely probabilistic terms interference can be represented as interference of probabilities of alternatives. Detailed analysis of this problem was performed by Khrennikov [3–6]. It was shown that interference of probabilities can be represented as violation of the *law of total probability* (also

*This work is supported by the profile of växjö university – "mathematical modelling and system collaboration."

called the *law of alternatives*) which is widely used in classical statistics. This effect was confirmed (at least preliminary) experimentally [3].

Recently a similar viewpoint to the role of the law of total probability was presented by Busemeyer [7] who described the well known disjunction effect [9, 8] (violating Savage sure thing principle - STP) by using the quantum formalism. We recall this law in the simplest case of dichotomous random variables, $a = \pm$ and $b = \pm$:

$$P(b = \pm) = P(a = +)P(b = \pm|a = +) + P(a = -)P(b = \pm|a = -). \quad (1)$$

This formula plays the fundamental role in modern science. It was pointed [3–6] that the quantum formalism induces a modification of this formula. An additional term appears in the right hand side of (1), so called *interference term.* Violation of the law of total probability can be considered as an evidence that the classical probabilistic description could not be applied. Our aim is to show that QL probabilistic descriptions could be applied. The terminology "quantum-like" and not simply "quantum" is used to emphasize that violations of (1) are not reduced to those which can be described by the conventional quantum model. Contexts which are nonclassical (in the sense of violation of (1)), but at the same time cannot be described by the conventional quantum formalism may appear outside quantum physics.

What are the sources of violation of the law of total probability?

The most natural explanation can be provided in so called contextual probabilistic framework [3–6]. The basic notion of this approach is *context.* In quantum mechanics it is a complex of experimental physical conditions. In the present paper it will be a complex of mental conditions [5]. In particular, we shall consider contexts corresponding to *Prisoner's Dilemma* (PD) as well as contexts for gambling experiments[9]. The crucial point is that probabilities in the law of total probability correspond to different contexts. A priori there is no reason to assume that all those (essentially different contexts) could be "peacefully combined." Therefore in the contextual framework one could not use *Boolean algebra* for contexts. We recall that Boolean algebra is used in classical probability theory. It is important to remark that in the latter conditioning is considered not with respect to a context, but with respect to an event.

Roughly speaking violation of the law of total probability is not surprising. It is surprising that we were able to find so many situations (in particular, in classical statistical physics, psychology and economics) in which it can be applied and that we were lucky to proceed so far by using classical probability. The latter can be explained if we consider this law as an *approximative law.* If the additional term which should appear in the general case in the right-hand side of (1), the "interference term", is relatively small, then one could neglect by it and proceed by applying (1) without problem. In fact, the fundamental contribution of Tversky and Shafir [9,8] is that they found statistical data which *violates essentially* the law of total probability.

Our contextual approach does not contradict Bayesian approach which nowadays is extremely popular in cognitive science and psychology. We just say that Bayesian analysis is an *approximative theory.* It has its domain of application. But (as any mathematical model) it has its boundaries of application. From our viewpoint the disjunction effect demonstrated that we have approached these boundaries.

Thus the formula of total probability which is the basis of Bayesian analysis is, in fact, not the precise equality (1), but it should be written as an approximative formula:

$$P(b = \pm) \approx P(a = +)P(b = \pm|a = +) + P(a = -)P(b = \pm|a = -). \quad (2)$$

2. Prisoner's Dilemma

In game theory, PD is a type of non-zero-sum game in which two players can cooperate with or defect (i.e. betray) the other player. In this game, as in all game theory, the only concern of each individual player (prisoner) is maximizing his/her own payoff, without any concern for the other player's payoff. In the classic form of this game, cooperating is strictly dominated by defecting, so that the only possible equilibrium for the game is for all players to defect. In simpler terms, no matter what the other player does, one player will always gain a greater payoff by playing defect. Since in any situation playing defect is more beneficial than cooperating, all rational players will play defect. This is the *principle of rational behavior* (a form of Savage's sure thing principle) which is basic for rational choice theory which is the dominant theoretical paradigm in microeconomics. It is also central to modern political science and is used by scholars in other disciplines such as sociology. However, Shafir and Tversky [8] found that players frequently behave irrationally.

Contexts in PD:

Each contextual model is based on a collection of contexts and a collection of observables. Such observables can be measured[a] for each of contexts

[a]By measurements we understand even self-measurements which are performed by e.g.

under consideration [6] for the general formalism. The following mental contexts are involved in PD:

Context C representing the situation such that a player has no idea about planned action of another player.

Context C_+^A representing the situation such that the B-player supposes that A will cooperate and context C_-^A – A will compete. We can also consider similar contexts $C_\pm^B$.

We define dichotomous observables a and b corresponding to *actions* of players A and $B : a = +$ if A chooses to cooperate and $a = -$ if A chooses to compete, b is defined in the same way.

A priory the law of total probability might be violated for PD, since the B-player is not able to combine contexts. If those contexts were represented by subsets of a so called space of "elementary events" as it is done in classical probability theory (based on Kolmogorov (1933) measure-theoretic axiomatics), the B-player would be able to consider the conjunction of the contexts C and e.g. C_+^A and to operate in the context $C \wedge C_+^A$ (which would be represented by the set $C \cap C_+^A$). But the very situation of PD is such that one could not expect that contexts C and $C_\pm^A$ might be peacefully combined. If the B-player obtains information about the planned action of the A-player (or even if he just decides that A will play in the definite way, e.g. the context C_+^A will be realized), then the context C is simply destroyed. It could not be combined with C_+^A.

We can introduce the following contextual probabilities:

$P(b = \pm|C)$ – probabilities for actions of B under the complex of mental conditions C.

$P_{\pm,+} \equiv P(b = \pm|C_+^A)$ and $P_{\pm,-} \equiv P(b = \pm|C_-^A)$ – probabilities for actions of B under the complexes of mental conditions C_+^A and C_-^A, respectively.

$P(a = \pm|C)$ – priory probabilities which B assigns for actions of A under the complex of mental conditions C.

As we pointed out, there are no priory reasons for the equality (1) to hold. And experimental statistics collected by Shafir and Tversky [8] demonstrated that this equality could be really violated (as we shall show!). From paper [8] for PD experiment we have: $P(b = -|C) = 0.63$ and hence $P(b = +|C) = 0.37$; $P_{-,-} = 0.97$, $P_{+,-} = 0.03$; $P_{-,+} = 0.84$, $P_{+,+} = 0.16$.

As always in probability theory it is convenient to introduce the ma-

---

the brain.

trix of transition probabilities $P = \begin{pmatrix} 0.16 & 0.84 \\ 0.03 & 0.97 \end{pmatrix}$. We point out that this matrix is *stochastic.* It is a square matrix each of whose rows consists of nonnegative real numbers, with each row summing to 1. This is the common property of all matrices of transition probabilities.

We now recall the definition of a *doubly stochastic matrix:* in a doubly stochastic matrix all entries are nonnegative and all rows and all columns sum to 1. It is clear that the *matrix obtained by Shafir and Tversky is not doubly stochastic.*

In the simplified framework the prisoner B considers (typically unconsciously) priory probabilities $p = P(a = +|C)$ and $1 - p = P(a = -|C)$ which B assigns for actions of A under the complex of mental conditions C. These probabilities are parameters of the model. In the simplest case B assigns some fixed value p to A-cooperation. The mental wave function depends on p.

However, in reality the situation is essentially more complicated. The B is not able to determine precisely p. He considers a spectrum of possible p which might be assigned to A-cooperation. Therefore, instead of a pure QL-state (mental wave function), the B-brain creates a *statistical mixture of mental wave functions* corresponding to some range of parameters p which could be assigned to A-cooperation. In this statistical mixture different wave functions are mixed with some weights. Instead of the wave function, B creates a von Neumann density matrix which describes B's state of mind. We emphasize that the latter operation of statistical mixing is purely classical. The crucial step is creation of the QL-representation for fixed value of the parameter p.

3. Contexts in Gambling Experiments

In paper[9] it was proposed to test disjunction effect for the following gambling experiment. In this experiment, you are presented with two possible plays of a gamble that is equally likely to win 200 USD or lose 100USD. You are instructed that the first play has completed, and now you are faced with the possibility of another play.

Here a gambling device, e.g., roulette, plays the role of A; B is a real player, his actions are $b = +$, to play the second game, $b = -$, not. Here the context C correspond to the situation such that the result of the first game is unknown for B; the contexts $C_{\pm}^{A}$ correspond to the situations such that the results $a = \pm$ of the first play in the gamble are known. From Tversky and Shafir [9] we have: $P(b = +|C) = 0.36$ and hence $P(b = -|C) = 0.64$; $P_{+,-} = 0.59$, $P_{-,-} = 0.41$; $P_{+.+} = 0.69$, $P_{-.+} = 0.31$.

We get the following matrix of transition probabilities: $P = \begin{pmatrix} 0.69 & 0.31 \\ 0.59 & 0.41 \end{pmatrix}$. This matrix of transition probabilities is not doubly stochastic either (cf. with the experiment in paper [8]).

In this experiment (in contrast to paper [8]) probabilities $P(a = \pm|C)$ are not subject of a priory consideration. They are fixed from the very beginning as 1/2.

4. Measure of Interference

Violation of the law of total probability implies that the left-hand and right-hand sides of (1) do not coincides. Therefore it is natural to consider the difference between them as a measure of incompatibility between contexts C and $C_A^{\pm}$. We denote it by the symbol $\delta_{\pm}$. It is the measure of impossibility to combine these contexts in a single space of elementary events. In PD C can be called uncertainty context – B has no information about planned actions of A. This context is incompatible with the contexts $C_A^{\pm}$ corresponding to definite actions of A. We propose to measure this incompatibility numerically by using δ. This number can be found if one have all probabilities involved in the law of total probability.

The next important question is the choice of normalization of δ. Here we proceed as in Khrennikov's article [3] . We are lucky that quantum mechanics has been already discovered. Its formalism implies that for quantum systems (e.g. photons) this coefficient of incompatibility has the form $2\cos\theta$ (where the angle θ is called phase) multiplied by the normalization factor which is equal to square root of the product Π of all probabilities in the right-hand side of (1). Thus

$$\delta = 2\cos\theta\,\sqrt{\Pi}.$$

We proposed to use the same normalization in the general case of any collection of contextual probabilities. Thus we introduce the normalized coefficient of incompatibility of mental contexts: $\lambda = \frac{\delta}{2\sqrt{\Pi}}$. As was mentioned, in the conventional quantum mechanics it is always bounded by one. Hence, it can be written as $\lambda = \cos\theta$, where $\theta = \arccos\lambda$.

However, it could as well be larger than one[6]. In such a case it can be written as $\lambda = \pm\cosh\theta$, where $\theta = \operatorname{arccosh}|\lambda|$.

Since in the conventional quantum mechanics the term $\delta = 2\cos\theta\,\sqrt{\Pi}$ describes interference, we can call δ the interference term even in the general contextual framework. The same terminology we use for the normalized coefficient λ : the coefficient of interference. It can be considered as a measure of *"interference of mental contexts."*

5. Interference in Disjunction Experiments

Since in gambling experiment [9] the A-probabilities are fixed, it is easier for investigation. Simple arithmetic calculations give $\delta_+ = -0.28$, and hence $\lambda_+ = -0.44$. Thus the probabilistic phase $\theta_+ = 2.03$. We recall [3] that $\delta_+ + \delta_- = 0$ (in the general case). Thus $\delta_- = 0.28$, and hence $\lambda_- = 0.79$. Thus the probabilistic phase $\theta_- = 0.66$.

In the case of PD-experiment[8] the B-player assigns probabilities of the A-actions, p and $1-p$ (in the simplest case). Thus coefficients of interference depend on p. We start with $\delta_- = -(0.21 + 0.13p)$ and $\lambda_- = -(0.12 + 0.07p)/\sqrt{p(1-p)}$. For example, if B would assume that A will act randomly with probabilities $p = 1 - p = 1/2$, then the interference between contexts is given by $\lambda_- = -0.31$ and hence the phase $\theta = 1.89$. We now find $\delta_+ = (0.21+0.13p)$ and $\lambda_+ = -(1.52+0.94p)/\sqrt{p(1-p)}$. For example, if B would assume that A will act randomly with probabilities $p = 1 - p = 1/2$, then the interference between contexts is given by $\lambda_+ = 3.98$. Thus interference is very high. It exceeds the possible range of the conventional trigonometric interference. This is *the case of hyperbolic interference!* Here the hyperbolic phase $\theta_+ = \text{arccosh}\ (3.98) = 2.06$.

6. Quantum-like Representation Algorithm – QLRA

This algorithm will produce a probability amplitude from contextual probabilities. We shall consider separately two cases:

Trigonometric Mental Interference:

The coefficients of interference are bounded by one.

In this case we can represent $\lambda_\pm$ in the form $\lambda_\pm = 2\cos\theta_\pm\sqrt{\Pi}$. Hence we obtain the following modification of the law of total probability:

$$P(b = \pm) = P(a = +)P_{\pm,+} + P(a = -)P_{\pm,-} + 2\cos\theta_\pm\sqrt{\Pi}, \qquad (3)$$

where $\Pi_\pm = P(a = +|C)P(a = -|C)P_{\pm,+}P_{\pm,-}$. In a special case – for a doubly stochastic matrix of transition probabilities – this law can be derived in the conventional quantum formalism.

We now recall elementary formula from algebra of complex numbers: $k = k_1 + k_2 + 2\sqrt{k_1 k_2}\cos\theta = |\sqrt{k_1} + e^{i\theta}\sqrt{k_2}|^2$, for real numbers $k_1, k_2 > 0, \theta \in [0, 2\pi]$. Thus $k = |\psi|^2$, where $\psi = \sqrt{k_1} + e^{i\theta}\sqrt{k_2}$. Let us compare this formula and the interference law of total probability (3). We set $k = P(b = \pm), k_1 = P(a = +)P_{\pm,+}, k_2 = P(a = -)P_{\pm,-}$. We introduce the complex probability amplitudes: $\psi(\pm) = \sqrt{P(a = +)P_{\pm,+}} + e^{i\theta_\pm}\sqrt{P(a = -)P_{\pm,-}}$. We call its mental wave function (it is defined on the set $\{+, -\}$ and takes complex values) representing the context C via observables a and b.

The crucial point is that Born's rule takes place: $P(b = \pm) = |\psi(\pm)|^2$. We speculate that the brain can apply such an algorithm to probabilistic data about contexts and construct the complex probability amplitude, the mental wave function. Then it operates only with such amplitudes and not with original probabilities.

Finally, we provide the interpretation of coefficients $k_1 = P(a = +)P_{\pm,+}$ and $k_2 = P(a = -)P_{\pm,-}$. For example, $P_{a\to b}(+, +) \equiv P(a = +)P_{+,+}$ is the probability to get values $a = +, b = +$ in "measurement" in that the a is measured at the first step and the b is measured at the second step; $P_{a\to b}(-, +) \equiv P(a = +)P_{-,+}$ is the probability to get values $a = +, b = -$.

Hyperbolic Mental Interference:

The coefficients of interference are larger than one.

Here mathematics is more complicated. One should use so called hyperbolic numbers, instead of complex numbers. We would not like to go in mathematical details. We just mention that one should change everywhere the imaginary unit i (such that $i^2 = -1$) to hyper-imaginary unit $j : j^2 = +1$ and usual trigonometric functions $\cos\theta$ and $\sin\theta$ to their hyperbolic analogues $\cosh\theta$ and $\sinh\theta$, see e.g. Khrennikov [6] for details. Here the probabilistic image of incompatible mental contexts is given by the hyperbolic probabilistic amplitude: $\psi(\pm) = \sqrt{P(a = +)P_{\pm,+}} \pm e^{j\theta_\pm}\sqrt{P(a = -)P_{\pm,-}}$.

We emphasize that some cognitive systems may exhibit (for some mental contexts) hyper-trigonometric interference: one coefficient, e.g., λ_+ is bounded by one and another is larger than one.

Mental Wave Function for Tversky-Shafir Experiment:

This experiment has simpler QL-representation. Both coefficients of interference are bounded by one. Thus we can represent incompatible contexts by the complex probability amplitude: $\psi(+) \approx 0.59 + e^{2.03i}\, 0.54$; $\psi(-) \approx 0.39 + e^{0.79i}\, 0.45$. We remind again that our algorithm for representing statistical data by complex probability amplitudes reproduces Born's rule (which is postulated in QM): $|\psi(+)|^2 = P(b = +)$, $|\psi(-)|^2 = P(b = -)$. Thus in the situation (context) C such that B

does not know about the result of his first game the probabilities of his actions are encoded in the mental wave function ψ. In this simplest case it is just a vector with two complex coordinates.

Finally, we provide the interpretation of coefficients $k_1 = P(a = +)P_{\pm,+}$ and $k_2 = P(a = -)P_{\pm,-}$ for this "gambling measurement." For example, $P_{a\to b}(+,+) \equiv P(a = +)P_{+,+}$ is the probability that the roulette (computer) produces the value $a = +1$ (so the first game is successful) and the B decides to play again. It should be pointed out that B was informed about the result of his first game. Thus: $P_{a\to b}(+,+) = P(B$ won $\to$ informed $\to$ decided to play again). In the same way we have: $P_{a\to b}(-,+) = P(B$ won $\to$ informed $\to$ decided not to play again).

Mental Wave Function for Shafir-Tversky experiment:

Here the B-player creates QL-representation by assigning the probabilities p and $1-p$ to possible actions of A. The wave function depends on p. For example, suppose that B assigned to the A-actions equal probabilities. Then the B-brain would represent the PD game by the following hyper-trigonometric amplitude: $\psi(+) \approx 0.28 + e^{2.06j}\ 0.12$; $\psi(-) \approx 0.65 + e^{1.89i}\ 0.7$ It is once again: $|\psi(+)|^2 = P(b = +)$, $|\psi(-)|^2 = P(b = -)$. Thus in the situation (context) C such that B does not know about the strategy chosen by A the probabilities of B's actions are encoded in the mental wave function ψ.

We remark that in general ψ depends on the prior probabilities $P(a = +) = p$ and $P(a = -) = 1 - p : \psi \equiv \psi_p$. In the experiment which was performed by Shafir-Tversky the prior probability p was not determined – participants were not asked about their priors. Therefore we cannot reconstruct ψ without an additional assumption, namely, on p.

Evidently this experiment can be generalized. Each B-participant should be asked: "Which prior probabilities do you assign to actions of A? However, there are no reasons to get one fixed probability value, say p. It is more natural to expect to find a mixture of prior probabilities: values $p_1,\ldots,p_N$ are realized with probabilities $q_1, \ldots q_N$. Here $q_1 + \ldots + q_N = 1$.

At the next step the refined version of the Shafir-Tversky experiment should be repeated. Denote by Ω_p the B-population of those who chosen the prior probability p. The for this population we shall find its own transition probabilities $P_{\pm,\pm}(p)$. Here p plays the role of a parameter. On the basis of p and these probabilities by applying QLRA we reconstruct the "mental wave function" ψ_p of the population Ω_p.

Finally, we reconstruct the quantum state of the total population $\Omega =$

$\cup_p \Omega_p$ of B players participated in the experiment. It is not a pure state (wave function), but a mixed state (density matrix): $\rho = \sum_j q_j \psi_{p_j} \otimes \psi_{p_j}$. Here $\psi \otimes \psi$ denotes the orthogonal projector π_ψ onto the vector $\psi : \pi_\psi \phi = (\psi, \phi)\phi$. Finally, we provide the interpretation of coefficients $k_1 = P(a = +)P_{\pm,+}$ and $k_2 = P(a = -)P_{\pm,-}$ for PD under the pure state assumption, i.e. for players B belonging to the population Ω_p.

For example, $P_{a\to b}(+,+) \equiv P(a = +)P_{+,+}$ is the probability the following process:

P1). B assumes that A would cooperate with the probability $p = P(a = +)$ and hence compete with the probability $1 - p = P(a = -)$;

P2). Under assumption P1 the B decides that in this concrete "PD-game" A will cooperate;

P3). Under assumption P2 the B also decides to cooperate.

Thus: $P_{a\to b}(+,+)$

$$= P(\text{priors } p, 1-p \to \ A \text{ would cooperate } \to \ B\text{cooperating}).$$

In the same way $P_{a\to b}(+,-)$

$$= P(\text{priors } p, 1-p \to \ A \text{ would compete } \to \ B\text{cooperating}).$$

We emphasize again the difference between the Shafir-Tversky experiment of the PD-type and Tversky-Shafir gambling experiments. In the latter experiment probabilities for roulette are known for players: p=1-p=1/2. These are "objective probabilities." In the PD-type experiment they are not known for B. These are "subjective probabilities." However, the Tversky-Shafir gambling experiment can be easily modified into the same subjective probabilistic framework. It is sufficient not inform players about probabilities for roulette. Then players would guess about these probabilities in the same way as in the PD-type game B-players guess about probabilities for possible actions of A-players.

7. Hilbert-space Representation of Mental Information and Decision Making

By using QLRA we can represent statistical data from the Tversky-Shafir and Shafir-Tversky experiments (and, in fact, from any psychological experiment) in the QL form. In this way we just reconstructed the "mental wave function" which was created in the brains of players in the context of gambling.[b]

---

[b]Existing of such a "mental wave" in the brain is the basic postulate of our QL model of brain's functioning. This wave need not to have a direct relation with real quantum

For example, in Tversky-Shafir gambling experiment the $\psi(+)$ is the complex amplitude of "intention" of a player B to continue to play in the absence of information about the result of the first game. However, the B is not totally ignorant. First of all, the B knows probabilities for roulette (computer). Thus he knows that he can win with probability 1/2 (as it was in Tversky-Shafir gambling experiment). Then he has already been trained (on the previous experience, may be in quite different gambling) to make decision to gamble or not depending on the result of the previous gambling. These probabilities, $P_{\pm,\pm}$, are also present in B's brain.

If B were used only these probabilities, his decision making would be performed on the basis of the classical formula of total probability. Thus the B would be a "purely classical cognitive system."

But a QL player also uses the phase angle θ which is a new parameter. It was absent in the classical decision making process. The presence of this new parameter provides new possibilities for decision making which were totally absent in classical statistical decision making.

Our algorithm QLRA produces θ on the basis of given probabilistic data. But it is only reconstruction of the "mental phase" which is present (by our conjecture) in B's brain. Thus by using the mathematical terminology we solved a special inverse problem. However, the direct (and great!!!) problem is still open: "How does brain produces phases θ_j?"

Our conjecture is that the brain can learn itself to create phases between possible alternatives ($a = \pm 1$) preceding decision and possible alternatives for decision ($b = \pm 1$).

The easiest way to proceed is to create a Hilbert space representation for alternatives, i.e., to represent contexts $C_{\pm}^{A}$ by vectors $e_{\pm}^{A}$ and $C_{\pm}^{B}$ by vectors $e_{\pm}^{B}$. The presence of QLRA justifies the conjecture on the existence of such a representation. It can be reconstructed on the basis of probabilistic data. However, creation of QLRA also lighted difficulties with application of the conventional quantum formalism. It is possible to choose both bases $e_{\pm}^{A}$ and $e_{\pm}^{B}$ as the orthogonal ones (and hence to represent a and b by self-adjoint operators $\hat{a}$ and $\hat{b}$) only in the case of the doubly stochastic matrix of transition probabilities. If it is not doubly stochastic then one should generalize the conventional quantum formalism by considering non self-adjoint operators.

We emphasize again the difference between Tversky-Shafir gambling experiment and the PD-type experiment. In the PD-game the B player does

wave functions of quantum physical systems composing the brain.

not know (at least precisely) the probabilities for the A-choices. The B considers a priori probabilities $P(a = \pm)$. In general the PD-type experiment is represented by a mixed quantum state, von Neumann density operator.

8. Concluding Remarks: Accardian Models, Ohya's Adaptive Dynamics

My conjecture is that the laws of classical probability theory can be violated in cognitive sciences, psychology, social sciences and economy. However, nonclassical statistical data is not covered completely by the conventional quantum model. As Luigi Accardi [10, 11] pointed out the main distinguishing feature of quantum probability is its *non-Kolmogorovness.* Accardi emphasized that in the same way as in geometry (where starting with Lobachevsky's model various non-Euclidean geometries were developed and widely applied e.g. in relativity theory) in probability theory various non-Kolmogorov models may be developed to serve applications. The QM probabilistic model was one of the first non-Kolmogorovian models which had important applications. Thus one may expect development of other types of models which would be neither Kolmogorovian nor quantum. We call such models *Accardian probabilistic models.*

My personal explanation is based on the evidence [3] that violation of the formula of total probability does not mean that we should obtain precisely the formula of total probability with the interference term which is derived in the conventional quantum formalism. Nevertheless, the conventional quantum formalism can be used as the simplest nonclassical model for mental and social modelling. My model with the hyperbolic probability interference is an interesting example of an Accardian probabilistic model.

It is important to mention that non-Kolmogorovness may appear as a consequence of a very special representation of probabilistic data – in my approach via QLRA. Kolmogorovian data may be represented in special ways (convenient for some applications). In this way new models can be induced. I think that QL models generated via QLRA would play an important role in applications. In fact, our construction provides a *linear representation of data of any origin.*

Finally, we point out to one possible application of the QLRA-representation, namely, to *adaptive dynamical systems,* see e.g. Ohya [12]. On the basis of two fundamental features of adaptive dynamics – "state adaptivity" and "observable adaptivity" – we could represent probabilistic data produced by an adaptive dynamical system by using QLRA. In this way in general nonlinear dynamical system would be represented in complex (or even may be hyperbolic?) Hilbert space. However, realization of such a program needs essential technical efforts.

Conclusion. *By using violation of the law of total probability as the starting point we created the QL-representation of mental contexts. We found that, besides the conventional trigonometric interference in cognitive science can be exhibited so called hyperbolic interference. Thus the probabilistic structure of cognitive science is not simply nonclassical, but it is even essentially richer than the probabilistic structure of quantum mechanics.*

Acknowledgments

I would like to thank L. Accardi and M. Ohya for numerous discussions on applications of the quantum formalism outside of physics.

References

1. N. Bohr, *The philosophical writings of Niels Bohr*, 3 vols., Ox Bow Press, 1987.
2. C. G. Jung, and W. Pauli, In: C. A. Meier (Ed), *Ein Briefweksel,* Spinger Verlag, 1992, 679-702.
3. A. Khrennikov, *Information Dynamics in Cognitive, Psychological and Anomalous Phenomena,* ser. *Fundamental Theories of Physics,* Kluwer Academic, 2004.
4. A. Khrennikov, Interference in the classical probabilistic framework, *Fuzzy Sets and Systems* **155**, 4-17 (2005).
5. A. Khrennikov, Quantum-like brain: Interference of minds, *BioSystems* **84**, 225–241 (2006).
6. A. Khrennikov, The principle of supplementarity: A contextual probabilistic viewpoint to complementarity, the interference of probabilities, and the incompatibility of variables in quantum mechanics, *Foundations of Physics* **35**(10), 1655 - 1693 (2005).
7. J. B. Busemeyer, Z. Wang, and J. T. Townsend, Quantum dynamics of human decision making, *J. Math. Psychology* **50**, 220-241 (2006).
8. E. Shafir and A. Tversky, Thinking through uncertainty: nonconsequential reasoning and choice, *Cognitive Psychology* **24**, 449-474 (1992).
9. A. Tversky E. and Shafir, The disjunction effect in choice under uncertainty, *Psychological Science* **3**, 305-309 (1992).
10. L. Accardi, *Urne e Camaleoni: Dialogo sulla realta, le leggi del caso e la teoria quantistica,* Rome, 1997.
11. L. Accardi, The probabilistic roots of the quantum mechanical paradoxes. *The wave–particle dualism. A tribute to Louis de Broglie on his 90th Birthday,* Edited by S. Diner, D. Fargue, G. Lochak and F. Selleri. D. Reidel Publ. Company, 1970, 47–55.

12. M. Ohya, Addaptive dynamics and its applications to chaos and NPC problem. *Quantum Bio-Informatics*, Edited by L. Accardi, W. Freudenberg, M. Ohya, ser. *Quantum Prob. and White Noise An.* **21**, 2007, 181-216.

ON COMPLETELY POSITIVE NON-MARKOVIAN EVOLUTION OF A D-LEVEL SYSTEM

ANDRZEJ KOSSAKOWSKI
Institute of Physics, Nicolaus Copernicus University,
87–100 Toruń, Poland
E-mail: kossak@fizyka.umk.pl

ROLANDO REBOLLEDO
Laboratorio de Análisis Estocático, Facultad de Matemáticas
Pontificia Universidad Católica de Chile
Casilla 306, Santiago 22, Chile
E-mail: rrebolle@puc.cl

A sufficient condition for non-Markovian master equation which ensures the complete positivity of the resulting time evolution is presented.

1. Introduction

An open system is one coupled to an external environment [1,2]. The interaction between the system and its environment leads to phenomena of decoherence and dissipation, and for this reason recently this class of problems receives intense consideration in quantum information, where decoherence is viewed as a fundamental obstacle to the construction of quantum information processors [3]. In principle, the von Neumann equation for the total density matrix of the system and the reservoir provides complete predictions for all the observables. However, this equation is in practice impossible to solve since all degrees of freedom of the reservoir have to be taken into account. Main efforts have focused in deducing the time evolution of the reduced state density matrix. This is the aim of the well-known exact theory of subsystem dynamics due to Nakajima-Zwanzig [4,5] which relies in a generalized (non-Markovian) master equation approach.

The Nakajima-Zwanzig projection operator method makes possible to derive an exact equation for the reduced density from the von Neumann equation of the composed system. The resulting generalized master equation — an integrodifferential equation — is mostly of formal interest since such an exact equation can hardly ever be merely written down explicitely in closed form. In contrast, when one makes the Markovian approximation, i.e., when one neglects the reservoir memory effects, the resulting Markovian master equation [6,7] takes a simple form and the required [8] complete positivity of the resulting time evolution is maintained. The main goal of the theory of open quantum systems is a non-Markovian description of the dynamics which at the same time include reservoir memory effects and retain complete positivity.

A variety of non-Markovian master equations has been proposed (cf. [2, 9 – 29]). However, the complete positivity of the resulting time evolution is still an important problem to be investigated.

In the present paper a sufficient condition for non-Markovian master equations is given, which ensures that the resulting time evolution is completely positive. It is shown that this condition is rather difficult to verify in practice. The main reason for that is related to the normalization condition of the time evolution. This difficulty can be overcome if one looks first for completely positive unnormalized solutions to non-Markovian master equations, while the normalization is imposed separately.

2. Notations

Let $\mathbb{C}^d$ be a d-dimensional Hilbert space with the scalar product $\langle \cdot, \cdot \rangle$ and elements $e, x, y, z, \ldots$ The C^*-algebra of linear operators on $\mathbb{C}^d$ will be denoted by M_d. Elements of M_d will be denoted by $a, b, c, \ldots$ and the unit of M_d is $\mathbb{1}_d$. The M_d is the Hilbert space under the scalar product $\langle a, b \rangle = \operatorname{tr} a^* b$.

The C^*-algebra of linear maps from M_d into M_d will be denoted by $\mathfrak{L}(M_d)$, its elements are $A, B, C, \ldots$ and the identity map in $\mathfrak{L}(M_d)$ will be denoted by id. The conjugation (duality) $^{\#}$ in $\mathfrak{L}(M_d)$ is defined by the relation:

$$\langle A^{\#} a, b \rangle = \langle a, Ab \rangle , \tag{1}$$

for all $a, b \in M_d$. This operation endows the following property: the relation $A \mathbb{1}_d = \mathbb{1}_d$ is equivalent to $\operatorname{tr}(A^{\#} a) = \operatorname{tr} a$, and similarly — $L \mathbb{1}_d = 0$ to $\operatorname{tr}(L^{\#} a) = 0$. The cone of all completely positive maps on M_d will be denoted by $\mathcal{B}^+(M_d)$. Finally, if $A_t \in \mathfrak{L}(M_d)$, $t \geq 0$, then the Laplace transform of A_t will be denoted by $\widehat{A}_p$.

3. Non-Markovian Master Equations

The reduced dynamics can be studied equivalently in the Schrödinger or the Heinsenberg pictures. Suppose that $A_t : M_d \to M_d$ describes the reduced dynamics in the Heisenberg picture, then it should satisfy the following conditions: $A_t \in \mathcal{B}^+(M_d)$, $A_t \mathbb{1}_d = \mathbb{1}_d$, for all $t \geq 0$, and $A_0 = \lim_{t\to 0} A_t =$ id. In the Schrödinger picture these relations are given in terms of $A_t^{\#}$, $t \geq 0$.

In the present section, the reduced dynamics is investigated under the assumption that A_t is the solution of a non-Markovian master equation of the form:

$$\frac{dA_t}{dt} = LA_t + \int_0^t ds L_{t-s} A_s \,, \tag{1}$$

with the initial condition $A_0 = \text{id}$, where

$$La = i[h, a] + Fa - \frac{1}{2}\{F(\mathbb{1}_d), a\} \,, \tag{2}$$

and $h = h^* \in M_d$, $F \in \mathcal{B}^+(M_d)$, that is, L is the generator of a completely positive semigroup. The normalization condition $A_t \mathbb{1}_d = \mathbb{1}_d$ implies the equality

$$L_t \mathbb{1}_d = 0 \,. \tag{3}$$

A non-Markovian master equation of a somewhat simpler form than (1) can be easily derived from the Heisenberg equation for the composed system by the Nakajima-Zwanzig method under the assumption of factorization of the initial state of the composed system and the invariance of the initial reservoir state under the reservoir free evolution, c.f. [2]. In this case, $La = i[h, a]$ only, with $h = h^* \in M_d$.

Taking the Laplace transform of (1) one finds:

$$(\text{id}\, p - L - \widehat{L}_p)\widehat{A}_p = \text{id} \,. \tag{4}$$

The equality before implies that both relations below:

$$\widehat{A}_p = (p\, \text{id} - L - \widehat{L}_p)^{-1} \tag{5}$$

and

$$\widehat{A}_p(\text{id} - L - \widehat{L}_p)^{-1} = \text{id} \,, \tag{6}$$

hold. It follows from (6) that equation (1) can also be written in the form

$$\frac{dA_t}{dt} = A_t L + \int_0^t ds A_s L_{t-s} \,, \tag{7}$$

and consequently, the dual dynamics becomes:

$$\frac{dA_t^{\#}}{dt} = -LA_t^{\#} + \int_0^t ds L_{t-s}{}^{\#} A_s^{\#} \,. \tag{8}$$

This means that in the case of non-Markovian master equations there is an analogy to the Markovian case.

To find conditions on L and L_t that ensure that the time evolution A_t resulting from (1) is completely positive for all $t \geq 0$ is the fundamental problem of non-Markovian master equations. The main result of the current paper can be summarized in the following theorem.

Let us suppose that A_t is the solution of the equation (1), where L_t has the form

$$L_t = B_t + Z_t \,, \tag{9}$$

where $B_t \in \mathcal{B}^+(M_d)$ for all $t \geq 0$,

$$Z_t a = -\frac{1}{2}\{B_t(\mathbb{1}_d), a\} + i[h_t, a] \,, \tag{10}$$

and $h_t = h_t^*$, then A_t is completely positive for all $t \geq 0$ if the solution of the normalization equation

$$\frac{dN_t}{dt} = LN_t + \int_0^t ds Z_{t-s} N_s \,, \tag{11}$$

with the initial condition $N_0 = \mathrm{id}$, is completely positive for all $t \geq 0$.

Proof. It follows from (1) that the Laplace transform $\widehat{A}_p$ of A_t is given by the formula

$$\widehat{A}_p = (\mathrm{id}\, p - L - \widehat{Z}_p - \widehat{B}_p)^{-1}, \tag{12}$$

and satisfies the equation

$$\widehat{A}_p = (\mathrm{id}\, p - L - \widehat{Z}_p)^{-1} + (\mathrm{id}\, p - L - \widehat{Z}_p)^{-1}\widehat{B}_p \widehat{A}_p \,. \tag{13}$$

It follows from (13) and (11) that (1) can be written in the form:

$$A_t = N_t + \int_0^t du \int_0^{t-u} ds N_{t-u-s} B_u A_s \,. \tag{14}$$

If N_t is completely positive for all $t \geq 0$, then iterating (14) it is easy to see that A_t is completely positive as well for all $t \geq 0$, since $B_t \in \mathcal{B}^+(M_d)$, provided the iteration procedure converges.

In order to analyze the problems related to the solution of the normalization equation let us consider the non-Markovian master equation of the form

$$\frac{dA_t}{dt} = \int_0^t ds k(t-s)(B_{t-s} - \mathrm{id})A_s\,, \tag{15}$$

where $B_t \in \mathcal{B}^+(M_d)$ and $B_t(\mathbb{1}_d) = \mathbb{1}_d$ for all $t \geq 0$, and $k(t) \geq 0$.

The normalization equation takes the form

$$\frac{dN_t}{dt} = -\int_0^t ds k(t-s)N_s\,, \tag{16}$$

with the initial condition $N_0 = \mathrm{id}$. The solution of (16) has the form

$$N_t = f(t)\mathrm{id}\,, \tag{17}$$

where $f(t)$ satisfies the equation

$$\frac{df(t)}{dt} = -\int_0^t ds k(t-s)f(s)\,, \tag{18}$$

and $f(0) = 1$.

As a particular case, let us choose $k(t)$ in the Lidar-Shabani form, cf. [27], i.e.,

$$k(t) = \kappa^2 e^{-2\kappa\gamma t}. \tag{19}$$

In this case one easily finds

$$f(t) = \begin{cases} e^{-\kappa\gamma t}\Big[\cos(\kappa t\sqrt{1-\gamma^2}) + \frac{\gamma}{\sqrt{1-\gamma^2}}\sin(\kappa t\sqrt{1-\gamma^2})\Big]\,, & \text{if } 0 \leq \gamma < 1, \\ e^{-\kappa t}(1+\kappa t)\,, & \text{if } \gamma = 1, \\ e^{-\kappa\gamma t}\Big[\cosh \kappa t\sqrt{\gamma^2-1}) + \frac{\gamma}{\sqrt{\gamma^2-1}}\sinh(\kappa t\sqrt{\gamma^2-1})\Big] & \text{,if } \gamma > 1. \end{cases} \tag{20}$$

It follows from (17) that N_t is completely positive if and only if $f(t) \geq 0$ for all $t \geq 0$ for all $t \geq 0$, and (20) shows that $f(t) \geq 0$ for all $t \geq 0$ if and only if $\gamma \geq 1$.

The above example clearly indicates that the structure of non-Markovian master equations is much more complicated than the Markovian ones.

4. Modified non-Markovian Master Equations

The time evolution (in the Heisenberg picture) is given by the family of maps $A_t : M_d \to M_d$, $t \geq 0$, such that $A_t \in \mathcal{B}^+(M_d)$, for all $t \geq 0$, (complete positivity condition), $A_t(\mathbb{1}_d) = \mathbb{1}_d$ for all $t \geq 0$, (normalization condition) and $A_0 := \lim_{t\downarrow 0} A_t = \mathrm{id}$. In Sect. 3 it has been shown that if A_t satisfies equation (1), then the normalization condition can be imposed with no trouble. Indeed, if (2), (9) and (10) are satisfied, then the normalization condition is trivially fulfilled. On the other hand, complete positivity of A_t leads to complete positivity of solutions to the normalization equation (10) but this is considerably more difficult to demonstrate. However, one can circumvent this in the following manner. Let V_t, $t \geq 0$ be the family of complete positive maps on M_d such that $\lim_{t\to 0} V_t = \mathrm{id}$. If $V_t(\mathbb{1}_d) > 0$ for all $t \geq 0$, then the maps A_t, $t \geq 0$, defined as

$$A_t(a) = V_t(\mathbb{1}_d)^{-1/2} V_t(a) V_t(\mathbb{1}_d)^{-1/2}, \tag{1}$$

are completely positive and normalized.

Let V_t, $t \geq 0$ be the solution of the following modified non-Markovian master equation:

$$\frac{dV_t}{dt} = PV_t + \int_0^t ds B_{t-s} V_s \,, \tag{2}$$

with the initial condition $\lim_{t\to 0} V_t = \mathrm{id}$, where P is a completely positive map and $B_t \in \mathcal{B}^+(M_d)$ for all $t \geq 0$. The resolvent of (2),

$$\widehat{V}_p = (\mathrm{id}\, p - P - \widehat{B}_p)^{-1}, \tag{3}$$

satisfies the equation

$$\widehat{V}_p = (\mathrm{id}\, p - P)^{-1} + (\mathrm{id}\, p - P)^{-1} \widehat{B}_p \widehat{V}_p \,, \tag{4}$$

which is the integral form of (2). Iteration of (4) yields that V_t is completely positive since $\exp(tP)$ and B_t are completely positive. If the solution of (2) satisfies the condition $V_t(\mathbb{1}_d) > 0$ for all $t \geq 0$, then A_t, $t \geq 0$, defined through (1) gives the correct time evolution, i.e., it is completely positive and normalized.

The above approach contains as a special case the semigroup form of the dynamics. Let us consider the equation

$$\frac{dV_t}{dt} = LV_t + \lambda^2 \int_0^t ds\, e^{(t-s)L} V_s\,, \tag{5}$$

where L is the generator of a completely positive semigroup. One easily finds the solution of (5) which is of the form:

$$V_t = \cosh(\lambda t) e^{tL}\,, \tag{6}$$

and $V_t(\mathbb{1}_d) = \mathbb{1}_d \cosh(\lambda t)$. The corresponding normalized evolution A_t has the form

$$A_t = e^{tL}\,, \tag{7}$$

that is, it is a semigroup.

Acknowledgment

A. K. has been supported by MNiSW grant No. NN202300433. R. R. acknowledges a partial support by PBCT-ACT13 and Dirección de Relaciones Internacionales-PUC.

References

1. R. Alicki, K. Lendi, *Quantum Dynamical Semigroups and Applications*, Lecture Notes in Physics **286**, Springer-Verlag, 1987.
2. H.-P. Breuer, F. Petruccione, *The theory of open quantum systems*, Oxford University Press, New York, 2002.
3. M. A. Nielsen, I. L. Chuang, *Quantum computation and quantum information*, Cambridge University Press, Cambridge, 2000.
4. S. Nakajima, Prog. Theor. Phys. **20**, 948 (1958).
5. R. Zwanzig, J. Chem. Phys. **33**, 1338 (1960).
6. V. Gorini, A. Kossakowski, E. C. G. Sudarshan, J. Math. Phys. **17**, 821 (1976).
7. G. Lindblad, Commun. Math. Phys. **48**, 119 (1976).
8. K. Kraus, *States, Effects and Operations, Fundamental notions of Quantum Theory*, Academics, Berlin, 1983.
9. N. Shibata, Y. Takahashi, J. Stat. Phys. **17**, 171 (1977).
10. A. Imamoglu, Phys. Rev. A **50**, 3650 (1994).
11. A. Royer, Phys. Rev. Lett. **77**, 3272 (1996).
12. A. Royer, Phys. Lett. A **315**, 335 (2003).
13. M. Barnett, S. Stenholm, Phys. Rev. A **64**, 033808 (2001).
14. H.-P. Breuer, B. Kappler, F. Petruccione, Ann. Phys. **291**, 36 (2001).

15. H.-P. Breuer, F. Petruccione, Phys. Rev. A **63**, 032102 (2001).
16. H.-P. Breuer, B. Kappler, F. Petruccione, Phys. Rev. A **59**, 1633 (1999).
17. H.-P. Breuer, Phys. Rev. A **69**, 022115 (2004).
18. H.-P. Breuer, Phys. Rev. A **70**, 012106 (2004).
19. H.-P. Breuer, J. Gemmer, M. Michel, Phys. Rev. E **73**, 016139 (2006).
20. S. Maniscalco, Phys. Rev. A **72**, 024103 (2005).
21. S. Maniscalco, F. Petruccione, Phys. Rev. A **73**, 012111 (2006).
22. J. Wilkie, Phys. Rev. E **62**, 8808 (2000).
23. J. Wilkie, J. Phys. A **114**, 7736 (2001).
24. H.-P. Breuer, Phys. Rev. A **75**, 022103 (2007).
25. A. Budini, Phys. Rev. A **69**, 042107 (2004).
26. A. Budini, Phys. Rev. A **74**, 053815 (2006).
27. A. Shabani, D. Lidar, Phys. Rev. A **71**, 020101 (2005).
28. J. Lee, I. A. D. M. A. H. Kim, M. Kim, Phys. Rev. A **70**, 024301 (2004).
29. J. Pülo, S. K. Maniscalco, quant-ph/07064438.

MEASURES OF ENTANGLEMENT - A HILBERT SPACE APPROACH

WŁADYSŁAW A. MAJEWSKI
Institute of Theoretical Physics and Astrophysics
Gdańsk University
Wita Stwosza 57
80-952 Gdańsk, Poland
E-mail: fizwam@univ.gda.pl
Joint work with M. Ohya and T. Matsuoka

We present the description of positive partially transposed (PPT) states. Our approach is based on Tomita-Takesaki theory. The discussion about effectiveness of this approach is given. The notion degrees of entanglement are introduced.

1. Definitions and notations

In Quantum Computing a characterization of states with positive partial transposition is important problem (see [13] and also [5]). Recently, some partial results in this direction were obtained (see [8], [9], [17], and [21]). The aim of this note, partially based on [2], [3], and [16], is to proceed with the analysis of entanglement and PPT states (see also Matsuoka's contribution[22] to this proceedings). The detailed proofs will appear in a forthcoming publication[15] .

For the sake of convenience, we provide all necessary preliminaries and set up the notation. Let $\mathcal{B}(\mathcal{H})$ be the set of all linear bounded operators on a Hilbert space $\mathcal{H}$. We denote the set of all positive elements of $\mathcal{B}(\mathcal{H})$ by $\mathcal{B}(\mathcal{H})^+$. A *state* on $\mathcal{B}(\mathcal{H})$ is a linear functional $\phi : \mathcal{B}(\mathcal{H}) \longrightarrow \mathbb{C}$ such that $\phi(A) \geq 0$ for every $A \in \mathcal{B}(\mathcal{H})^+$ and $\phi(\mathbb{1}) = 1$, where $\mathbb{1}$ is the unit of $\mathcal{B}(\mathcal{H})$. The set of all states on $\mathcal{B}(\mathcal{H})$ is denoted by $\mathcal{S}_{\mathcal{B}(\mathcal{H})}$.

A linear map $\Psi : \mathcal{B}(\mathcal{H}_1) \longrightarrow \mathcal{B}(\mathcal{H}_2)$ is called *positive* if $\Psi(\mathcal{B}(\mathcal{H}_1)^+) \subset \mathcal{B}(\mathcal{H}_2)^+$. For $k \in \mathbb{N}$ we consider a map $\Psi_k : M_k \otimes \mathcal{B}(\mathcal{H}_1) \longrightarrow M_k \otimes \mathcal{B}(\mathcal{H}_2)$ where M_k denotes the algebra of $k \times k$-matrices with complex entries and $\Psi_k = \mathrm{id}_{M_k} \otimes \Psi$. We say that Ψ is *k-positive* if the map Ψ_k is positive. The map Ψ is said *completely positive* when Ψ is k-positive for every $k \in \mathbb{N}$. Let us recall that for a finite dimensional Hilbert space $\mathcal{L}$ every state ϕ on

$\mathcal{B}(\mathcal{L})$ has the form of $\phi(A) = \mathrm{Tr}\,(\varrho A)$, where ϱ is a uniquely determined *density matrix*, i.e. an element of $\mathcal{B}(\mathcal{L})^+$ such that $\mathrm{Tr}\,\varrho = 1$.

Throughout this note $\mathcal{H}$ and $\mathcal{K}$ will be fixed finite-dimensional Hilbert spaces. We also fix orthonormal bases $\{e_i\}_{i=1}^n$ and $\{f_j\}_{j=1}^m$ of the spaces $\mathcal{H}$ and $\mathcal{K}$ respectively, where $n = \dim \mathcal{H}$, $m = \dim \mathcal{K}$. For simplicity we will write $\mathcal{S}$, $\mathcal{S}_\mathcal{H}$, $\mathcal{S}_\mathcal{K}$ instead of $\mathcal{S}_{\mathcal{B}(\mathcal{H})\otimes\mathcal{B}(\mathcal{K})}$, $\mathcal{S}_{\mathcal{B}(\mathcal{H})}$, $\mathcal{S}_{\mathcal{B}(\mathcal{K})}$, respectively. By $\tau_\mathcal{H}$, $\tau_\mathcal{K}$, $\tau_{\mathcal{H}\otimes\mathcal{K}}$ we denote transposition maps on $\mathcal{B}(\mathcal{H})$, $\mathcal{B}(\mathcal{K})$, $\mathcal{B}(\mathcal{H}\otimes\mathcal{K})$, respectively, associated with bases $\{e_i\}$, $\{f_j\}$, $\{e_i \otimes f_j\}$, respectively. Let us note that for every finite dimensional Hilbert space $\mathcal{L}$ the transposition $\tau_\mathcal{L} : \mathcal{B}(\mathcal{L}) \longrightarrow \mathcal{B}(\mathcal{L})$ is a positive map but not completely positive (in fact it is not even 2-positive).

A positive map $\Psi : \mathcal{B}(\mathcal{H}) \longrightarrow \mathcal{B}(\mathcal{K})$ is called *decomposable* if there are completely positive maps $\Psi_1, \Psi_2 : \mathcal{B}(\mathcal{H}) \longrightarrow \mathcal{B}(\mathcal{K})$ such that $\Psi = \Psi_1 + \Psi_2 \circ \tau_\mathcal{H}$. Let $\mathcal{P}$, $\mathcal{P}_\mathrm{C}$ and $\mathcal{P}_\mathrm{D}$ denote the set of all positive, completely positive and decomposable maps from $\mathcal{B}(\mathcal{H})$ to $\mathcal{B}(\mathcal{K})$, respectively. Note that

$$\mathcal{P}_\mathrm{C} \subset \mathcal{P}_\mathrm{D} \subset \mathcal{P}(0.0) \tag{1.1}$$

(see also [5]- [7]).

A state $\varphi \in \mathcal{S}$ is said to be *separable* if it can be written in the form

$$\varphi = \sum_{n=1}^{N} a_n \varphi_n^\mathcal{H} \otimes \varphi_n^\mathcal{K}$$

where $N \in \mathbb{N}$, $\varphi_n^\mathcal{H} \in \mathcal{S}_\mathcal{H}$, $\varphi_n^\mathcal{K} \in \mathcal{S}_\mathcal{K}$ for $n = 1, 2, \ldots, N$, a_n are positive numbers such that $\sum_{n=1}^N a_n = 1$, and the state $\varphi_n^\mathcal{H} \otimes \varphi_n^\mathcal{K}$ is defined as $\varphi_n^\mathcal{H} \otimes \varphi_n^\mathcal{K}(A \otimes B) = \varphi_n^\mathcal{H}(A)\varphi_n^\mathcal{K}(B)$ for $A \in \mathcal{B}(\mathcal{H})$, $B \in \mathcal{B}(\mathcal{K})$. The set of all separable states on the algebra $\mathcal{B}(\mathcal{H}) \otimes \mathcal{B}(\mathcal{K})$ is denoted by $\mathcal{S}_\mathrm{sep}$. A state which is not in $\mathcal{S}_\mathrm{sep}$ is called *entangled* or *non-separable.*

Finally, let us define the family of *PPT* (transposable) states on $\mathcal{B}(\mathcal{H} \otimes \mathcal{K})$

$$\mathcal{S}_\tau = \{\varphi \in \mathcal{S} : \varphi \circ (\, id_{\mathcal{B}(\mathcal{H})} \otimes \tau_\mathcal{K}) \in \mathcal{S}\}.$$

Note that due to the positivity of the transposition $\tau_\mathcal{K}$ every separable state φ is transposable, so

$$\mathcal{S}_\mathrm{sep} \subset \mathcal{S}_\tau \subset \mathcal{S}.(0.0) \tag{1.2}$$

2. Tomita-Takesaki scheme for transposition

In this Section we wish to indicate how Tomita-Takesaki techniques may be used to describe a transposition (for details see [14]). In particular, we will examine how transposition is related to the modular conjugation and modular operator. Moreover, this brief description of Tomita-Takesaki theory presents a sufficient preparation for the next section where a characterization of PPT states will be given. For a comprehensive account of Tomita-Takesaki theory addressed to physicists we refer Haag's book [12], while mathematical description can be found in [1], and [24].

Again, let $\mathcal{H}$ be a finite dimensional (say n-dimensional) Hilbert space. Define $\omega \in \mathcal{S}_{\mathcal{B}(\mathcal{H})}$ as $\omega(a) = \mathrm{Tr}\,\varrho a$, where ϱ is an invertible density matrix, i.e. the state ω is a faithful one. Denote by $(\mathcal{H}_\pi, \pi, \Omega)$ the GNS triple associated with $(\mathcal{B}(\mathcal{H}), \omega)$. Then, one has:

- $\mathcal{H}_\pi$ is identified with $\mathcal{B}(\mathcal{H})$ where the inner product $(\cdot\,,\cdot)$ is defined as $(a,b) = \mathrm{Tr}\,a^*b$, $a, b \in \mathcal{B}(\mathcal{H})$;
- With the above identification: $\Omega = \varrho^{1/2}$;
- $\pi(a)\Omega = a\Omega$;
- The modular conjugation J_{m} is the hermitian involution: $J_{\mathrm{m}} a\varrho^{1/2} = \varrho^{1/2}a^*$;
- The modular operator Δ is equal to the map $\varrho \cdot \varrho^{-1}$;

As the next step let us define two conjugations: J_c on $\mathcal{H}$ and J on $\mathcal{H}_\pi$. To this end we note that the eigenvectors $\{x_i\}$ of $\varrho = \sum_i \lambda_i |x_i\rangle\langle x_i|$ form an orthonormal basis in $\mathcal{H}$ (due to the faithfulness of ω). Hence we can define

$$J_c f = \sum_i \overline{\langle x_i, f\rangle} x_i \tag{1}$$

for every $f \in \mathcal{H}$. Due to the fact that $\{E_{ij} \equiv |x_i\rangle\langle x_j|\}$ form an orthonormal basis in $\mathcal{H}_\pi$ we can define in the similar way a conjugation J on $\mathcal{H}_\pi$

$$Ja\varrho^{1/2} = \sum_{ij} \overline{(E_{ij}, a\varrho^{1/2})} E_{ij} \tag{2}$$

Obviously, $J\varrho^{1/2} = \varrho^{1/2}$.

Now let us define a transposition on $\mathcal{B}(\mathcal{H})$ as the map $a \mapsto a^t \equiv J_c a^* J_c$ where $a \in \mathcal{B}(\mathcal{H})$. By τ_0 we will denote the map induced on $\mathcal{H}_\pi$ by the transposition, i.e.

$$\tau_0 a\varrho^{1/2} = a^t \varrho^{1/2} \tag{3}$$

where $a \in \mathcal{B}(\mathcal{H})$. The main properties of the notions introduced above are the following

Proposition 2.1. *(see [14]) Let $a \in \mathcal{B}(\mathcal{H})$ and $\xi \in \mathcal{H}_\pi$. Then*

$$a^t \xi = Ja^* J\xi.$$

To present another property we define the unitary operator U on $\mathcal{H}_\pi$ by

$$U = \sum_{ij} |E_{ji}\rangle\langle E_{ij}| \tag{4}$$

Clearly, $UE_{ij} = E_{ji}$. We have the following

Proposition 2.2. *(see [14]) Let J and $J_{\rm m}$ be the conjugations introduced above and U be the unitary operator defined by (4). Then we have:*

(1) $U^2 = \mathbb{1}$ and $U = U^*$
(2) $J = UJ_{\rm m}$;
(3) J, $J_{\rm m}$ and U mutually commute.

Now, we are ready to describe a polar decomposition of the map τ_0.

Theorem 2.1. *(see [14]) If τ_0 is the map introduced in (3), then*

$$\tau_0 = U\Delta^{1/2}.$$

To present some properties of U which are analogous to that of the modular conjugation $J_{\rm m}$ we need[1]

$$V_\beta = \text{closure}\left\{\Delta^\beta a \varrho^{1/2} : a \geq 0,\ \beta \in \left[0, \frac{1}{2}\right]\right\}.$$

Recall that $V_{1/4}$ is nothing but **the natural cone** $\mathcal{P}$ associated with the pair $(\pi(\mathcal{B}(\mathcal{H})), \Omega)$. Finally, let us define an automorphism α on $\mathcal{B}(\mathcal{H}_\pi)$ by

$$\alpha(a) = UaU^* = UaU, \quad a \in \mathcal{B}(\mathcal{H}_\pi), \tag{5}$$

as U is self-adjoint. Then we have

Proposition 2.3. *(see [14])*

(1) $U\Delta = \Delta^{-1}U$
(2) α maps $\pi(\mathcal{B}(\mathcal{H}))$ onto $\pi(\mathcal{B}(\mathcal{H}))'$, where $\pi(\mathcal{B}(\mathcal{H}))'$ stands for the commutant of $\pi(\mathcal{B}(\mathcal{H}))$.
(3) J commutes with Δ.

Summarizing, this section establishes a close relationship between the Tomita-Takesaki scheme and transposition and we have the following:

Proposition 2.4. *(see [14]) Let $\xi \mapsto \omega_\xi$ be the homeomorphism between the natural cone $\mathcal{P}$ and the set of normal states on $\pi(\mathcal{B}(\mathcal{H}))$ described in [10], [4] , i.e. such that*

$$\omega_\xi(a) = (\xi, a\xi), \quad a \in \mathcal{B}(\mathcal{H}).$$

For every state ω define $\omega^{\tau}(a) = \omega(a^t)$ where $a \in \mathcal{B}(\mathcal{H})$. If $\xi \in \mathcal{P}$ then the unique vector in $\mathcal{P}$ mapped into the state ω_{ξ}^{τ} by the homeomorphism described above, is equal to $U\xi$

3. PPT states on the Hilbert-space level

Let us begin with a preliminary observation concerning separable states. We consider a composite system $A + B$ where a subsystem $i = A, B$ is described by $(\mathcal{B}(\mathcal{K}_i), \mathcal{P}_i, \varrho_i)$ where $\mathcal{P}_i$ denotes the natural cone associated with $(\mathcal{B}(\mathcal{K}_i), \varrho_i)$ (cf [17] and the previous Section). In [17], using Tomita-Takesaki approach, we have derived the one-to-one correspondence between the set of normalized vectors in $\mathcal{P}_A \otimes \mathcal{P}_B$ and the set of all separable states, where

$$\mathcal{P}_A \otimes \mathcal{P}_B \equiv \text{closure}\{\sum_k a_k x_k^{(1)} \otimes x_k^{(2)}, a_k \geq 0, \sum_k a_k = 1, x_k^{(i)} \in \mathcal{P}_i\}$$

Here we wish to extend this result and to get an analogous characterization of PPT states. So, again, we will consider a composite system A plus B. Moreover, again, to simplify the exposition we assume that the Hilbert spaces $\mathcal{K}_A$ and $\mathcal{K}_B$ are finite dimensional. Suppose that the subsystem A is described by a C^* algebra $\mathcal{A} \equiv \mathcal{B}(\mathcal{K}_A)$ equipped with a faithful state ω_A (so, of the form $\omega_A(a) \equiv Tr\{\varrho_A a\}$ where ϱ_A is an invertible density matrix). Similarly, let $\mathcal{B} \equiv \mathcal{B}(\mathcal{K}_B)$ for some Hilbert space $\mathcal{K}_B$, ϱ_B be an invertible density matrix in $\mathcal{B}(\mathcal{K}_B)$ and ω_B be a state on $\mathcal{B}$ such that $\omega_B(b) = \text{Tr}\,(b\varrho_B)$ for $b \in \mathcal{B}$. By $(\mathcal{H}, \pi, \Omega)$, $(\mathcal{H}_A, \pi_A, \Omega_A)$ and $(\mathcal{H}_B, \pi_B, \Omega_B)$ we denote the GNS representations of $(\mathcal{A} \otimes \mathcal{B}, \omega_A \otimes \omega_B)$, $(\mathcal{A}, \omega_A)$ and $(\mathcal{B}, \omega_B)$ respectively. We observe that we can make the following identifications (cf [11], [17]):

(1) $\mathcal{H} = \mathcal{H}_A \otimes \mathcal{H}_B$,
(2) $\pi = \pi_A \otimes \pi_B$,
(3) $\Omega = \Omega_A \otimes \Omega_B$.

With these identifications we have $J_{\text{m}} = J_A \otimes J_B$ and $\Delta = \Delta_A \otimes \Delta_B$ where J_{m}, J_A, J_B are modular conjugations and Δ, Δ_A, Δ_B are modular operators for $(\pi(A \otimes B)'', \Omega)$, $(\pi_A(A)'', \Omega_A)$, $(\pi_B(B)'', \omega_B)$ respectively ($\mathcal{M}''$ stands for the bicommutant of $\mathcal{M}$). Ω_A and Ω_B are separating vectors and we will write $a\Omega_A$ and $b\Omega_B$ instead of $\pi_A(a)\Omega_A$ and $\pi_B(b)\Omega_B$ for $a \in A$ and $b \in B$ when no confusion can arise. Moreover, as finite dimensionality of the corresponding Hilbert spaces was assumed, we will also identify $\pi_A(A)''$ with $\pi(A)$, etc. Furthermore, as $\mathcal{K}_B$ is a finite dimensional Hilbert space, we denote it dimension by n. Thus $\mathcal{B}(\mathcal{K}_B) \equiv \mathcal{B}(\mathbb{C}^n) \equiv M_n(\mathbb{C})$. To put some emphasis on the dimensionality of the "reference" subsystem B, by $\mathcal{P}_n$ we denote the natural cone for $(\pi(A) \otimes \mathcal{B}(\mathbb{C}^n), \omega \otimes \omega_0)$ where ω_0 is a faithful state on $\mathcal{B}(\mathbb{C}^n)$.

Finally, the partial transposition $(\mathrm{id} \otimes \tau)$ on $M_n^\pi(\mathcal{A}) \equiv \pi(A \otimes B)$ induces an operator at the Hilbert space level, but for the sake of simplicity we will where convenient retain the notation $(\mathrm{id} \otimes \tau)$ for this operator.

In order to achieve the desired characterization of PPT states we introduce the notion of the "transposed cone" $\mathcal{P}_n^\tau \equiv (\mathbf{I} \otimes U)\mathcal{P}_n$, where τ is transposition on $M_n(\mathbb{C})$ while the operator U was defined in the previous Section (we have used the following identification: for the basis $\{e_i\}_i$ in $\mathbb{C}^n$ consisting of eigenvectors of ϱ_{ω_0} $(\omega_0(\cdot) = Tr\{\varrho_{\omega_0} \cdot\}$, we have the basis $\{E_{ij} \equiv |e_i >< e_j|\}_{ij}$ in the GNS Hilbert space associated with $(\mathcal{B}(\mathbb{C}^n), \omega_0)$ with U defined in terms of that basis). Note that in the same basis one has the identification $\mathcal{B}(\mathbb{C}^n)$ with $M_n(\mathbb{C})$.

Now the natural cone $\mathcal{P}_n$ for $\pi(A \otimes \mathcal{B}(\mathbb{C}^n)) \equiv M_n^\pi(\mathcal{A})$ may be realized as

$$\mathcal{P}_n = \overline{\Delta_n^{1/4}\{[a_{ij}]\Omega_{:}[a_{ij}] \in M_n^\pi(\mathcal{A})^+\}}.$$

Obviously, $\overline{X}$ stands for the closure of X. We observe:

$$\begin{aligned}
&\{(\mathbf{I} \otimes U)\Delta^{1/4}[a_{ij}]\Omega : \; [a_{ij}] \in M_n^\pi(\mathcal{A})^+\} \\
&= \{(\Delta_A^{1/4} \otimes U\Delta_B^{1/4}) \circ \sum_{i,j} \pi_A(a_{ij}) \otimes \pi_B(E_{ij})\Omega : \; [a_{ij}] \in M_n^\pi(\mathcal{A})^+\} \\
&= \{(\Delta_A^{1/4} \otimes \Delta_B^{1/4}U\Delta_B^{1/2}) \circ \sum_{i,j} \pi_A(a_{ij}) \otimes \pi_B(E_{ij})\Omega : \; [a_{ij}] \in M_n^\pi(\mathcal{A})^+\} \\
&= \{\Delta^{1/4}[a_{ji}]\Omega : \; [a_{ij}] \in M_n^\pi(\mathcal{A})^+\}.
\end{aligned}$$

where in the last equality, Theorem 2.1 was used. Thus

$$\mathcal{P}_n^\tau = \overline{\Delta^{1/4}\{[a_{ji}]\Omega : [a_{ij}] \in M_n^\pi(\mathcal{A})^+\}},$$

and

Lemma 3.1. *(see [14])*

1. *For each n, $\mathcal{P}_n \cap \mathcal{P}_n^\tau$ and $\overline{co}(\mathcal{P}_n \cup \mathcal{P}_n^\tau)$ are dual cones.*
2. *Let n be given. For any $[a_{ij}] \in M_n^\pi(\mathcal{A})^+$, $\Delta^{1/4}[a_{ij}]\Omega \in \mathcal{P}_n \cap \mathcal{P}_n^\tau$ implies $[a_{ji}] \in M_n^\pi(\mathcal{A})^+$.*

Hence we arrived at

Theorem 3.1. *(see [14]) In the finite dimensional case* $\{\Delta^{1/4}[a_{ij}]\Omega : [a_{ij}] \geq 0, [a_{ji}] \geq 0\} = \mathcal{P}_n \cap \mathcal{P}_n^{\tau}$.

and

Corollary 3.2.

(1) *There is one-to-one correspondence between the set of PPT states and* $\mathcal{P}_n \cap \mathcal{P}_n^{\tau}$.
(2) *There is one-to-one correspondence between the set of separable states and* $\mathcal{P}_A \otimes \mathcal{P}_B$.

Remark 3.3.

(1) As U is nontrivial the inclusion $\mathcal{P}_n \cap \mathcal{P}_n^{\tau} \subseteq \mathcal{P}_n$ should be, in general, the proper one.
(2) As $\mathcal{P}_n^{\tau}$ and $\mathcal{P}_n$ contains $\mathcal{P}_A \otimes \mathcal{P}_B$, PPT states which are not separable are characterized by vectors in $\mathcal{P}_n \cap \mathcal{P}_n^{\tau} \setminus \mathcal{P}_A \otimes \mathcal{P}_B$. Thus, Corollary 3.2 gives a quite effective recipe for a construction of PPT state which is not a separable one (see the next section).
(3) Similarly, non-PPT states are characterized by vectors $\mathcal{P}_n \setminus \mathcal{P}_n \cap \mathcal{P}_n^{\tau}$. Again, this gives a recipe for a construction of non-PPT states.
(4) The above characterization of PPT states can be considered as dual approach to characterization obtained by Belavkin and Ohya (see [2], [3], [22], and [15]).

4. Effectiveness of the description of PPT states

Now we are in position to pose the following two problems:

(1) *to clarify effectiveness of the description of* $\mathcal{P}_n \cap \mathcal{P}_n^{\tau}$,
(2) *go one step further and classify various entanglements.*

Here we will be concerned with the first one while, in the next section, degrees of entanglement will be presented.

We begin with Størmer's characterization of decomposable maps[23] : a positive map T is decomposable if and only if for $\{a_{ij}\} \geq 0$ and $\{a_{ji}\} \geq 0$, $\{T(a_{ij})\} \geq 0$. It should be absolutely obvious that the form of $\mathcal{P}_n \cap \mathcal{P}_n^{\tau}$, given by Theorem 3.1, reflects the Størmer condition.

On the other hand, Tomiyama[26] gave a nice characterization of positivity for transposition. Namely, let $\mathcal{A}$ be a C^*-algebra. The transposition τ of $\{a_{ij}\}$ with $a_{ij} \in \mathcal{A}$ is a positive map if and only if $\mathcal{A}$ is abelian.

This result suggests that the condition $f \in \mathcal{P}_n \cap \mathcal{P}_n^\tau$ reflects a kind of "local commutativity".

Let us elaborate this supposition. Firstly we recall that any positive defined matrix $[x_{ij}]$ (with operator entries) can be written as[25]

$$[x_{ij}] = \sum_k ((v_i^{(k)})^* v_j^{(k)} \tag{6}$$

On the other hand, the positivity of the matrix $[\phi(x_{ij})]$ (with operator entries!) is equivalent to (cf [25])

$$\sum_{ij} y_i^* \phi(x_{ij}) y_j \geq 0 \tag{7}$$

where $\{y_i\}$ are arbitrary elements of $B(\mathcal{H})$.

Hence, applying condition (7) to matrices of the form $[a_i^* a_j]$ with the choice of x_i such that all $x_i = 0$ except for i_0 and j_0, then changing the numeration in such way that $x_{i_0} = x_1$ and $x_{j_0} = x_2$ we arrive to study the positivity of the following matrix

$$\begin{pmatrix} a_1^* a_1 & a_1^* a_2 \\ a_2^* a_1 & a_2^* a_2 \end{pmatrix} \geq 0 \tag{8}$$

and its transposition. However, block matrix techniques leads to the following result (for details see [18]): for any matrix $\begin{pmatrix} a_1^* a_1 & a_1^* a_2 \\ a_2^* a_1 & a_2^* a_2 \end{pmatrix}$ satisfying the Størmer condition, a_2 is of the form

$$a_2 = \sum_i \lambda_i |e_i >< a_1^* e_i| \tag{9}$$

so

$$\begin{pmatrix} a_1^* a_1 & a_1^* a_2 \\ a_2^* a_1 & a_2^* a_2 \end{pmatrix} = \sum_i \begin{pmatrix} 1 & \lambda_i \\ \bar{\lambda}_i & |\lambda_i|^2 \end{pmatrix} \begin{pmatrix} |a_1^* e_i >< a_1^* e_i| & 0 \\ 0 & |a_1^* e_i >< a_1^* e_i| \end{pmatrix} \tag{10}$$

where $\lambda_i \in \mathbb{C}$. To rewrite the above equality in more compact form, let us denote the norm of the vector $|a_1^* e_i >$ by α_i and the normalized vector $\frac{1}{\alpha_i}|a_1^* e_i >$ by φ_i. Then

$$\begin{pmatrix} a_1^* a_1 & a_1^* a_2 \\ a_2^* a_1 & a_2^* a_2 \end{pmatrix} = \sum_i \alpha_i^2 \begin{pmatrix} 1 & \lambda_i \\ \bar{\lambda}_i & |\lambda_i|^2 \end{pmatrix} \begin{pmatrix} |\varphi_i >< \varphi_i| & 0 \\ 0 & |\varphi_i >< \varphi_i| \end{pmatrix} \tag{11}$$

or symbolically

$$\begin{pmatrix} a_1^* a_1 & a_1^* a_2 \\ a_2^* a_1 & a_2^* a_2 \end{pmatrix} = \sum_i \alpha_i^2 \cdot \Lambda_i \cdot R_i \tag{12}$$

where Λ_i are "matrix" coefficients (matrices with complex entries!) while R_i are "matrix" projectors (not mutually orthogonal!). It is worth pointing out that (11) implies "separability" for any matrix $[a_i^* a_j]$ satisfying the Størmer condition. Therefore, it is important to realize that non-triviality of Størmer condition follows from the fact that when a positive matrix $[x_{i,j}]$ $(= \sum_k ((v_i^{(k)})^* v_j^{(k)}$) satisfies the Størmer condition some of its summand(s) $(v_i^{(k)})^* v_j^{(k)}$ may not.

We are now ready for explanation the phrase: *reflects a kind of "local commutativity"*, which was made at the beginning of this Section. Firstly, we note : Λ_i is a matrix with complex entries and the transposition on such matrices is a positive map. Secondly, R_i is a matrix with operator entries but this matrix is diagonal. Thus, for transposition, R_i is a fix point. Furthermore, Λ_i commutes with R_i. We emphasize that all these remarks stem from (11) - so this is a "local" property as we singled out two indices only. Nevertheless we can conclude : any summand of a positive matrix $[x_{ij}]$ in (6) satisfying the Størmer condition has "local-commutativity" which guarantees the nice behavior (positivity) of the transposition.

Finally, we are able to discuss the question of effectiveness of the description of PPT states. To this end we recall that Theorem 3.1 says: PPT states are characterized (uniquely) by vectors of the form $[a_{ij}]\Omega = \sum_k [a_{i,k}^* a_{j,k}]\Omega$ with $[a_{ji}] \geq 0$. However, the important point to note here is (cf the discussion following (12)) that some summands $[a_{i,k}^* a_{j,k}]\Omega$ may not be in $\mathcal{P}_n \cap \mathcal{P}_n^\tau$. Consequently, some vectors in the subcone $\mathcal{P}_n \cap \mathcal{P}_n^\tau$ which represent non-trivial (that is non-separable) PPT states can be obtained as a convex hull of vectors in such way that some summand(s) is (are) not necessary in this subcone. This suggests that the universally effective prescription for a vector representing PPT state is not available.

5. Measures of entanglement

In [19], [20] we have introduced the degree of quantum correlations. The basic idea is to describe how a given quantum system is close to "classical" world. We wish to repeat this idea but now in the context of Hilbert spaces. In this way we can employ the geometry of Hilbert spaces.

Definition 5.1. Let ξ be a vector in the natural cone $\mathcal{P}$ corresponding to a normal state of a composite system $1+2$ (cf Proposition 2.4). Then

(1) Degree of entanglement (or *quantum correlations*) is given by:

$$D_e(\xi) = inf\{||\xi - \eta||; \eta \in \mathcal{P}_1 \otimes \mathcal{P}_2\} \tag{13}$$

(2) Degree of genuine entanglement (or *genuine quantum correlations*) is defined as

$$D_{ge}(\xi) = inf\{||\xi - \eta||; \eta \in \mathcal{P} \cap \mathcal{P}^{\tau}\} \tag{14}$$

We will briefly discuss the geometric idea behind this definitions. The key to the argument is convexity (in Hilbert spaces). Namely, we observe

(1) $\mathcal{P} \supset \mathcal{P}_1 \otimes \mathcal{P}_2$ is a convex subset,
(2) $\mathcal{P} \supset \mathcal{P} \cap \mathcal{P}^{\tau}$ is a convex subset,
(3) the theory of Hilbert spaces says: $\exists \xi_0 \in \mathcal{P}_1 \otimes \mathcal{P}_2$, unique vector , such that $D_e(\xi) = ||\xi - \xi_0||$,
(4) analogously, $\exists \eta_0 \in \mathcal{P} \cap \mathcal{P}^{\tau}$, unique vector , such that $D_{ge}(\xi) = ||\xi - \eta_0||$,

The important point to note here is that we used the well known property of convex subsets in a Hilbert space: a closed convex subset W in a Hilbert space $\mathcal{H}$ contains the unique vector with the smallest norm. This ensures the existence of vectors ξ_0 and η_0 introduced in (3) and (4) respectively.

Concluding, for an entangled (non-PPT state) we are able to find *the best approximation* among separable states (PPT states, respectively). Moreover, this approach offers a classification of entanglement (genuine entanglement, respectively). More detailed discussion will appear in a forthcoming publication[15]

Acknowledgments

The author would like to express a deep gratitude to Masanori Ohya for kind hospitality. He would like also to acknowledge the supports of QBIC grant making his stay in Tokyo possible and BW grant 5400-5-0089-8.

References

1. H. Araki, Some properties of modular conjugation operator of a von Neumann algebra and non-commutative Radon-Nikodym theorem with a chain rule, Pac, J. Math. **50** (1974), 309.
2. V. P. Belavkin, M. Ohya, Quantum entropy and information in discrete entangled states, *Infinite analysis, quantum probability and related topics*, **4** 137 (2001)
3. V. P. Belavkin, M. Ohya, Quantum entanglement and entangled mutual entropy, *Proc. Royal. Soc. London A*, **458** 209 (2002)
4. Bratteli O., and Robinson D.W., *Operator Algebras ans Quantum Statistical Mechanics I*, Springer Verlag, 1987
5. Choi M.-D., Positive linear maps, Proc. Sympos. Pure. Math. **38**, 583-590 (1982)
6. Choi M.-D., Completely Positive Maps on Complex Matrices *Lin. Alg. Appl.* **10** (1975), 285–290.
7. Choi M.-D., Some assorted inequalities for positive linear maps on C^*-algebras, *J. Operator Th.* **4** (1980), 271–285.
8. D. Chruscinski, A. Kossakowski, *Phys. Rev. A* **74** 022308 (2006); quant-phys/0602076
9. D. Chruscinski, A. Kossakowski, On circulant states with positive partial transpose, quant-ph/0705.3534
10. A. Connes, Characterization des espaces vectoriels ordonnés sous-jacents aux algébres de von Neumann, *Ann. Inst. Fourier, Grenoble* **24** 121-155 (1974)
11. I. Cuculescu, Some remarks on tensor products of standard forms of von Neumann algebras, *Bolletino U. M. I.*, **7-B**, 907-919 (1993)
12. Haag, R. *Local Quantum Mechanics; Fields, Particles, Algebras*, Springer Verlag, 1992.
13. Horodecki M., Horodecki P. and Horodecki R., Separability of mixed states: necessary and sufficient conditions, *Phys. Lett. A* **223**, 1-8 (1996)
14. L. E. Labuschagne, W. A. Majewski, M. Marciniak, On k-decomposability of positive maps, *Expo. Math.* **24**, 103-125 (2006); math-ph/0306017
15. W. A. Majewski, T. Matsuoka, M. Ohya, Characterization of PPT states and measures of entanglement, in preparation,
16. W. A. Majewski, On a characterization of PPT states; arXiv:0708.3980
17. W. A. Majewski, Separable and entangled states of composite quantum systems - rigorous description, *Open Sys. & Inf. Dyn.* **6**, 79-86 (1999)
18. W. A. Majewski, A note on Størmer condition for decomposability of positive maps, arXiv:0806.3235
19. W. A. Majewski, "On entanglement of states and quantum correlations",in *Operator algebras and Mathematical Physics*, Eds. J.M. Combes, J. Cuntz, G.A. Elliott, G. Nenciu, H. Siedentop, S. Stratila; pp. 287-297, *Theta*, Bucharest, 2003. e-print, LANL math-ph/0202030;
20. W. A. Majewski, "On quantum correlations and positive maps", *Lett. Math. Physics,* **67**, 125-132 (2004); e-print, LANL math-ph/0403024
21. W. A. Majewski, M. Marciniak, On a characterization of positive maps, *J.*

Phys. A **34** 5863-5874 (2001)
22. T. Matsuoka, Some characterization of PPT states and their relations, in this proceedings
23. Størmer E., Decomposable positive maps on C^*-algebras, *Proc. Amer. Math. Soc.* **86** (1980), 402–404.
24. Takesaki M., *Tomita's theory of modular Hilbert algebras and its applications*, Lecture Notes in Mathematics **128** Springer Verlag, Berlin, 1970.
25. M. Takesaki, Theory of operator algebras I, Springer Verlag, 1979
26. Tomiyama J., On the transpose map of matrix algebras, *Proc. Amer. Math. Soc.* **88** (1983), 635–638.

SOME CHARACTERIZATIONS OF PPT STATES AND THEIR RELATION

TAKASHI MATSUOKA

Faculty of Management of Administration and Information
Tokyo University of Science, Suwa
Chino City, Nagano 391-0292, Japan
E-mail: matsuoka@rs.suwa.tus.ac.jp
Joint work with W. A. Majewski and M. Ohya

Recently some descriptions to characterize partially transposed (PPT) sates were obtained. In this note we focus the two of them and discuss their relation.

1. Introduction

One of the computable measures to detect the entanglement is partially transposed (PPT) criterion proposed by Peres[17], which is necessary condition but not sufficient one in general. However, in the lower dimensional case, Horodeckis showed that PPT criterion gives a necessary and sufficient condition of separability of compound state[7]. Since their papers there have been a lot of works investigating properties of entanglement in the terms of PPT criterion, but no one succeeded to introduce the computable measure to characterize the entangled state in general scheme (any finite dimension and also infinite dimension).

In order to find such a measure a characterization of states with positive partial transpose is one of important problems. Recently some results concerning the problem were obtained (see [4],[5],[9],[11]). In this note we focus the two of them, namely one is Majewski's approach (see [11] and also Majewski's contribution to this proceeding [12]) and another one is the characterization of PPT state via entanglement mappings discussed by Belavkin and Ohya[1,2,9]. It is shown that such two descriptions of PPT states are equivalent to each other. The detailed proofs will appear in a forthcoming publication[13].

2. Entanglement mapping and PPT condition

Belavkin and Ohya studied a rigorous mathematical structure of compound state by means of Hilbert Schmidt operator (it is called the entangling operator) and gave a classification of quantum composite system in terms of entanglement mapping[1,2]. In this section we review their approach and show how their classification corresponds to PPT condition[9].

2.1. *Pure state case*

Let $\mathcal{K}$ be a (separable) Hilbert space and $\mathcal{B}(\mathcal{K})$ be the set of all bounded linear operators on $\mathcal{K}$. For a normal state φ on $\mathbf{B}(\mathcal{K})$ there exists a density operator σ such that

$$\varphi(B) = tr_{\mathcal{K}} B\sigma, \quad B \in \mathbf{B}(\mathcal{K}). \tag{1}$$

Let $\mathcal{H}$ be an another Hilbert space and H be a Hilbert-Schmidt operator from $\mathcal{H}$ to $\mathcal{K}$ (i.e. $\sum_k \|Hx_k\|^2 < +\infty$ for any complete orthogonal system (CONS for short) $\{x_k\}$ in $\mathcal{H}$) defining the decomposition $\sigma = HH^*$ of the density with the adjoint operator H^* from $\mathcal{K}$ to $\mathcal{H}$. Then the normal state φ can be also expressed as

$$\varphi(B) = tr_{\mathcal{H}} H^* B H. \tag{2}$$

This H is called the amplitude operator, which can always be considered on $\mathcal{H} = \mathcal{K}$ as the square root of the operator σ (i.e. $H = \sigma^{\frac{1}{2}}$), and it is called just the amplitude if $\mathcal{H}$ is one dimensional space $\mathbb{C}$, corresponding to the pure state $\varphi(B) = \langle h, Bh\rangle$ here H is given as a vector $h \in \mathcal{K}$ with $\|h\|^2 = 1$.

The auxiliary Hilbert space $\mathcal{H}$ and the amplitude operator H are not unique; however H is defined uniquely up to unitary equivalence in $\mathcal{H}$. $\mathcal{H}$ can always be taken to be minimal by identifying it with the support $\mathcal{K}_\sigma \equiv E_\sigma \mathcal{K}$ for σ, defined as the closure of $\sigma\mathcal{K}$ (E_σ is the minimal orthoprojection in $\mathbf{B}(\mathcal{K})$ such that $\sigma E_\sigma = \sigma$). In this case the amplitude operator H can be represented as

$$H = \sum_n |h_n\rangle \langle e_n| \tag{3}$$

where $\{e_n\}$ is a CONS in $\mathcal{H}$ and h_n is the orthogonal eigen-amplitudes such that $h_n \in \mathcal{K}$, $\langle h_n, h_m\rangle = \lambda_n \delta_{n,m}$ with the eigenvalues λ_n of the density operator σ. Note that $\sigma = \sum_n |h_n\rangle\langle h_n|$ is a Schatten decomposition, i.e., a one-dimensional orthogonal decomposition of σ.

We can always equip $\mathcal{H}$ with an isometric involution $J_{\mathcal{H}} = J_{\mathcal{H}}^*$, $J_{\mathcal{H}}^2 = 1$, having the properties of complex conjugation in an eigen-representation of $\rho \equiv H^*H$ in (3) (i.e. $J_{\mathcal{H}}; \sum c_n|e_n\rangle \mapsto \sum \overline{c_n}|e_n\rangle$). Using $J_{\mathcal{H}}$ we can define a transposition ($\mathcal{H}$-transposition) $a^t \equiv J_{\mathcal{H}} a^* J_{\mathcal{H}}$ and a complex conjugation $\overline{a} \equiv J_{\mathcal{H}} a J_{\mathcal{H}}$ (i.e. it is easy to check the followoing equations: $\langle e_m, a^t e_n\rangle = \langle e_n, a e_m\rangle$, $\langle e_m, \overline{a} e_n\rangle = \overline{\langle e_m, a e_n\rangle}$.) on $\mathcal{H}$ such that $\rho = \overline{\rho} = \widetilde{\rho}$. The operator ρ is also recognized as a density operator defining a normal state $\psi(a) \equiv tr_{\mathcal{H}} a\rho$ on $\mathbf{B}(\mathcal{H})$ with the symmetry property above.

Given the amplitude operator H from $\mathcal{H}$ to $\mathcal{K}$, one can define not only the state φ and ψ but also a *pure entangled* state ω on $\mathbf{B}(\mathcal{H}\otimes\mathcal{K})$ by

$$\omega(a\otimes b) \equiv tr_{\mathcal{H}} a^t H^* b H \left(= tr_{\mathcal{H}} a (H^* b H)^t\right) = tr_{\mathcal{K}} H a^t H^* b. \tag{4}$$

The above bilinear form is uniquely expressed by the vector $\Omega \in \mathcal{H}\otimes\mathcal{K}$, which is given by $\langle \zeta\otimes\eta, \Omega\rangle = \langle \eta, H J_{\mathcal{H}} \zeta\rangle$ for all $\zeta \in \mathcal{H}$ and $\eta \in \mathcal{K}$. Then we have[1,2]

Lemma 2.1. *For any $a \in \mathcal{B}(\mathcal{H})$ and $b \in \mathcal{B}(\mathcal{K})$,*

$$\omega(a\otimes b) = tr_{\mathcal{K}} H a^t H^* b = tr_{\mathcal{H}} a^t H^* b H = \langle \Omega, (a\otimes b)\Omega\rangle \tag{5}$$

with its marginals given by $\varphi(b) = \omega(1\otimes b) = tr_{\mathcal{K}} b\sigma$, $\psi(a) = \omega(a\otimes 1) = tr_{\mathcal{H}} a\rho$.

Now let us move to the entangled mixed state and its entangling operator expression.

2.2. *Characterization of PPT states by entanglement mapping*

We can extend the argument of the normal state φ on $\mathcal{B}(\mathcal{K})$ to any normal compound state on $\mathcal{B}(\mathcal{H}\otimes\mathcal{K})$. Let ω be a normal compound state on $\mathcal{B}(\mathcal{H}\otimes\mathcal{K})$, then

$$\omega(a\otimes b) = tr_{\mathcal{F}} V^*(a\otimes b) V = tr_{\mathcal{H}\otimes\mathcal{K}}(a\otimes b)\theta \tag{6}$$

is described by an amplitude operator V from an another Hilbert space $\mathcal{F}$ to $\mathcal{H}\otimes\mathcal{K}$, satisfying the condition

$$VV^* \in \mathcal{B}(\mathcal{H}\otimes\mathcal{K}),\ tr_{\mathcal{H}\otimes\mathcal{K}}VV^* = 1, \tag{7}$$

where θ is the density operator corresponding to ω. $\mathcal{F}$ can always be taken as $\mathcal{F} \cong (\mathcal{H}\otimes\mathcal{K})_\theta = E_\theta(\mathcal{H}\otimes\mathcal{K})$ for $\theta = VV^*$ with an isometric involution $J_\mathcal{F}$ in a fixed CONS $\{f_k\} \subset \mathcal{F}$. Here $\mathcal{H}$ is also equipped with an isometric involution $J_\mathcal{H}$ in a CONS $\{e_n\}$ corresponding to the eigenvectors of the marginal density $\rho = tr_\mathcal{K}\theta$ (i.e., $\rho = \sum_n \lambda_n |e_n\rangle\langle e_n|$, $\lambda_n \geq 0$, $\sum_n \lambda_n = 1$).

Let $H_\mathcal{F}$ be a Hilbert-Schmidt operator from $\mathcal{F}$ to $\mathcal{H}\otimes\mathcal{K}$ given by

$$H_\mathcal{F} = |\Psi_k\rangle\langle f_k|, \tag{8}$$

where $|\Psi_k\rangle \in \mathcal{H}\otimes\mathcal{K}$ is the orthogonal eigen-amplitudes giving the Schatten decomposition of $\theta = \sum_k |\Psi_k\rangle\langle\Psi_k|$. This $H_\mathcal{F}$ is unitary equivalent to any amplitude operator V in (6) by $H_\mathcal{F}U = V$ where U is a unitary transformation $U = \sum_n |f_n\rangle\left\langle\widehat{f}_n\right|$ for the eigenvectors $\left\{\widehat{f}_n\right\}$ of V^*V. Then we can define an entangling operator $H_\mathcal{H} : \mathcal{H} \mapsto \mathcal{F}\otimes\mathcal{K}$ uniquely in the above sense by

$$\langle\zeta\otimes\eta, H_\mathcal{F}J_\mathcal{F}\xi\rangle = \langle\xi\otimes\eta, H_\mathcal{H}J_\mathcal{H}\zeta\rangle, \forall\zeta\in\mathcal{H}, \eta\in\mathcal{K}, \xi\in\mathcal{F}. \tag{9}$$

Theorem 2.1. *The normal compound state ω in (6) can be achieved as an entanglement*

$$\omega(a\otimes b) = tr_\mathcal{H} a^t H_\mathcal{H}^*(1\otimes b)H_\mathcal{H} = tr_{\mathcal{F}\otimes\mathcal{K}} H_\mathcal{H} a^t H_\mathcal{H}^*(1\otimes b) \tag{10}$$

with its marginal's densities $\rho = (H_\mathcal{H}^* H_\mathcal{H})^t$ on $\mathcal{H}$ and $\sigma = tr_\mathcal{F} H_\mathcal{H} H_\mathcal{H}^*$ on $\mathcal{K}$.

Note that the lemma 2.1 is corresponding to the case of $\mathcal{F} = \mathbb{C}$ in the theorem 2.1. Therefore in the following we will denote the entangling operator defined in (9) by the indifferently symbol H.

We introduce the linear map ϕ from $\mathcal{B}(\mathcal{K})$ to the predual $\mathcal{B}(\mathcal{H})_*$ by $\phi(b) = (H^*(1\otimes b)H)^t$ $\left(= J_\mathcal{H}H^*(1\otimes b)^* HJ_\mathcal{H}\right)$ for any $b\in\mathcal{B}(\mathcal{K})$ and its dual map ϕ^* from $\mathcal{B}(\mathcal{H})$ to the predual $\mathcal{B}(\mathcal{K})_*$ by $\phi^*(a) = tr_\mathcal{F} Ha^tH$ for any $a\in\mathcal{B}(\mathcal{H})$. Then the normal compound state ω is written as

$$\omega(a\otimes b) = tr_\mathcal{H} a\phi(b) = tr_\mathcal{K} b\phi^*(a). \tag{11}$$

Note that the map; $b(\in\mathcal{B}(\mathcal{K})) \mapsto (t\circ\phi)(a) \equiv \phi(b)^t$:

$$(t\circ\phi)(b) = H^*(1\otimes b)H \ \ (\in\mathcal{B}(\mathcal{H})_*) \tag{12}$$

is the normal completely positive (CP for short) map written in the Stein-spring form, and the map $a\,(\in \mathcal{B}(\mathcal{H})) \mapsto (\phi^* \circ t)(a) \equiv \phi^*(a^t)$:

$$(\phi^* \circ t)(a) = \sum_k \left(\langle k| \otimes 1\right) HaH^* \left(|k\rangle \otimes 1\right) \quad (\in \mathbf{B}(\mathcal{K})_*) \tag{13}$$

is also CP map written in the Kraus-Sudarshan form.

A linear map $\pi : \mathcal{B}(\mathcal{H}) \mapsto \mathcal{B}(\mathcal{K})_*$ is n co-positive (completely co-positive, or CCP for short) if the operator matrix

$$\pi_n^t([a_{ij}]) \equiv [\pi(a_{ji})] \tag{14}$$

is positive for every $n \times n$ positive-definite operator matrix $[a_{ij}]$ where $a_{ij} \in \mathcal{B}(\mathcal{H})$ (respectively for any $n \in \mathbb{N}$). Every n co-positive map is positive but not necessarily complete positive even if it is complete co-positive, unless $\mathcal{B}(\mathcal{K})$ is Abelian, in which case a positive map is both complete positive and complete co-positive. From (12) and (13) it is easy to check that both maps ϕ and ϕ^* are complete co-positive, but they are not necessarily complete positive. Using complete positivity of such maps one can show the necessary condition of separability of compound state θ as follows[1,2].

Lemma 2.2. *If θ is separable, then its entanglement ϕ^* is completely positive.*

Definition 2.1. The dual map $\phi^* : \mathcal{B}(\mathcal{H}) \to \mathcal{B}(\mathcal{K})_*$ of a CCP map $\phi : \mathcal{B}(\mathcal{K}) \to \mathcal{B}(\mathcal{H})_*$, normalized as $\mathrm{Tr}_{\mathcal{H}}\phi(1) = 1$, is called the (generalized) entanglement of the density $\rho \equiv \phi(1)$ to the density $\sigma \equiv \phi^*(1)$. We denote the set of all (generalized) entanglements by $\mathcal{E}$. The entanglement ϕ^* is called q-entanglement if ϕ^* is not CP. We denote q-entanglement by ϕ_q^* and the set of all q-entanglements by $\mathcal{E}_q$.

In [9], applying our entanglement to Choi's and Jamiołkowski's isomorphisms we showed that PPT states can be characterized by the complete positivity of entanglement mapping.

Let M_n denote the algebra of complex $n \times n$ matrices and let $\mathcal{L}[M_n, M_m]$ be the complex space of linear maps from M_n to M_m. In the following argument we focus the isomorphism $\mathcal{W} : \mathcal{L}[M_n, M_m] \mapsto M_n \otimes M_m$ as follows:

For each $\Lambda \in \mathcal{L}[M_n, M_m]$

$$\mathcal{W}_{\mathbb{P}_+}(\Lambda) = \sum_{k,l}^{n} |e_k\rangle\langle e_l| \otimes \varphi\left(|e_k\rangle\langle e_l|\right) = (\mathbf{1} \otimes \Lambda)\,\mathbb{P}_+ \tag{15}$$

$$\mathcal{W}_{\mathbb{J}}(\Lambda) = \sum_{k,l}^{n} |e_l\rangle\langle e_k| \otimes \varphi(|e_k\rangle\langle e_l|) = (\mathbf{1} \otimes \Lambda)\,\mathbb{J} \tag{16}$$

where $\{|e_k\rangle\langle e_l|\}$ is an orthonormal base of M_n. The map (15) is called the Choi's isomorphism[3] with $\mathbb{P}_+ = \sum_{k,l}^{n} |e_k\rangle\langle e_l| \otimes |e_k\rangle\langle e_l|$ which gives the maximally entanglement state (not normalized), and the map (16) is called the Jamiołkowski isomorphism[8] with $\mathbb{J} = \sum_{k,l}^{n} |e_l\rangle\langle e_k| \otimes |e_k\rangle\langle e_l|$. Note that n-positivity of $\Lambda \in \mathcal{L}[M_n, M_m]$ means complete positivity of it. Then, one has:

Lemma 2.3. [3,14,18] *For $\Lambda \in \mathcal{L}[M_n, M_m]$ the following statements hold.*

(1) Λ is a CP map iff $\mathcal{W}_{\mathbb{P}_+}(\Lambda) \geq 0$.
(2) Λ is a CCP map iff $\mathcal{W}_{\mathbb{J}}(\Lambda) \geq 0$.

Let ω be a compound state on $M_n \otimes M_m$ with its corresponding density matrix given by θ. Applying (16) to the quantum entanglement ϕ^* associated with ω we have

$$\mathcal{W}_{\mathbb{J}}(\phi^*) = \theta \geq 0, \tag{17}$$

The equation (17) shows that ϕ^* reconstructs its density θ via Jamiołkovski isomorphism. Moreover the lemma 3.1. says that ϕ^* is a CCP map and the complete positivity of ϕ^* can be characterized by the positivity of $\mathcal{W}_{\mathbb{P}_+}(\phi^*)$. Then we can prove that $\mathcal{W}_{\mathbb{P}_+}(\phi^*)$ is positive iff $\theta^{T_{\mathcal{K}}}$ is positive where $\theta^{T_{\mathcal{K}}}$ is a partically transpose of θ[9]. This fact means the following theorem .

Theorem 2.2. *θ is the PPT state iff $\phi^* \notin \mathcal{E}_q$.*

Thus, in the lower dimensional case, the separability condition can be characterized by the entanglement mapping.

Corollary 2.1. *In $M_2 \otimes M_2$ or $M_2 \otimes M_3$ case θ is separable iff $\phi^* \notin \mathcal{E}_q$.*

Note that in terms of the compound density operator $\theta = VV^*$ for the compound state $\omega(a \otimes b) = tr(a \otimes b)\,\theta$ the above maps are written simply as

$$\phi^*(a) = tr_{\mathcal{H}}(a \otimes 1)\,\theta, \quad \phi(b) = tr_{\mathcal{K}}(1 \otimes b)\,\theta. \tag{18}$$

In the following sections the above descriptions (18) of entanglement maps are available.

3. Characterization of CP and CCP map via Størmer's isomorphism

In the previous section the lemma 2.3. plays a important role of showing the equivalence between PPT condition and non q-entanglement condition. We can extend the lemma 2.3. to characterization of CP and CCP map from $\mathcal{B}(\mathcal{H})$ to $\mathcal{B}(\mathcal{K})_*$[13].

Lemma 3.1. *(1) There is an isometric isomorphism* $\psi \longmapsto \Psi$ *between* $\mathcal{L}[\,\mathcal{B}(\mathcal{H}), \mathcal{B}(\mathcal{K})_*]$ *and* $(\mathcal{B}(\mathcal{H}) \otimes \mathcal{B}(\mathcal{K}))^*$ *given by*

$$\Psi\left(\sum_i a_i \otimes b_i\right) = \sum_i tr_{\mathcal{K}}\psi(a_i)\, b_i^t, \quad a_i \in \mathcal{B}(\mathcal{H}), \ b_i \in \mathcal{B}(\mathcal{K}). \tag{19}$$

Furthermore Ψ *is positive iff* ψ *is complete positive.*

(2) There is an isometric isomorphism $\phi \longmapsto \Phi$ *between* $\mathcal{L}[\,\mathcal{B}(\mathcal{H}), \mathcal{B}(\mathcal{K})_*]$ *and* $(\mathcal{B}(\mathcal{H}) \otimes \mathcal{B}(\mathcal{K}))^*$ *given by*

$$\Phi\left(\sum_i a_i \otimes b_i\right) = \sum_i tr_{\mathcal{K}}\phi(a_i)\, b_i, \quad a_i \in \mathcal{B}(\mathcal{H}), \ b_i \in \mathcal{B}(\mathcal{K}). \tag{20}$$

Furthermore Φ *is positive iff* ϕ *is complete co-positive.*

The isomorphism (19) is so-called Størmer's isomorphism given in [15], [16]. We assume the normality of Ψ. Then ψ is given by B-O construction in (18) as follows:

$$\begin{aligned}
\Psi(a \otimes b) &= tr_{\mathcal{H}\otimes\mathcal{K}}(a \otimes b)\,\theta \\
&= tr_{\mathcal{K}}\left(tr_{\mathcal{H}}(a \otimes 1)\,\theta\right) b \\
&= tr_{\mathcal{K}}\left(tr_{\mathcal{H}}(a \otimes 1)\,\theta\right)^{tot} b \\
&= tr_{\mathcal{K}}\left(tr_{\mathcal{H}}(a \otimes 1)\,\theta\right)^{t} b^t \\
&= tr_{\mathcal{K}}\psi(a)\, b^t.
\end{aligned} \tag{21}$$

In the similar way we have:

$$\begin{aligned}
\Phi(a \otimes b) &= tr_{\mathcal{H}\otimes\mathcal{K}}(a \otimes b)\,\theta \\
&= tr_{\mathcal{K}}\left(tr_{\mathcal{H}}(a \otimes 1)\,\theta\right) b \\
&= = tr_{\mathcal{K}}\phi(a)\, b,
\end{aligned} \tag{22}$$

where θ is a trace class operator on $\mathcal{H} \otimes \mathcal{K}$ corresponding to the positive functional Φ. Now we apply the lemma 3.1. to a normal compound state ω. From the positivity of ω we have

$$0 \le \omega\left(a_i^* a_j \otimes b_i^* b_j\right) = \sum_{i,j} tr_{\mathcal{K}}\phi^*\left(a_i^* a_j\right) b_i^* b_j. \tag{23}$$

This means complete co-positivity of any entanglement $\phi^* \in \mathcal{E}$. Applying ϕ^* to (19), we define $\omega^{\tau\mathcal{K}}$ as:

$$\begin{aligned}\omega^{\tau\mathcal{K}}\left(a_i^* a_j \otimes b_i^* b_j\right) &\equiv \sum_{i,j} tr_{\mathcal{K}}\phi^*\left(a_i^* a_j\right)\left(b_i^* b_j\right)^t \\ &= \sum_{i,j} tr_{\mathcal{H}\otimes\mathcal{K}}\left(a_i^* a_j \otimes \left(b_i^* b_j\right)^t\right)\theta \\ &= \sum_{i,j} tr_{\mathcal{H}\otimes\mathcal{K}}\left(a_i^* a_j \otimes b_i^* b_j\right)\theta^{\tau\mathcal{K}},\end{aligned} \tag{24}$$

where $\theta^{\tau\mathcal{K}}$ denotes the partially transpose of θ. From the lemma 3.1. we know that the positivity of $\omega^{\tau\mathcal{K}}$ is equivalent to the complete positivity of ϕ^*, and this property also means that θ satisfies the PPT condition. Therefore we can show the theorem 2.2. by using Størmer's isomorphism.

4. Majewski's description of PPT states on Tomita-Takesaki scheme (see [11], [12])

Majewski gave a characterization of PPT states by revealing the how Tomita-Takesaki scheme may be used to describe a transposition. In this section we review his approach (see also Majewski's contribution to this proceeding [12]).

Let $\mathcal{H}$ be a finite Hilbert space. Using an invertible density matrix ρ we can define a faithful state ω on $\mathcal{B}(\mathcal{H})$ as $\omega(a) = tr\rho a$ for $a \in \mathcal{B}(\mathcal{H})$. Let us consider the GNS triple $(\mathcal{H}_\pi, \pi, \Omega)$ associated with $(\mathcal{B}(\mathcal{H}), \omega)$. Such triple are given by:

- GNS Hilbert space: $\mathcal{H}_\pi = \overline{\{a\Omega\,; a \in \mathcal{B}(\mathcal{H})\}}^{(\cdot,\cdot)}$ with $(a, b) = tra^*b$ for $a, b \in \mathcal{B}(\mathcal{H})$.
- cyclic vector: $\Omega = \rho^{1/2}$.
- representation: $\pi(a)\Omega = a\Omega$.

In this case the modular conjugation J_m is the hermitian involution as $J_m a\rho^{1/2} = \rho^{1/2}a^*$, and the modular operator Δ is equal to the map $\rho \cdot \rho^{-1}$. In order to discuss the transposition on $\pi(\mathcal{B}(\mathcal{H}))$ we introduce the following two conjugations: J_c on $\mathcal{H}$ and J on $\mathcal{H}_\pi$. Thanks to the faithfulness of ω the eigenvectors $\{e_i\}$ of ρ form an orthogonal basis in $\mathcal{H}$. Hence we can define

$$J_c x = \sum_i \overline{\langle e_i, x\rangle} e_i \tag{25}$$

for every $x \in \mathcal{H}$. Due to the fact that $\{E_{ij} = |e_i\rangle \langle e_j|\}$ form an orthogonal basis in $\mathcal{H}_\pi$ we can also define a conjugation J on $\mathcal{H}_\pi$

$$Ja\Omega = \sum_i \overline{(E_{ij}, a\Omega)} E_{ij} \tag{26}$$

with $J\Omega = \Omega$.

According to [11] and [12] let us define a transposition on $\mathcal{B}(\mathcal{H})$ as the map $a \in \mathcal{B}(\mathcal{H}) \mapsto a^t \equiv J_c a^* J_c$. By τ_0 we will denote the map induced on $\mathcal{H}_\pi$ by the transposition, i.e.

$$\tau_0 a\Omega = a^t \Omega.$$

The main properties of τ_0 are the following.

Proposition 4.1. *(see [11]) (1) Let $a \in \mathcal{B}(\mathcal{H})$ and $\xi \in \mathcal{H}_\pi$. Then*

$$a^t \xi = Ja^* J\xi. \tag{27}$$

(2) The map τ_0 has its polar decomposition, i.e.

$$\tau_0 = U\Delta^{1/2} \tag{28}$$

where U is the unitary operator on $\mathcal{H}_\pi$ defined by $U = \sum_{ij} |E_{ij})(E_{ij}|$.

On the above setting we can introduce the natural cone $\mathcal{P}$ associated with $(\pi(\mathcal{B}(\mathcal{H})), \Omega)$:

$$P = \overline{\left\{\Delta^{1/4} a\Omega : a \geq 0, a \in \pi(\mathcal{B}(\mathcal{H}))\right\}}^{(\cdot,\cdot)}.$$

Majewski has revealed a close relationship between the Tomita-Takesaki scheme and transposition, and he has the following:

Proposition 4.2. *(see [11]) Let $\xi \mapsto \omega_\xi$ be the homeomorphism between the natural cone $\mathcal{P}$ and the set of normal states on $\pi(\mathcal{B}(\mathcal{H}))$, such that*

$$\omega_\xi(a) = (\xi, a\xi), \quad a \in \mathcal{B}(\mathcal{H}).$$

For every state ω define $\omega^\tau(a) = \omega(a^t)$. If $\xi \in \mathcal{P}$ then the unique vector in $\mathcal{P}$ mapped into the state ω_ξ^τ by the homeomorphism described above, is equal to $U\xi$, i.e.

$$\omega_\xi^\tau(a) = (U\xi, aU\xi), \quad a \in \mathcal{B}(\mathcal{H}).$$

Notice that Belavkin and Ohya also discussed the transposition on the algebra $\mathcal{B}(\mathcal{H})$ by using an isometric involution J defined on the eigenvectors of marginal density operator ρ associated with the compound density operator θ in question (see section 2). In the following section we will show that their construction can be recognized as the dual description of Majewski's construction of PPT state. Before coming to our result we review the Majewski's characterization of PPT states.

Let $\mathcal{H}_A$ and $\mathcal{H}_B$ be a finite Hilbert space. Again we will consider a composite system $A+B$. Suppose that a subsystem A is described by $\mathcal{A}=\mathcal{B}(\mathcal{H}_A)$ equipped with a faithful state ω_A given by an invertible density matrix ρ_A as $\omega_A(a) \equiv tr\rho_A a$. Similarly, let $\mathcal{B}=\mathcal{B}(\mathcal{H}_B)$ as another subsystem B, ρ_B be an invertible density matrix in $\mathcal{B}(\mathcal{H}_B)$ and ω_B be a state on $\mathcal{B}$ such that $\omega_B(b) \equiv tr\rho_B b$ for $b \in \mathcal{B}$. By $(\mathcal{K},\pi,\Omega)$, $(\mathcal{K}_A,\pi_A,\Omega_A)$ and $(\mathcal{K}_B,\pi_B,\Omega_B)$ we denote the GNS representations of $(\mathcal{A}\otimes\mathcal{B},\ \omega_A\otimes\omega_B)$, $(\mathcal{A},\ \omega_A)$ and $(\mathcal{B},\ \omega_B)$ respectively. Then the triple $(\mathcal{K},\pi,\Omega)$ can be given by following identifications (cf [6], [10]):

$$\mathcal{K}=\mathcal{K}_A\otimes\mathcal{K}_B,\ \pi=\pi_A\otimes\pi_B,\ \Omega=\Omega_A\otimes\Omega_B.$$

With these identification we have

$$J_m=J_A\otimes J_B,\ \Delta=\Delta_A\otimes\Delta_B$$

where J_m, J_A, J_B are modular conjugation and Δ, Δ_A, Δ_B are modular operator for $\left(\pi\left(\mathcal{A}\otimes\mathcal{B}\right)'',\ \Omega\right),\left(\pi_A\left(\mathcal{A}\right)'',\ \Omega_A\right)$, $\left(\pi_B\left(\mathcal{B}\right)'',\ \Omega_B\right)$ respectively. Due to the finite dimensionality of the corresponding Hilbert spaces we will identify $\pi_A(\mathcal{A})''=\pi_A(\mathcal{A})$, etc. Moreover we will also write $a\Omega_A$ and $b\Omega_B$ instead of $\pi_A(a)\Omega_A$ and $\pi_B(b)\Omega_B$ for $a\in\mathcal{A}$, $b\in\mathcal{B}$ without confusion. Furthermore we denote the finite dimension of $\mathcal{H}_B$ by n. Thus $\mathcal{B}(\mathcal{H}_B)\equiv\mathcal{B}(\mathbb{C}^n)\equiv M_n(\mathbb{C})$. To put some emphasis on the dimensionality of the "reference" subsystem B, by $\mathcal{P}_n$ we denote the natural cone for $(M_n^\pi(\mathcal{A}),\omega_A\otimes\omega_0)$ where $\pi(\mathcal{A}\otimes M_n(\mathbb{C}))$ is denoted by $M_n^\pi(\mathcal{A})$ and ω_0 is a faithful state on $M_n(\mathbb{C})$.

In order to characterize the set of PPT state Majewski introduced the notion of the "transposed cone" $\mathcal{P}_n^\tau=(I\otimes U)\mathcal{P}_n$, where τ is transposition on $M_n(\mathbb{C})$ and U is the unitary operator given in (28) with the eigenvectors of density matrix ρ_0 corresponding to ω_0.

In his construction $\mathcal{P}_n$ and $\mathcal{P}_n^\tau$ may be realized as follows:

$$\mathcal{P}_n=\overline{\left\{\Delta^{1/4}\left[a_{ij}\right]\Omega:\left[a_{ij}\right]\in M_n^\pi\left(\mathcal{A}\right)^+\right\}},$$

$$\mathcal{P}_n^\tau=\overline{\left\{\Delta^{1/4}\left[a_{ji}\right]\Omega:\left[a_{ij}\right]\in M_n^\pi\left(\mathcal{A}\right)^+\right\}}.$$

Then he arrived at

Theorem 4.1. *(see [11]) In the finite dimensional case*

$$\mathcal{P}_n^{\tau} \cap \mathcal{P}_n = \left\{ \Delta^{1/4} \left[a_{ij}\right] \Omega : \left[a_{ij}\right] \geq 0, \left[a_{ji}\right] \geq 0 \right\}.$$

Corollary 4.1. *(1) There is one to one correspondence between the set of PPT states and $\mathcal{P}_n^{\tau} \cap \mathcal{P}_n$.*

(2) There is one to one correspondence between the set of separable states and $\mathcal{P}_A \otimes \mathcal{P}_B$

5. Equivalence between two types of characterization of PPT states

Let us discuss the relation between Majewski's description of PPT states and Belavkin-Ohya's one via Størmer's isomorphism. In the lemma 3.1. the condition of positivity is crucial. So, applying the scheme of lemma 3.1. the compound state ω can be represented by

$$\omega \left(\sum_i a_i \otimes b_i \right) = \sum_i \varphi_{\xi, a_i} \left(b_i\right)$$

where $\varphi_{\xi,a_i}(b_i) \equiv (\xi, (a_i \otimes b_i)\, \xi)$. If we fix a_i, then $\varphi_{\xi,a_i}(\cdot)$ can be recognized as a functional on $\mathcal{B} = M_n(\mathbb{C})$. Hence we have

$$\begin{aligned} 0 \leq \omega &\left(\sum_{i,j} a_i^* a_j \otimes b_i^* b_j \right) \\ &= \sum_i \varphi_{\xi, a_i^* a_j} \left(b_i^* b_j\right) = \left(\xi, \left(a_i^* a_j \otimes b_i^* b_j\right) \xi\right). \end{aligned}$$

The positivity of ω means that the entanglement mapping ϕ_ξ^* associated with ω is a CCP map and $\xi \in \mathcal{P}_n$. Now we can define $\omega^{\tau} \in (\mathcal{A} \otimes \mathcal{B})$ by using the same vector $\xi \in \mathcal{P}_n$ as

$$\omega^{\tau} \left(\sum_i a_i \otimes b_i \right) \equiv \sum_i \varphi_{\xi, a_i}^{\tau} \left(b_i\right)$$

where $\varphi_{\xi,a_i}^{\tau}(b_i) \equiv (\xi, (a_i \otimes b_i^t)\, \xi)$. Then we have

$$\begin{aligned} \omega^{\tau} \left(\sum_{i,j} a_i^* a_j \otimes b_i^* b_j \right) &= \sum_i \varphi_{\xi, a_i^* a_j}^{\tau} \left(b_i^* b_j\right) \\ &= \left(\xi, \left(a_i^* a_j \otimes \left(b_i^* b_j\right)^t\right) \xi\right) \\ &= \left(\xi, \left(a_i^* a_j \otimes b_j^t \left(b_i^*\right)^t\right) \xi\right). \end{aligned}$$

From the lemma 3.1. and the theorem 4.1. the positivity of ω^τ means that $\phi_\xi^* \notin \mathcal{E}_q$ and $\xi \in \mathcal{P}_n^\tau \cap \mathcal{P}_n$. So we conclude that

Theorem 5.1. *The description of PPT states by $\mathcal{P}_n^\tau \cap \mathcal{P}_n$ can be recognized as the dual description of PPT states by $\mathcal{E}/\mathcal{E}_q$.*

In the base of the above equivalence of two types of description of PPT states we may discuss the effectiveness of such characterizations from different points of view. One is the possibility to apply the geometry of positive cone to measures of entanglement (see Majewski's contribution to this proceeding [12]). In the other hand we can discuss how entangled state achieve the maximum channel capacity in quantum encoding via entanglement mapping (see [1], [2]). More detailed discussion will appear in a forthcoming publication[13].

References

1. V. P. Belavkin, M. Ohya, *Infinite Dimensional Analysis, Quantum Probability and Related Topics*, **4**, No.2, 137 (2001).
2. V. P. Belavkin, M. Ohya, *Proc. R. Soc. Lond.*, **A 458**, 209 (2002).
3. M. D. Choi, *Lin. Alg. Appl.*, **10**, 285 (1975).
4. D. Chruscinski, A. Kossakowski, *Phy. Rev.* **A 74,** 022308 (2006), quant-ph/0602076.
5. D. Chruscinski, A. Kossakowski, "On circulant states with positive partial transpose", quant-ph/0705.3534.θ
6. I. Cuculescu, *Bolletino U. M. I.,* **7-B,** 907 (1993).
7. M. Horodecki, P. Horodecki and R. Horodecki, *Phy. Lett.*, **A 223**, 1 (1996).
8. A. Jamiołkowski, *Rep. Math. Phys.*, **3**, 275 (1972).
9. A. Jamiołkowski, T. Matsuoka and M. Ohya, "Entangling operator and PPT condition", TUS preprint (2007).
10. W. A. Majewski, *Open Sys. & Inf. Dyn.* **6,** 79 (1999).
11. W. A. Majewski, "On a characterization of PPT states", arXiv:0708.3980
12. W. A. Majewski, "Measures of Enanglement - a Hilbert space approach -" in this proceeding.
13. W. A. Majewski, T. Matsuoka and M. Ohya, Characterization of PPT states and measures of entanglement, in preparation.
14. G. Kimura, A. Kossakowski, *Open Sys. & Information Dyn.*, **12**, 1 (2005).
15. E. Størmer, *J. Func. Ana.,* **66**, 235 (1986).
16. E. Størmer, *Contemporary Math.,* **62**, 345 (1987).
17. A. Peres, *Phys.Rev.Lett.*, **77,** 1413 (1996).
18. S. L. Woronowicz, *Rep. Math. Phys.*, **10**, 165 (1976).

ON THE DYNAMICS OF ENTANGLEMENT AND CHARACTERIZATION OF ENTANGLING PROPERTIES OF QUANTUM EVOLUTIONS

MIŁOSZ MICHALSKI

Institute of Physics, Nicolaus Copernicus University
Grudziądzka 5, 87–100 Toruń, Poland
milosz@fizyka.umk.pl

We study patterns of creation, temporal evolution, and transport of entanglement in quantum spin chains. The model used is an isotropic Heisenberg chain with periodic boundary conditions, and we obtain analytic form of time dependence of concurrence and negativity of various spin pairs in the system. The objective of the present study is to assess the usefulness of various averaged entanglement measures in the characterization of entangling properties of quantum evolutions.

Keywords: *Entanglement evolution; isotropic Heisenberg chains; entangling power of operators; time-averaged entanglement.*

1. Introduction

In the last decade, quantum information theory and its most promising applications — quantum communication and computing have experienced a remarkable development, feedbacked by the fast growing interest of scientific community. Quantum entanglement, the yesterday's "spooky action at a distance", turned out to be an invaluable and manageable resource lying at the heart of intriguing applications, like teleportation of quantum states, spectacular power of quantum computation or ultimately safe cryptographic systems, to name just a few.

Despite the continuing research effort, our understanding and characterization of entanglement for general quantum states remains far from being complete. While it is entirely understood for pure states of bipartite systems or — in the case of mixed states — for low-dimensional ones, testing for or measuring entanglement in larger, in particular multipartite, systems continues to be a serious research challenge.

Entanglement emerges as a result of interaction among parts of a quantum system. It is thus of immediate interest to ask about entangling properties of physical interactions or, more generally, of quantum operations. Namely, given a completely positive, trace preserving map Λ, it is natural to ask how much entanglement is produced by acting with Λ on a state ρ, relative to its initial entanglement.

In order to characterize the entangling capability of Λ in more global terms, one may look for a maximum [10] or average amount of entanglement generated by Λ on an ensemble of states,

$$\mathcal{E}(\Lambda) = \int \big[E(\Lambda\rho) - E(\rho)\big]\, d\mu(\rho)\,, \tag{1}$$

with E being some fixed entanglement measure. This is, in particular, the scheme adopted and developed by Zanardi *et al.* [21] for so-called *entangling power* of unitary operators acting on pure bipartite states, $\Lambda \in \mathrm{U}(\mathcal{H}_{AB})$. In Zanardi's approach, μ is taken to be the uniform measure on pure product states $|\Psi\rangle = |\psi_A\rangle \otimes |\psi_B\rangle$, and E is the linear entropy of the corresponding partial trace, denoted here by E_L,

$$E_L(|\Phi\rangle) = 1 - \mathrm{Tr}_A\, \sigma^2\,, \qquad \sigma = \mathrm{Tr}_B\, |\Phi\rangle\langle\Phi|\,. \tag{2}$$

Due to its computational simplicity, linear entropy is frequently used to replace von Neumann entropy $E_{\mathrm{vN}} = -\,\mathrm{Tr}_A\, \sigma \log \sigma$, much more appropriate in this context.

We want to stress that the integration in (1) is in general troublesome, even in simplified settings like that of Ref. [21]. Actually, using some symmetry arguments, Zanardi and Wang [22,19,11] managed to avoid this integration completely by expressing the entangling power of a unitary operator U by the *entanglement of* U and of its various partial transposes and realignments. The operator entanglement mentioned here is yet another measure characterizing entangling properties of matrices: it is defined by making simple use of the natural Hilbert space isomorphism between the Hilbert-Schmidt space $\mathcal{H}_{HS}$ of linear operators on finite-dimensional $\mathcal{H}$ and the space $\mathcal{H} \otimes \mathcal{H}$,

$$\Gamma : A \longmapsto (A \otimes I)|\Phi^+\rangle\,, \qquad |\Phi^+\rangle = \sum_i |e_i\rangle \otimes |e_i\rangle\,, \tag{3}$$

where $\{|e_i\rangle\}$ is an orthonormal basis in $\mathcal{H}$. The entanglement of an operator A is then defined as $\tilde{E}(A) = E_L(\Gamma A)$, with E_L as in (2). As we have just mentioned above, in [11] Wang gave the following relation between the entangling power $\mathcal{E}$ of a unitary operation U and its entanglement $\tilde{E}(U)$,

$$\begin{aligned}\mathcal{E}(U) &= \left(\frac{d}{d+1}\right)^2 \left[\tilde{E}(U) + \tilde{E}(US_{AB}) - \tilde{E}(S_{AB})\right] \\ &= \left(\frac{d}{d+1}\right)^2 \left[2 - \tilde{E}(S_{AB})\right] \\ &\qquad - \frac{1}{(d+1)^2 d^2} \operatorname{Tr}\left(\left[U^R (U^R)^\dagger\right]^2 + \left[U^{T_A} (U^{T_A})^\dagger\right]^2\right) ,\end{aligned}$$

where S_{AB} is the swap operator, R denotes the matrix realignment, $(U^R)_{ij,kl} = U_{ik,jl}$ and T_A — the partial transpose.

Let us also mention here that the definition of operator entanglement based on (3) can immediately be lifted up to apply to arbitrary completely positive maps [3,22] by using various guises of the Jamiołkowski isomorphism [9], i.e. by expressing the entanglement of a *quantum operation* by that of the corresponding *state*.

Although the operator entanglement and its entangling power are visibly related, this relation is certainly not trivial. A manifest example is the swap operator S, whose image by Γ is maximally entangled, while its entangling power is null as S maps product states onto product states.

For the completeness of this overview, let us also mention a few other approaches like that of Nielsen *et al.* [14], where the *entangling strength* of an operation related to its nonlocality is an axiomatically introduced measure, without any reference to state entanglement. Some authors [5,17,2] concentrate their study on entangling abilities and nonlocality of *Hamiltonians* rather than unitaries. Entangling capacity (or nonlocal strength) of interactions is considered there in the context of their efficiency in implementing various 2-qubit quantum gates. Given nonlocal H, assisted by possible local control operations on subsystems, one characterizes its entangling power by the rate of entanglement creation, e.g.

$$\tilde{E}(H) \;=\; \sup_{|\psi\rangle} \lim_{t\to 0} \frac{E(e^{-iHt}|\psi\rangle) - E(|\psi\rangle)}{t} \, .$$

Yet another group of papers deals with time evolution of entanglement under the action of a quantum dynamical (semi)group $\{\Lambda_t\}$ with a generator H, namely $\tilde{E}(t) = E(\rho_t) = E\big(\Lambda_t \rho_0\big)$, for some fixed entanglement measure E and initial state ρ_0. Then *time-averages* of $\tilde{E}$ can enter the stage,

$$\mathcal{E}(H) = \lim_{T\to\infty} \frac{1}{T} \int_0^T E\big(\Lambda_t \rho_0\big)\, dt \, . \tag{4}$$

Some models studied along similar lines include: kicked quantum tops [4], the Dicke model [7], Bose-Einstein condensates [20], vibration of triatomic molecules [8], to name just a few. Often the limit $T \to \infty$ is replaced with a finite time average, where the value of T is taken to be some characteristic time scale of the system in question. Our model studied below shows that such a cut-off may lead to results differing significantly from the true time-average.

It is natural to ask about possible relation between ensemble- (1) and time-averages (4), following the arguments of classical ergodic theory. Although quantum unitary dynamics are in principle nonergodic, Naudts, Van der Straeten and Verhulst [13,16] have recently discussed the possibility of using the two averages interchangeably by introducing generalized quantum microcanonical ensembles of rather specific form.

To sum up, a substantial body of results and tools characterizing the entangling capabilities of quantum operations is now available, yet natural quantitative measures taking the form of averages (1) and (4), are in principle hard to compute. We study below a simple, analytically solvable model of an isotropic Heisenberg chain with periodic boundary conditions to get an insight into the mechanism of formation of temporal entanglement patterns among interacting particles. To obtain a "realistic" image of the dynamics, in our computational experiments we use a 30-particle chain, where all but a pair of particles are traced out, leaving the two-particle mixed state whose time-varying entanglement is assessed by means of concurrence. The remainder of the chain is thus treated as a reservoir establishing and transmitting entanglement between various parts of the system in the course of dynamics. We observe a rather clear pattern of entanglement evolution along the chain in the short-time regime, and a very complicated long-time behaviour. Then we test numerically the possibility of replacing a complicated time-averaging of entanglement measure with an appropriate ensemble mean along the lines of [13,16].

2. The Heisenberg model

We consider a standard 1-dimensional isotropic XX Heisenberg model [18,1,15] of N interacting spins, whose Hamiltonian acting on $\mathcal{H} = (\mathbf{C}^2)^{\otimes N}$ is given by

$$H = \frac{J}{2} \sum_{i=1}^{N} \left(\sigma_i^x \sigma_{i+1}^x : + : \sigma_i^y \sigma_{i+1}^y \right), \tag{5}$$

and where we have written $\sigma_i^\alpha = I \otimes \cdots \otimes \sigma^\alpha \otimes \cdots \otimes I$, $\alpha = x, y$. Moreover, the index value $i = N+1$ is identified with 1 (periodic boundary conditions).

Since H commutes with $S^z = \sum \sigma_i^z$, the total excitation number is conserved. Therefore, invariant subspaces of H are spanned by vectors with precisely m spins up. We confine ourselves to the dynamics on $m = 1$ subspace Span$\{|j\rangle\}$, with $|j\rangle = |0 \dots 1 \dots 0\rangle$ at j^{th} site. The eigenvalues and eigenvectors of H are:

$$E_k = J\cos(2\pi k/N)\,, \qquad |\psi_k\rangle = \frac{1}{\sqrt{N}}\sum_{j=1}^{N} e^{2\pi i jk/N}\,|j\rangle\,. \tag{6}$$

Choosing the initial state $|\Phi_0\rangle = |\ell\rangle = \frac{1}{\sqrt{N}}\sum_{k=1}^{N} e^{-2\pi i\ell k/N}\,|\psi_k\rangle$ we get

$$|\Phi(t)\rangle = e^{-iHt}|\Phi_0\rangle = \sum_{j=0}^{N} \alpha_j(t)\,|j\rangle\,,$$

with $\alpha_j(t) = \frac{1}{N}\sum_{k=1}^{N} e^{2\pi i(j-\ell)k/N - itJ\cos(2\pi k/N)}$.

By tracing out all but a pair of spins at i and j, we obtain the corresponding two-particle mixed state in the form

$$\varrho_{ij}(t) = \begin{bmatrix} 1-|\alpha_i|^2-|\alpha_j|^2 & 0 & 0 & 0 \\ 0 & |\alpha_i|^2 & \alpha_i\overline{\alpha_j} & 0 \\ 0 & \overline{\alpha_i}\alpha_j & |\alpha_j|^2 & 0 \\ 0 & 0 & 0 & 0 \end{bmatrix}. \tag{7}$$

From this, one can easily compute the concurrence and negativity of ϱ_{ij},

$$C_{ij} = 2|\alpha_i\alpha_j|\,, \qquad \mathcal{N}_{ij} = \frac{1}{2}\left(\beta - \sqrt{\beta^2 + C_{ij}^2}\right),$$

where $\beta = 1 - |\alpha_i|^2 - |\alpha_j|^2$.

Using the above formulae we have studied the time dependence of C_{ij} and $\mathcal{N}_{ij}$ for various pairs (ij) of spins in the chain and different initial data. As it can be expected, the qualitative images of the time evolution of both the concurrence and the negativity are similar. One can observe a characteristic pattern of the occurrence of entanglement in different parts of the chain when the initial state is a product one, e.g. $|\Phi_0\rangle = |10\dots0\rangle$ — the phenomenon of *entanglement creation*. As the excitation of the first spin is propagated in time along the chain, more and more distant pairs of neighbouring spins become entangled, and likewise for pairs of nonadjacent

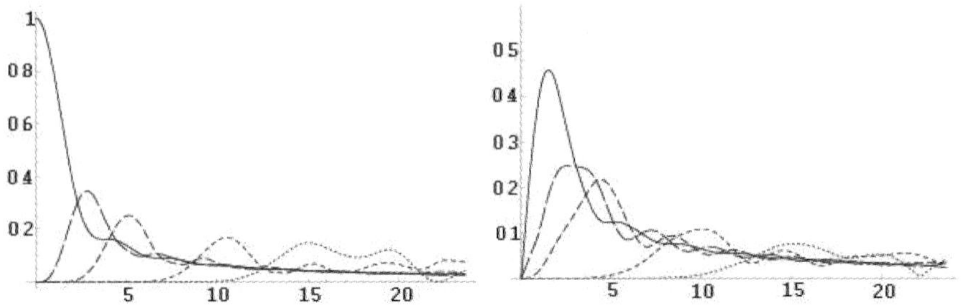

Figure 1. Entanglement transport – concurrence dynamics, Ci $j(t)$, for various pairs of spins with locally entangled initial state $|\Phi_0\rangle$, cf. text. Left plot –neighbouring pairs of spins(i,j), respectively: (1,2), (3,4), (5,6), (10,11), (14,15), Right plot (note the y-scale change) – distant spins, respectively: (1,3), (1,4). (1,5), (1,10), (1,15).

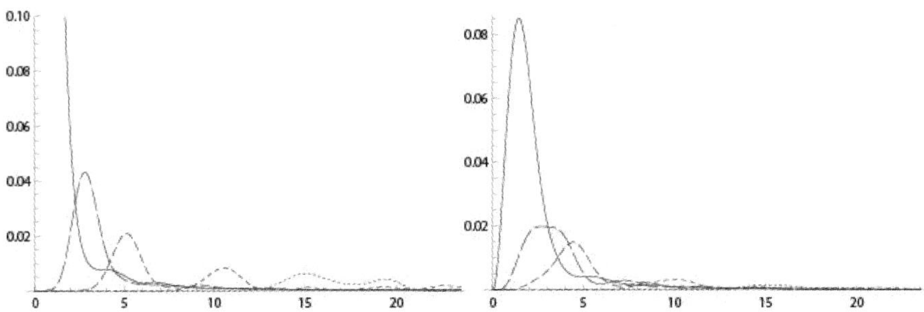

Figure 2. The corresponding negativity dynamics, $Nij(t)$, for various pairs of spins (for details see the caption of Fig.1)

spins. Locally, this "wave" of entanglement has oscillatory character, however it is highly nonperiodic due to the fact that the eigenvalues (6) are, in general, rationally incommensurate.

Note that due to the circular shift symmetry of the Hamiltonian, changing initial conditions to $|\ell\rangle$ induces simply the respective cyclic change in (ij) among C_{ij}.

Fig. 1 presents the patters of time dependence of C_{ij} in a chain of length $N = 30$ for locally entangled initial state $|\Phi_0\rangle = \frac{1}{\sqrt{2}}(|10\rangle + |01\rangle) \otimes |0\ldots0\rangle$, which may be loosely termed *entanglement transport*. Fig. 2 shows similar dependence of the negativity $\mathcal{N}_{ij}$. The general character of these patterns is similar to that of entanglement creation discussed above, we therefore do

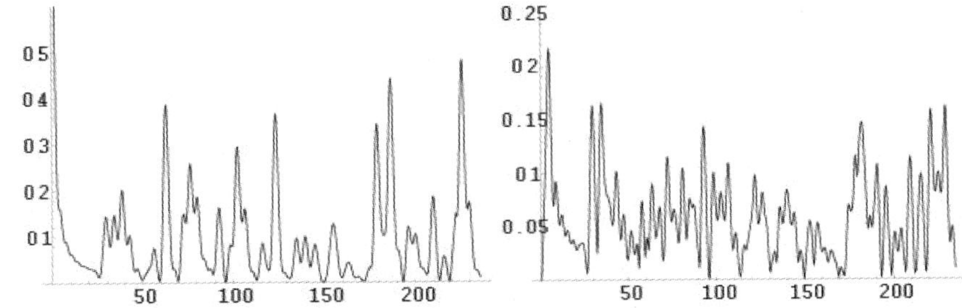

Figure 3. Long-time behaviour of C_{12} (left) and C_{15} (right) for initial $|\Phi_0\rangle$ as in Fig.1.

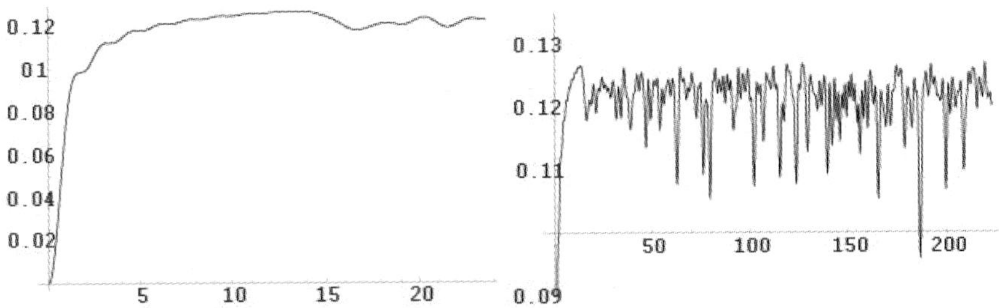

Figure 4. Short-(left) and long-time (right) behaviour of global entanglement for $|\Phi(0)\rangle = |10\ldots0\rangle$.

not reproduce the images of the latter here.

The regular relaxation pattern which can be observed in these plots turns out to be deceiving when one goes to longer time scales. Fig. 3 shows how the regular initial transient evolves into a complicated, nonperiodic pattern resulting from the closedness of our system — the interference of entanglement carrying waves, recurring back and forth along the chain. The characteristic time scale for this regular behaviour can be seen to change with N as $T_0 \approx \frac{N}{4J}\pi$ (in our numerical example we have set the coupling strength $J = 1$ for simplicity).

In order to assess the average entangling power of the Heisenberg interaction (5) one may attempt to compute the time average (4) for C_{ij} and, subsequently, subject it to averaging over different admissible initial conditions. There exists a simple entanglement measure which, applied to our model, resembles such an averaging procedure. It is the so called *global entanglement* [12], one of the earliest multipartite entanglement measures in the literature, which in the present case evaluates to

$$Q(t) = Q(|\Phi(t)\rangle) = \frac{4}{N}\sum_{j=1}^{N} |\alpha_j|^2\left(1 - |\alpha_j|^2\right) \tag{8}$$

and can be easily seen to equal, up to a constant, the sum of squared pair concurrences. Taking into account the cyclic shift symmetry of (5) mentioned above, summing the concurrences over different (ij) is the same as the summation over initial states $|\ell\rangle$. Therefore, the time average (4) of Q above can be regarded as a good candidate for a measure of mean entangling capability of H. However, Fig. 4 showing short- (T_0 scale) and long-time behaviour of $Q(t)$, suggests that even numerical time-averaging of Q may not be an easy task.

Following Refs. [13] and [16] we can form now an ensemble of states

$$\left\{|\Phi\rangle,\ d\beta\right\} = \left\{\frac{1}{\sqrt{N}}\sum_{k=1}^{N} e^{-2\pi i\ell k/N} e^{i\beta_k}|\psi_k\rangle,\ d\beta\right\},$$

where $d\beta = \frac{1}{(2\pi)^N}\prod_k d\beta_k$ on $(0, 2\pi)^N$. Here $|\ell\rangle$ denotes an arbitrary fixed initial state. The value of $Q(|\Phi\rangle)$ is clearly independent of $|\ell\rangle$. Then the ensemble average of Q is computed as

$$\int Q(|\Phi\rangle)\, d\beta = \frac{4}{(2\pi)^N N}\iint\cdots\int\Big(\sum_{j=1}^{N}|\alpha_j|^2\left(1 - |\alpha_j|^2\right)\Big)\, d\beta_1\ldots d\beta_N\,, \tag{9}$$

where $\alpha_j = \frac{1}{N}\sum_{k=1}^{N} e^{2\pi i(j-\ell)k/N + i\beta_k}$. Although our system is nonergodic due to 2-fold degeneracy of eigenvalues (6) and rational proportionality of some of them, one can still expect the ensemble average to correctly reproduce the time one as the measure $d\beta$ is invariant with respect to the e^{-iHt} action [13].

We have performed a series of tests computing the time average of (8) for a few values of N by direct numerical integration and by a Monte Carlo method. The direct integration showed slow convergence with growing T. In particular, the results of integration over the characteristic time scale T_0 were off by more than 3% from the results obtained for longer intervals. Monte Carlo methods produced satisfactory results in good agreement with the latter with much more modest computational effort. Lastly, Monte Carlo results obtained for the ensemble averages (9) exhibited very good stability and agreement with time averages as above, the deviations being less than 10^{-4} or 0.1%. The table below shows exemplary numerical results obtained for $N = 30$ and $T \sim 20\,T_0$.

Time average (4) of $Q(t)$, direct integration	0.120378
Time average (4) of $Q(t)$, Monte Carlo integration	0.1204...
Ensemble average (9), Monte Carlo integration	0.1204...
Time average (4) of $Q(t)$ for $T = T_0$, direct integration	0.116713

References

1. D. Bruss *et al.*, *Phys. Rev. A* **72** (2005) 014301.
2. A.M. Childs *et al.*, *Quant. Inf. Comput.* **3** (2003) 97.
3. J.I. Cirac *et al.*, *Phys. Rev. Lett.* **86** (2001) 544.
4. R. Demkowicz-Dobrzański, M. Kuś, *Phys. Rev. E* **70** (2004) 066216.
5. W. Dür *et al.*, *Phys. Rev. Lett.* **87** (2001) 137901.
6. A. Hamma, P. Zanardi, *Phys. Rev. A* **69** (2004) 062319, quant-ph/03081131v2.
7. X-W. Hou, B. Hu, *Phys. Rev. A* **69** (2004) 042110.
8. X-W. Hou, M-F. Wan, Z-Q. Ma, *Chem. Phys. Lett.* **426** (2006) 469.
9. A. Jamiołkowski, *Rep. Math. Phys.* **3** (1972) 275.
10. B. Kraus, J.I. Cirac, *Phys. Rev. A* **63** (2001) 062309.
11. Z. Ma, X. Wang, *Phys. Rev. A* **75** (2007) 014304, quant-ph/0702103.
12. D.A. Meyer, N.R. Wallach, *J. Math. Phys.* **43** (2002) 4273, quant-ph/0108104.
13. J. Naudts, E. Van der Straeten, *J. Stat. Mech.* (2006) P06015, quant-ph/0602039.
14. M.A. Nielsen *et al.* *Phys. Rev. A* **67** (2003) 052301.
15. V. Subrahmanyam, A. Lakshminarayan, *Phys. Lett. A* **349** (2006) 164.
16. T. Verhulst, J. Naudts, *J. Phys. A: Math. Theor.* **40** (2007) 2475.
17. G. Vidal, K. Hammerer, J.I. Cirac, *Phys. Rev. Lett.* **88** (2002) 237902.
18. X. Wang, *Phys. Rev. A* **64** (2001) 012313, quant-ph/0101013v2.
19. X. Wang, P. Zanardi, *Phys. Rev. A* **66** (2002) 044303.
20. Q. Xie, W. Hai, *Eur. Phys. J.* D **33** (2005) 265.
21. P. Zanardi, Ch. Zalka, L. Faoro, *Phys. Rev. A* **62** (2000) 030301(R).
22. P. Zanardi, *Phys. Rev. A* **63** (2001) 040304.

PERSPECTIVE FROM MICRO-MACRO DUALITY – TOWARDS NON-PERTURBATIVE RENORMALIZATION SCHEME –

IZUMI OJIMA

RIMS, Kyoto University, Japan

The problem of renormalization procedure is re-examied from the viewpoint of Micro-Macro duality.

1. Micro-Macro duality

"Micro-Macro duality" is one of the basic features found between the invisible microscopic nature and its visible macroscopic manifestations, which can be understood in parallel with the Fourier duality between an abstract group and its concrete representations. This viewpoint has played crucial roles in our analysis of the mutual relations between virtual dynamical levels and specific geometric ones in various contexts (see, Ref.1). Using this general notion, we can provide the heuristic idea of "Quantum-Classical Correspondence" with precise mathematical formulations in which Micro and Macro are mutually and closely related with each other; the latter, Macro, emerges from the former, Micro, through the processes of condensation of infinitely many quanta and the essential features of the former can be determined and re-constructed to certain extent from the data structure at the levels of Macro, in close analogy with the above-mentioned duality in the context of groups and representations. From this viewpoint of Micro-Macro duality, we try here to sketch the essential ingredients for a natural reformulation of the traditional theory of renormalization procedures commonly adopted in the physical applications of quantum field theory (QFT for short). For this purpose, the most relevant notions in what follows are the group of scale transformations and the associated aspects of broken symmetry which is not unitarily implementable within a sector (defined by a quasi-equivalence class of factor representations of the algebra of observables) but which generates a family of (mutually disjoint) sectors along an orbit of symmetry transformations.

2. Broken scale invariance: imaginary-time vs. real-time

Here we briefly summarize some consequences of broken scale invariance in relativistic QFT.

1) Imaginary-time version = "temperature as order parameter of broken scale invariance":

Theorem 2.1. *(IO04[2]) In the standard setting of algebraic QFT, the inverse temperature $\beta := (\beta^\mu \beta_\mu)^{1/2}$ is a macroscopic* ***order parameter*** *for parametrizing mutually disjoint sectors in the thermal situation arising from the* ***broken scale invariance*** *under the renormalization-group transformations, where β^μ is an inverse temperature 4-vector of a relativistic KMS state ω_{β^μ} describing a thermal equilibrium in its rest frame.*

This result is based on the notion of a scaling algebra due to Buchholz-Verch[3] in combination with Takesaki's theorem[4] on the disjointness of KMS states at different temperatures valid for a system with physical observables constituting a von Neumann algebra of type III.

2) What should be the corresponding "real-time" version to the above?: Renormalization Theory (at $T = 0K$).

2.1. *How to formulate broken scale invariance*

Theorem 2.2. *(Takesaki'70[4]) For a quantum C^*-dynamical system with type III representations in its KMS states, any pair of KMS states for different (inverse) temperatures $\beta_1 \neq \beta_2$ are mutually disjoint $\omega_{\beta_1} \overset{|}{\circ} \omega_{\beta_2}$.*

The claim of the first theorem (Theorem 2.1) due to myself is that the above disjointness allows us to interpret the inverse temperature β as an order parameter of broken scale invariance. In the usual situation, this kind of symmetry breakdown arises as a spontaneous breakdown of a symmetry described by a group acting on the algebra of physical quantities by automorphisms. In contrast, the present case of broken scale invariance usually involves explicit breaking terms such as mass, which seem to prevent scale transformations from being treated as automorphisms. However, the results on scaling algebra in algebraic QFT due to Ref.3, 2 shows that the above negative anticipation can be avoided.

The results can be summarized as follows. Let the following requirements be imposed on all the possible renormalization-group transformations R_λ:

(i) R_λ should map the given net $\mathcal{O} \to \mathcal{A}(\mathcal{O})$ of local observables at spacetime scale 1 onto the corresponding net $\mathcal{O} \to \mathcal{A}_\lambda(\mathcal{O}) \doteq \mathcal{A}(\lambda\mathcal{O})$ at scale λ, i.e.,

$$R_\lambda : \mathcal{A}(\mathcal{O}) \to \mathcal{A}_\lambda(\mathcal{O})$$

for every region $\mathcal{O} \subset \mathbb{R}^4$, through which the light velocity c is kept unchanged: $(\lambda x)^i/(\lambda x)^0 = x^i/x^0$.

(ii) In the Fourier-transformed picture, the subspace $\widetilde{\mathcal{A}}(\widetilde{\mathcal{O}})$ of all (quasi-local) observables carrying energy-momentum in the set $\widetilde{\mathcal{O}} \subset \mathbb{R}^4$ is transformed as

$$R_\lambda : \widetilde{\mathcal{A}}(\widetilde{\mathcal{O}}) \to \widetilde{\mathcal{A}}_\lambda(\widetilde{\mathcal{O}})$$

for $\forall\widetilde{\mathcal{O}}$, where $\widetilde{\mathcal{A}}_\lambda(\widetilde{\mathcal{O}}) := \widetilde{\mathcal{A}}(\lambda^{-1}\widetilde{\mathcal{O}})$, through which the Planck constant $\hbar$ is unchanged: $(\lambda^{-1}p)_\mu(\lambda x)^\mu/\hbar = p_\mu x^\mu/\hbar$.

(iii) For scale invariant theories R_λ may not be isomorphisms but are maps continuous and bounded uniformly in λ.

Then, the scaling net $\mathcal{O} \to \widehat{\mathcal{A}}(\mathcal{O})$ corresponding to the original local net $\mathcal{O} \to \widehat{\mathcal{A}}(\mathcal{O})$ of observables is defined as the local net consisting of scale-changed observables under the action of all the possible choice of R_λ satisfying (i)-(iii). Mathematically the algebra $\widehat{\mathcal{A}}(\mathcal{O})$ can be understood as the algebra $\Gamma(\mathbb{R}^+ \times \mathcal{A}(\mathcal{O}))$ of sections $\mathbb{R}^+ \ni \lambda \longmapsto \hat{A}(\lambda) \in \mathcal{A}_\lambda(\mathcal{O})$ of algebra bundle $\amalg_{\lambda\in\mathbb{R}^+}\mathcal{A}_\lambda(\mathcal{O}) \twoheadrightarrow \mathbb{R}^+$ over the multiplicative group $\mathbb{R}^+$ of scale changes. This algebra $\widehat{\mathcal{A}}(\mathcal{O})$ is seen to be identical to the augmented algebra $\mathcal{A}(\mathcal{O}) \rtimes_\alpha \widehat{\mathbb{R}^+}$ defined in Ref.5. Then, the scaling algebra $\hat{\mathcal{A}}$ is defined by the C*-inductive limit of all local algebras $\hat{\mathcal{A}}(\mathcal{O})$. Algebraic structures making $\hat{\mathcal{A}}(\mathcal{O})$ a unital C*-algebra are defined in a *pointwise manner*, for instance, by $(\hat{A} \cdot \hat{B})(\lambda) := \hat{A}(\lambda)\hat{B}(\lambda)$, $(\hat{A}^*)(\lambda) := \hat{A}(\lambda)^*$, etc., and $\|\hat{A}\| := \sup_{\lambda\in\mathbb{R}^+} \|\hat{A}(\lambda)\|$.

From the scaled actions $\mathcal{A}_\lambda \underset{\alpha^{(\lambda)}}{\curvearrowleft} \mathcal{P}_+^\uparrow$ of Poincaré group on $\mathcal{A}_\lambda$ with $\alpha^{(\lambda)}_{x,\Lambda} = \alpha_{\lambda x,\Lambda}$, an action of $\mathcal{P}_+^\uparrow$ is induced on $\widehat{\mathcal{A}}$ by

$$(\hat{\alpha}_{x,\Lambda}(\hat{A}))(\lambda) := \alpha_{\lambda x,\Lambda}(\hat{A}(\lambda)).$$

Then the conditions (ii), (iii) are expressed simply as the continuity of Poincaré-group action: $\|\hat{\alpha}_{x,\Lambda}(\hat{A}) - \hat{A}\| \underset{(x,\Lambda)\to(0,1)}{\longrightarrow} 0$. Then, the scaling net $\mathcal{O} \to \widehat{\mathcal{A}}(\mathcal{O})$ is shown to satisfy all the properties to characterize a relativistic local net of observables if the original one $\mathcal{O} \to \mathcal{A}(\mathcal{O})$ does.

Now scale transformations can be defined by an automorphic action $\hat{\sigma}_{\mathbb{R}^+}$ of the $\mathbb{R}^+$ on the scaling algebra $\hat{\mathcal{A}}$, given for $\forall \mu \in \mathbb{R}^+$ by

$$(\hat{\sigma}_\mu(\hat{A}))(\lambda) := \hat{A}(\mu\lambda), \quad \lambda > 0,$$

satisfying

$$\hat{\sigma}_\mu(\hat{\mathcal{A}}(\mathcal{O})) = \hat{\mathcal{A}}(\mu\mathcal{O}), \quad \mathcal{O} \subset \mathbb{R}^4,$$
$$\hat{\sigma}_\mu \circ \hat{\alpha}_{x,\Lambda} = \hat{\alpha}_{\mu x,\Lambda} \circ \hat{\sigma}_\mu, \quad (x,\Lambda) \in \mathcal{P}_+^\uparrow.$$

Remark 2.1. Scaling transformations $\hat{\sigma}_{\mathbb{R}^+}$ play the role of renormalization group transformations to relate observables at different scales.

Remark 2.2. Since a broken symmetry can always be restored by taking all breaking parameters as variables undergoing the broken symmetry transformations, there is no miracle in the results due to Buchholz and Verch through their complicated analysis: it can naturally be accommodated as a special case into the general definition of an augmented algebra[5] $\hat{\mathcal{F}} := \Gamma(G \times_H \mathcal{F})$ with the choice of $H := \mathcal{P}_+^\uparrow$, $G = H \rtimes \mathbb{R}^+$ (semidirect product) and together with slight modifications due to spacetime dependence $\mathcal{F} \Longrightarrow \mathcal{A}(\mathcal{O})$ (which is affected by the action of $\mathbb{R}^+$) (and the intervention of the centre due to SSB: $SO(3)\backslash L_+^\uparrow \cong \mathbb{R}^3$ at $T \neq 0°K)^2$.
Scaled actions $\alpha_{x,\Lambda}^{(\lambda)} = \alpha_{\lambda x,\Lambda}$ of Poincaré group on $\mathcal{A}_\lambda$ can also be naturally understood as the conjugacy change of the stability group $H \to gHg^{-1}$ from the point He to Hg^{-1} on the base space $H\backslash G = \mathbb{R}^+$: $s_\mu(x,\Lambda)s_\mu^{-1} = (\mu x, \Lambda)$.

2.2. *Scale changes on states*

Corresponding to each probability measure μ on the centre $\mathfrak{Z}(\widehat{\mathcal{A}}) = \mathfrak{Z}(\widehat{\mathcal{A}}(\mathcal{O})) = C_b(\mathbb{R}^+)$ due to the broken scale invariance, we have a conditional expectation $\hat{\mu}$ from $\widehat{\mathcal{A}}$ to $\mathcal{A}$:

$$\hat{\mu} : \widehat{\mathcal{A}} \ni \hat{A} \longmapsto \int_{\mathbb{R}^+} d\mu(\lambda)\hat{A}(\lambda) \in \mathcal{A}. \tag{1}$$

Instead of $d\mu(\lambda)$, it is also possible to take the Haar measure $d\lambda/\lambda$ of $\mathbb{R}^+$. As it is a positive unbounded measure but not a probability one with the total mass one, however, the map corresponding to $\hat{\mu}$ becomes an *operator-valued weight* whose images are not guaranteed to be finite. Any state $\omega \in E_\mathcal{A}$ can be lifted onto $\widehat{\mathcal{A}}$ through $\hat{\mu}$ by

$$E_\mathcal{A} \ni \omega \longmapsto \hat{\mu}^*(\omega) = \omega \circ \hat{\mu} = \omega \otimes \mu \in E_{\widehat{\mathcal{A}}}, \tag{2}$$

where we have used $\widehat{\mathcal{A}} \subset C_b(\mathbb{R}^+, \mathcal{A}) \cong \mathcal{A} \otimes C_b(\mathbb{R}^+)$.

In Ref.3 the case $\mu = \delta_{\lambda=1}$(: Dirac measure at the identity of $\mathbb{R}^+$) is called a *canonical lift* $\hat{\omega} := \omega \circ \hat{\delta}_1$. The scale transformed state defined by

$$\hat{\omega}_\lambda := \hat{\omega} \circ \hat{\sigma}_\lambda = \omega \circ \hat{\delta}_\lambda \tag{3}$$

describes the situation at scale λ due to the renormalization-group transformation of scale change λ.

Conversely, starting from a state $\hat{\omega}$ of $\widehat{\mathcal{A}}$, we can obtain its central decomposition: first, we call two natural embedding maps $\iota : \mathcal{A} \hookrightarrow \widehat{\mathcal{A}}$ $[[\iota(A)](\lambda) \equiv A]$ and $\kappa : C_b(\mathbb{R}^+) \simeq \mathfrak{Z}(\widehat{\mathcal{A}}) \hookrightarrow \widehat{\mathcal{A}}$. Pulling back $\hat{\omega}$ by $\kappa^* : E_{\widehat{\mathcal{A}}} \to E_{C_b(\mathbb{R}^+)}$, we can define a probability measure $\rho_{\hat{\omega}} := \kappa^*(\hat{\omega}) = \hat{\omega} \circ \kappa = \hat{\omega} \upharpoonright_{C_b(\mathbb{R}^+)}$ on $\mathbb{R}^+$, namely, $\hat{\omega} \upharpoonright_{C_b(\mathbb{R}^+)} (f) = \int_{\mathbb{R}^+} d\rho_{\hat{\omega}}(\lambda) f(\lambda)$ for $\forall f \in C_b(\mathbb{R}^+)$.

For any positive operator $\hat{A} = \int a d\hat{E}_{\hat{A}}(a) \in \widehat{\mathcal{A}}$, we can consider the central supports $c(\hat{E}_{\hat{A}}(\Delta)) \in Proj(\mathfrak{Z}_{\hat{\pi}_{\hat{\omega}}}(\widehat{\mathcal{A}}))$ of $\hat{E}_{\hat{A}}(\Delta) \in Proj(\hat{\pi}_{\hat{\omega}}(\widehat{\mathcal{A}})'')$ with a Borel set Δ in $Sp(\hat{A}) \subset \mathbb{R}^+$ satisfying $c(\hat{E}_{\hat{A}}(\Delta))\hat{E}_{\hat{A}}(\Delta) = \hat{E}_{\hat{A}}(\Delta)$. From this we see that $\rho''_{\hat{\omega}}(c(\hat{E}_{\hat{A}}(\Delta))) = 0$ implies $\hat{\omega}''(\hat{E}_{\hat{A}}(\Delta)) = 0$, where $\hat{\omega}''$ and $\rho''_{\hat{\omega}}$ are the extensions of $\hat{\omega}$ and $\rho_{\hat{\omega}}$ to $\hat{\pi}_{\hat{\omega}}(\widehat{\mathcal{A}})''$ and $L^\infty(\mathbb{R}^+, d\rho_{\hat{\omega}})$, respectively. Thus we can define the Radon-Nikodym derivative $\omega_\lambda := \frac{d\hat{\omega}}{d\rho_{\hat{\omega}}}(\lambda)$ of $\hat{\omega}$ w.r.t. $\rho_{\hat{\omega}}$ as a state on $\hat{\pi}_{\hat{\omega}}(\widehat{\mathcal{A}})''$ so that

$$\hat{\omega}(\hat{A}) = \int d\rho_{\hat{\omega}}(\lambda)\omega_\lambda(\hat{A}(\lambda)) = \int d\rho_{\hat{\omega}}(\lambda)\omega_\lambda(\hat{\delta}_\lambda(\hat{A})) = \int d\rho_{\hat{\omega}}(\lambda) \left[\omega_\lambda \otimes \hat{\delta}_\lambda\right](\hat{A}). \tag{4}$$

Then the pull-back $\iota^*(\hat{\omega}) = \hat{\omega} \circ \iota \in E_{\mathcal{A}}$ of $\hat{\omega} \in E_{\widehat{\mathcal{A}}}$ by $\iota^* : E_{\widehat{\mathcal{A}}} \to E_{\mathcal{A}}$ is given by

$$\iota^*(\hat{\omega}) = \int d\rho_{\hat{\omega}}(\lambda)\omega_\lambda, \tag{5}$$

owing to the relation

$$\iota^*(\hat{\omega})(A) = \hat{\omega}(\iota(A)) = \int d\rho_{\hat{\omega}}(\lambda)\omega_\lambda(A) = \left[\int d\rho_{\hat{\omega}}(\lambda)\omega_\lambda\right](A). \tag{6}$$

Using this relation to the scaled canonical lift, $\hat{\omega}_\lambda := \hat{\omega} \circ \hat{\sigma}_\lambda = (\omega \circ \hat{\delta}_1) \circ \hat{\sigma}_\lambda = \omega \circ \hat{\delta}_\lambda$, of a state $\omega \in E_{\mathcal{A}}$, we can easily see $\iota^*(\omega \circ \hat{\delta}_\lambda) = \iota^*(\hat{\omega}_\lambda) = \omega_\lambda [= \frac{d\hat{\omega}_\lambda}{d\delta_\lambda}(\lambda)] = \phi_\lambda(\omega)$, where ϕ_λ is the isomorphism introduced in Ref.3 between ω and the canonical lift $\hat{\omega}_\lambda \in E_{\widehat{\mathcal{A}}}$ projected onto $\widehat{\mathcal{A}}/\ker(\hat{\pi}_{\hat{\omega}} \circ \hat{\sigma}_\lambda)$.

Thus we can lift any state $\omega \in E_{\mathcal{A}}$ canonically from $\mathcal{A}$ to $\hat{\omega} \in E_{\widehat{\mathcal{A}}}$, and, after the scale shift $\hat{\sigma}_\lambda$ on $\widehat{\mathcal{A}}$, return $\hat{\omega} \circ \hat{\sigma}_\lambda$ back onto $\mathcal{A}$: $\phi_\lambda(\omega) = \omega_\lambda = \iota^*(\omega \circ \hat{\delta}_\lambda)$, as result of which we obtain the scale-shifted state $\omega_\lambda \in E_{\mathcal{A}}$ from $\omega \in E_{\mathcal{A}}$ *in spite of the absence of scale invariance on* $\mathcal{A}$.

Now applying this procedure to $\omega = \omega_\beta$ (: any state belonging to the family of relativistic KMS states with the same $(\beta^2)^{1/2}$), we have a genuine KMS state by going to their rest frames. Then we have $\hat{\omega}_\lambda = (\widehat{\omega_\beta})_\lambda = \omega_\beta \circ \widehat{\delta_\lambda}$ which is shown to be a KMS state at β/λ:

$$
\begin{aligned}
(\omega_\beta \circ \widehat{\delta_\lambda})(\hat{A}\hat{\alpha}_t(\hat{B})) &= \omega_\beta(\hat{A}(\lambda)\alpha_{\lambda t}(\hat{B}(\lambda))) = \omega_\beta(\alpha_{\lambda t - i\beta}(\hat{B}(\lambda))\hat{A}(\lambda)) \\
&= \omega_\beta(\alpha_{\lambda(t-i\beta/\lambda)}(\hat{B}(\lambda))\hat{A}(\lambda)) \\
&= (\omega_\beta \circ \widehat{\delta_\lambda})(\hat{\alpha}_{t-i\beta/\lambda}(\hat{B})\hat{A}),
\end{aligned} \tag{7}
$$

and hence, $(\widehat{\omega_\beta})_\lambda \in \hat{K}_{\beta/\lambda}$, $\phi_\lambda(\omega_\beta) \in K_{\beta/\lambda}$.

As already remarked, the above discussion is seen to apply equally to the spontaneous as well as explicitly broken scale invariance with explicit breaking parameters such as mass terms. The actions of scale transformations on such variables as x^μ, β^μ and also conserved charges are just straightforward, which is justified by such facts that the first and the second ones are of kinematical nature and that the second and the third ones exhibit themselves in the state labels for specifying the relevant sectors in the context of the superselection structures[6,5]. This gives an alternative verification to the so-called non-renormalization theorem of conserved charges. In sharp contrast, other such variables as coupling constants (to be read off from the data of correlation functions or Green's functions) are affected by the scaled dynamics, and hence, may show non-trivial scaling behaviours with deviations from the canonical (or kinematical) dimensions, in such forms as the running couplings or anomalous dimensions. Thus, the transformations $\hat{\sigma}_\lambda$ (as "exact" symmetry on the augmented algebra $\widehat{\mathcal{A}}$) are understood to play the roles of the renormalization-group transformations (as broken symmetry on the original algebra $\mathcal{A}$). As a result, we see that *classical macroscopic observable* β naturally emerging from a microscopic quantum system is verified to be an order parameter of broken scale invariance involved in the renormalization group.

In the present context of the scale transformations in real version, we can use these scale changes of states to compare different theories renormalized by renormalization conditions imposed at different scale points.

3. Nuclearity Condition & Renormalizability

For the purpose of controlling the phase space properties in algebraic QFT, the nuclearity condition is formulated as follows: the map $\Phi_{\mathcal{O},E} : \mathcal{A}(\mathcal{O})_1 \ni A \longmapsto P_E A\Omega \in \mathfrak{H}$ with P_E the spectral projection on state vectors having energy below E is required to be *nuclear*, admitting such a decomposition as

$$\Phi_{\mathcal{O},E}(A) = \sum_{i=1}^{\infty} \varphi_i(A)\xi_i \quad \text{for } \forall A \in \mathcal{A}(\mathcal{O})_1$$

$$\text{with } \varphi_i \in \mathcal{A}(\mathcal{O})^* \text{ and } \xi_i \in \mathfrak{H} \text{ s.t. } \sum_{i=1}^{\infty} ||\varphi_i||\,||\xi_i|| < \infty,$$

on the unit ball $\mathcal{A}(\mathcal{O})_1 := \{A \in \mathcal{A}(\mathcal{O});\ ||A|| \leq 1\}$ of any local subalgebra $\mathcal{A}(\mathcal{O})$ of observables. This condition excludes such "unphysical" fields as generalized free fields without discrete mass spectrum admitting no particle picture to be detected in scattering experiments. The nuclearity condition and the assumption of the approximate scale invariance are known[8] to imply that the local subalgebra $\mathcal{A}(\mathcal{O})$ is a factor von Neumann algebra of *type III* with *no minimal projections*, i.e., any projection operator $E \in \mathcal{A}(\mathcal{O})\diagdown\{0\}$ is equivalent to the identity operator $I = id_{\mathfrak{H}}$: $\exists v \in \mathcal{A}(\mathcal{O})$ s.t. $v^*v = I, vv^* = E$.

3.1. *Point-like fields as idealized local observables*

On the basis of the nuclearity condition[7] and the energy-bound, the notion of point-like field operators[9,10] has been established satisfying the operator-product expansion (OPE) in a non-perturbative way in algebraic QFT by Bostelmann[12]. The energy bound means the requirement that observed values of quantum fields $\hat{\phi}(f)$ can become large only with large energy: for any $l > 0$, there is a sufficiently large $k > 0$ that the inequality

$$||(1+H)^{-k}\,\hat{\phi}(f)\,(1+H)^{-k}|| \leq c \int dx\,|(1-\Delta)^{-l} f(x)|,$$

holds with a (positive) Hamiltonian H, operator norm $||\cdot||$ in the vacuum sector $\mathcal{H}$ and Δ: Laplacian on $\mathbb{R}^4$. When this holds, there exist a sequence of test functions $\{f_i\}$ tending to Dirac measure at x: $f_i \underset{i\to\infty}{\longrightarrow} \delta_x$, and a sufficiently large integer $k > 0$ such that

$$\lim_{i\to\infty} (1+H)^{-k}\,\hat{\phi}(f_i)\,(1+H)^{-k} =: (1+H)^{-k}\,\hat{\phi}(x)\,(1+H)^{-k}.$$

Then a *field* $\hat{\phi}(x)$ *at a point* x is well-defined as a linear form on such states ω in the vacuum sector as $\omega((1+H)^{2k}) < \infty$. Hermitian elements in the sets $\mathcal{Q}_{k,x} := \{\hat{\phi}(x);\ ||(1+H)^{-k}\hat{\phi}(x)(1+H)^{-k}|| < \infty\}$ of point-like fields are idealized observables at spacetime points x meaningful for such states ω as $\omega((1+H)^{2k}) < \infty$. The set $\mathcal{Q}_{k,x}$ of such point-like fields are, in general, finite-dimensional linear spaces satisfying $\mathcal{Q}_{k,x} \subset \mathcal{Q}_{k',x}$ for $k \leq k'$ and are invariant under the stability group of x in $\mathcal{P}_+^\uparrow$. The meaningless notion of *product of fields at a point* x is replaced in $\mathcal{Q}_{k,x}$ by *normal products* defined by the following OPE: for instance, ill-defined square $\hat{\phi}(x)^2$ is replaced by the subspaces $\mathcal{N}(\hat{\phi}^2)_{q,x} \subset \mathcal{Q}_{n,x}$ generated by normal products $\hat{\Phi}_j(x)$, $j = 1, \cdots, J(q)$, appearing in OPE of $\hat{\phi}(x + \frac{\xi}{2})\hat{\phi}(x - \frac{\xi}{2})$:

$$||(1+H)^{-n}\left[\hat{\phi}(x+\frac{\xi}{2})\hat{\phi}(x-\frac{\xi}{2}) - \sum_{j=1}^{J(q)} c_j(\xi)\,\hat{\Phi}_j(x)\right](1+H)^{-n}|| \leq c\,|\xi|^q,$$

which is satisfied for any $\hat{\phi} \in \mathcal{Q}_{k,x}$ for spacelike $\xi(\in \mathbb{R}^4) \to 0$ with arbitrary $q > 0$, by choosing a finite number of fields $\hat{\Phi}_j(x) \in \mathcal{Q}_{n,x}$ and sufficiently large n, and some analytic functions $\xi \mapsto c_j(\xi)$, $j = 1, \cdots, J(q)$. Using this definition, the spaces $\mathcal{N}(\hat{\phi}^p)_{q,x}(\subset \mathcal{Q}_{n,x})$ of normal products of higher powers p can similarly be defined. While the linear spaces $\mathcal{Q}_{k,x}$ of point-like fields lack the multiplication structure, the validity of OPE allows us to provide them with a structure generalizing a *product system of Hilbert modules* $\mathcal{Q}_{n,x}$. It is also possible for the partial derivatives ∂_ξ of spacetime coordinates ξ to act on these spaces through the "*balanced derivatives*" $\partial_\xi\,\hat{\phi}(x + \frac{\xi}{2})\hat{\phi}(x - \frac{\xi}{2})$ (Ref.14) which are contained in $\mathcal{N}(\hat{\phi}^2)_{q,x}$ (for large q) as shown by the relation,

$$||(1+H)^{-n}[\partial_\xi\,\hat{\phi}(x+\frac{\xi}{2})\hat{\phi}(x-\frac{\xi}{2}) - \sum_{j=1}^{J(q)} \partial_\xi\, c_j(\xi)\;\hat{\Phi}_j(x)](1+H)^{-n}|| \leq c\,|\xi|^r,$$

valid for $\forall r > 0\ \exists q$ and $\exists n$ sufficiently large.

3.2. *Comparison between OPE & Wigner-Eckhart theorem*

To take advantage of the above OPE structure, we compare it with the basic feature of the Wigner-Eckhart theorem for an irreducible family of tensor operators $\{F^{(\gamma_1)}_{m_1}; m_1 = -\gamma_1, -\gamma_1+1, \cdots, \gamma_1 - 1, \gamma_1\}$ under the action of a (compact) group G (such as $SU(2)$, typically):

$$\langle \gamma m | F^{(\gamma_1)}_{m_1} | \gamma_2 m_2 \rangle = \langle \gamma || F^{(\gamma_1)} || \gamma_2 \rangle \langle \gamma m | (\gamma_1 m_1), (\gamma_2 m_2) \rangle,$$

where $\langle \gamma m | (\gamma_1 m_1), (\gamma_2 m_2) \rangle$ are the Clebsch-Gordan coefficients describing a branching rule from the Kronecker tensor product $[\gamma_1 \hat{\otimes} \gamma_2](g) = \gamma_1(g) \otimes \gamma_2(g)$ of representations (γ_i, V_{γ_i}) $(i = 1, 2)$ into irreducible ones $\{(\gamma, V_\gamma)\} \in Rep(G)$ $(|\gamma, m\rangle \in V_\gamma)$ of G. Note that the Kronecker tensor product $\gamma_1 \hat{\otimes} \gamma_2$ of G is the restriction of the tensor product representation $(\gamma \boxtimes \gamma_2)(g_1, g_2) =$

$\gamma(g_1) \otimes \gamma_2(g_2)$ of $G \times G$ onto a subgroup G embedded via the diagonal map $\delta_G : G \ni g \longmapsto \delta_G(g) = (g, g) \in G \times G$:

$$[\gamma_1 \hat{\otimes} \gamma_2](g) = [(\gamma_1 \boxtimes \gamma_2) \circ \delta_G](g) = \gamma_1(g) \otimes \gamma_2(g).$$

According to this formula, the matrix elements of the tensor operator $\{F^{(\gamma_1)}_{m_1}; m_1 = -\gamma_1, -\gamma_1 + 1, \cdots, \gamma_1 - 1, \gamma_1\}$ are decomposed into two factors, *G-invariant dynamical* one $\langle\gamma||F^{(\gamma_1)}||\gamma_2\rangle$ and purely *kinematical* one $\langle\gamma m|(\gamma_1 m_1), (\gamma_2 m_2)\rangle$ determined completely by the G-transformation property of $F^{(\gamma_1)}$.

In the case of OPE,

$$\varphi_1(x + \frac{\xi}{2})\varphi_2(x - \frac{\xi}{2}) \underset{\xi \to 0}{\sim} \sum_i N(\varphi_1\varphi_2)_i(x) C_i(\xi) + \cdots ,$$

the dual map δ^* of δ given by $\delta^*(\varphi_1 \boxtimes \varphi_2)(x) = (\varphi_1 \boxtimes \varphi_2)(\delta(x)) = \varphi_1(x) \otimes \varphi_2(x)$ is *ill-defined* for operator-valued distributions φ_i. In this context, therefore, the diagonal map $\delta(x) = (x, x)$ should be understood in the limit: $(x + \frac{\xi}{2}, x - \frac{\xi}{2}) \underset{\xi \to 0}{\sim} \delta(x) = (x, x)$ after "removing" such divergent terms as $C_i(\xi)$. Except for this difference, the essence of OPE formula is just in parallel with the above Wigner-Eckhart case: *factorization* of the product $\varphi_1(x + \frac{\xi}{2})\varphi_2(x - \frac{\xi}{2})$ into two components, *dynamical non-singular* factors $N(\varphi_1\varphi_2)_i(x)$ depending only on the "centre of mass" $[(x + \frac{\xi}{2}) + (x - \frac{\xi}{2})]/2 = x$ and *c-number kinematical singular functions* $C_i(\xi)$ of the relative coordinates $(x + \frac{\xi}{2}) - (x - \frac{\xi}{2}) = \xi$. Note that the *singularity of the product* $\varphi_1(x + \frac{\xi}{2})\varphi_2(x - \frac{\xi}{2})$ *in the limit of* $\xi \to 0$ is isolated into these kinematical c-number factors $C_i(\xi) = N_i(\lambda) C_i^{reg}(\xi)$, where $\lambda := |\xi|^{-1}$ represents the *cutoff momentum* to regularize the *UV divergences* in a *non-perturbative* way and $N_i(\lambda)$ can be taken as *counter terms* to define *renormalized field operators* (formally) by

$$\varphi_{ren}(x) := \Pi_i N_i(\lambda)^{-1/2} \varphi(x).$$

It may be instructive to find the analogy of the present structure with the time-localization scale Δt of Hida derivatives a_t, a_t^* in the White-Noise Analysis[15].

Since the limit $\underset{\xi \to 0}{\sim}$ means

$$||(1 + H)^{-n} \left[\hat{\phi}(x + \frac{\xi}{2})\hat{\phi}(x - \frac{\xi}{2}) - \sum_{j=1}^{J(q)} c_j(\xi)\, \hat{\Phi}_j(x) \right] (1 + H)^{-n}|| \leq c\, |\xi|^q,$$

the convergence $\hat{\phi}(x+\frac{\xi}{2})\hat{\phi}(x-\frac{\xi}{2}) \to \sum_{j=1}^{J(q)} c_j(\xi)\,\hat{\Phi}_j(x)$ is *state-dependent* so that

$$\omega(\left[\hat{\phi}(x+\frac{\xi}{2})\hat{\phi}(x-\frac{\xi}{2}) - \sum_{j=1}^{J(q)} c_j(\xi)\,\hat{\Phi}_j(x)\right]) \underset{\xi\to 0}{\to} 0$$

holds only for those states ω which satisfy

$$\omega((1+H)^{2n}) < constant.$$

Thus, states ω for which OPE is valid *cannot be localized*, and hence, to such an extent, the spacetime point x in $\varphi(x)$ is actually extended!

In the above situations the common essence can be found in the relevance of some selective filters depending on the choice of states ω. We note here that the (approximate) *diagonal maps* $\delta(x) = (x, x)$,

$$(x+\frac{\xi}{2}, x-\frac{\xi}{2}) \underset{\xi\to 0}{\sim} (x, x) = \delta(x);$$
$$[(\gamma_1 \boxtimes \gamma_2)\circ\delta](g) = [\gamma_1 \hat{\otimes} \gamma_2](g) = \gamma_1(g)\otimes\gamma_2(g),$$

play essential roles in the definition of *Hopf algebra* structures with the harmonic-analytic dualities controlled by the Kac-Takesaki operator (of the so-called duality transformations), which should play crucial roles in extending the above relations for two-point functions to arbitrary n-point functions.

By the above condition $\omega((1+H)^{2n}) <$ *constant*, the selective filter on the initial state ω is related with the nuclearity condition $\Phi_{\mathcal{O},E}(A) = P_E A\Omega = \sum_{i=1}^{\infty}\varphi_i(A)\xi_i$ whose energy scale E can be related to the above *cutoff* λ. In spite of the sharp contrast between *almost finite-dimensitonality* as nuclearity and ∞-*dimensionality* inherent to *type III*, both properties are closely related with the *nuclearity condition* and are crucial for *renormalizability* and for shifts of the *renormalization points* by *renormalization-group* transformations:

1) *renormalizability* = finiteness of the number of graph types of divergent "1-particle irreducible (1PI)" diagrams is expected to follow from the very nuclearity condition (= *intra-sectorial* structure);

2) the *absence of minimal projection in type III* von Neumann factors (due to approximate scale invariance) allows the *shifts of renormalization points* by scale transformations = *renormalization-group transformations*. This gives the *inter-sectorial* relations among *"sectors parametrized by renormalization conditions" at different renormalization points* (on the centre $\mathfrak{Z}(\widehat{\mathcal{A}}) = \mathfrak{Z}(\widehat{\mathcal{A}}(\mathcal{O})) = C_b(\mathbb{R}^+)$ of the scaling algebra).

In this sense, the *nuclearity condition* can be regarded as the *mathematical version of the renormalizability condition* and *broken scale invariance* inherent to local subalgebras $\mathcal{A}(\mathcal{O})$ of *type III* with no minimal projection requires the *renormalization condition to be specified at some renormalization point* which can, however, be *chosen arbitrarily.*

Here we present some new perspectives for understanding the conceptual and mathematical meaning of *renormalization scheme* in relation with such key notions as the nuclearity condition, broken scale invariance and the type III nature of local subalgebras of quantum fields in close relation with algebraic QFT. What remains to be clarified is the following:

(1) *Counter terms* $N_i(\lambda)$ are expected to be *factors of automorphy* associated to the fractional linear transformations of (approximate) *conformal* symmetry $SO(2,4)(\simeq SU(2,2))$ associated with (approximate) scale invariance. Along this line, the Callan-Symanzik type equation for $N_i(\lambda)$ involving running coupling constants and anomalous dimensions should be established.

(2) *In the opposite direction* to the conventional renormalization scheme based on *perturbative expansion method* starting from a *"Lagrangian"* (along such a flow chart as "Lagrangian" $\rightarrow$ perturbative expansion $\rightarrow$ renormalization + OPE), the perturbation expansion itself should be derived and justified as a kind of asymptotic analysis within the non-perturbative formulation of renormalization based on OPE: namely, we advocate such a flow chart as starting from OPE $\rightarrow$ renormalization $\rightarrow$ perturbative method as asymptotic expansion $\rightarrow$ "Lagrangian" determined by Γ_{1PI} & renormalizability (= finite generation property).

(3) More detailed mathematical connections should be exhibited among *nuclearity condition, renormalizability, renormalization conditions, renormalization group to shift renormalization point* and *broken scale invariance inherent to local subalgebras* $\mathcal{A}(\mathcal{O})$ *of type III* from the viewpoint of *non-standard analysis.*

Acknowledgement

I would like to express my sincere thanks to Prof. M. Ohya for the invitation to the pioneering and inspiring International Conference, QBIC2008, and to Prof. L. Accardi and to Prof. T. Hida for their encouragements.

References

1. Ojima, I., Micro-macro duality in quantum physics, pp.143–161 in Proc. Intern. Conf. on Stochastic Analysis, Classical and Quantum, World Scientific, 2005; Ojima, I. and Takeori, M, How to observe quantum fields and recover them from observational data? – Takesaki duality as a Micro-Macro duality –, Open Sys. & Inf. Dyn. **14**, 307 – 318 (2007) (math-ph/0604054 (2006)).
2. Ojima, I., Temperature as order parameter of broken scale invariance, Publ. RIMS **40**, 731-756 (2004) (math-ph0311025).
3. Buchholz, D. and Verch, R., Scaling algebras and renormalization group in algebraic quantum field theory, Rev. Math. Phys. **7** (1995), 1195-1240.
4. Takesaki, M., Disjointness of the KMS states of different temperatures, Comm. Math. Phys. **17** (1970), 33-41.
5. Ojima, I., A unified scheme for generalized sectors based on selection criteria –Order parameters of symmetries and of thermality and physical meanings of adjunctions–, Open Systems and Information Dynamics, **10** (2003), 235-279 (math-ph/0303009).
6. Ojima, I., How to formulate non-equilibrium local states in QFT?–General characterization and extension to curved spacetime–, pp.365-384 in "*A Garden of Quanta*", World Scientific (2003) (cond-mat/0302283).
7. Buchholz, D., Wichmann, E. H.: Causal independence and the energy-level density of states in local quantum field theory. Commun. Math. Phys. **106**, 321–344 (1986); Buchholz, D., Porrmann, M.: How small is the phase space in quantum field theory? Ann. Inst. Henri Poincaré - Physique théorique **52**, 237–257 (1990).
8. Fredenhagen, K., On the modular structure of local algebras of observables, Commun. Math. Phys. **97** (1985), 79–89; Buchholz, D., D'Antoni, C. and Fredenhagen, K., The universal structure of local algebras, Comm. Math. Phys. **111** (1987), 123-135.
9. Fredenhagen, K. and Hertel, J., Comm. Math. Phys. **80**, 555(1981).
10. Haag, R. and Ojima, I., On the problem of defining a specific theory within the frame of Local Quantum Physics, Ann. Inst. H. Poincaré **64**, (1996) 385 – 393.
11. Wilson, K. G. and Zimmermann, W., Operator product expansions and composite field operators in the general framework of quantum field theory, Commun. Math. Phys., **24**, 87–106, 1972.
12. Bostelmann, H., Lokale Algebren und Operatorprodukte am Punkt, PhD Thesis, Universität Göttingen, 2000; Operator product expansions as a consequence of phase space properties. J. Math. Phys., **46**, 082304, 2005; Phase space properties and the short distance structure in quantum field theory. J. Math. Phys., **46**, 052301, 2005.
13. Bhat, B.V.R. and Skeide, M., Tensor product systems of Hilbert modules and dilations of completely positive semigroups, Infin. Dimens. Anal. Quantum Probab. Relat. Top. **3** (2000), 519–575; Skeide, M., E_0–semigroups for continuous product systems, Infin. Dimens. Anal. Quantum Probab. Relat. Top. **10** (2007), 381–395,

14. Buchholz, D., Ojima, I. and Roos, H., Thermodynamic properties of non-equilibrium states in quantum field theory, Ann. Phys. (N.Y.) **297** (2002), 219 - 242.
15. Hida, T., Kuo, H.-H., Potthoff, J. and Streit, L., *White Noise*, Kluwer (1993).

A SIMPLE SYMMETRIC ALGORITHM USING A LIKENESS WITH INTRONS BEHAVIOR IN RNA SEQUENCES

MASSIMO REGOLI

Centro Vito Volterra,
Università di Roma "Tor Vergata",
Roma I-00133, Italy
E-mail: regoli@uniroma2.it

The RNA-Crypto System (shortly RCS) is a symmetric key algorithm to cipher data. The idea for this new algorithm starts from the observation of nature. In particular from the observation of RNA behavior and some of its properties. The RNA sequences has some sections called Introns. Introns, derived from the term "intragenic regions", are non-coding sections of precursor mRNA (pre-mRNA) or other RNAs, that are removed (spliced out of the RNA) before the mature RNA is formed. Once the introns have been spliced out of a pre-mRNA, the resulting mRNA sequence is ready to be translated into a protein. The corresponding parts of a gene are known as introns as well. The nature and the role of Introns in the pre-mRNA is not clear and it is under ponderous researches by Biologists but, in our case, we will use the presence of Introns in the RNA-Crypto System output as a strong method to add chaotic non coding information and an unnecessary behaviour in the access to the secret key to code the messages. In the RNA-Crypto System algoritnm the introns are sections of the ciphered message with non-coding information as well as in the precursor mRNA..

1. Introduction

In this text a new cryptographic algorithm is described.

As described before the algorithm is symmetric and uses a pre shared secret key. The length of the key (as shown in the following) has only the role to prevent the massive attacks techniques: it does not affect the speed of the encryption or decryption phases. So it is possible to use a very long key without loosing the speed peculiarity of the method.

Before introducing the core algorithm it is necessary to explain some symbols used in the text:

- m is the clear message
- o is the ciphered message
- k is the secret shared key

- KP is the secret key actual pointer (pointer to the next bit to be used in the secret key)
- MP is the message text actual pointer (pointer to the next bit to be processed in the message)

The algorithm works using two different actions (the boxes). These boxes act in different way on the message and on the secret key and produce different behavior.
The boxes are labeled as follows:

- Inserting Introns (box B_1)
- Coding (box B_2)

During all the ciphering phase these two *boxes* are selected at random for each step of the algorithm (for example using a random number from a linear congruential generator). Then, for each step the selected box acts on message, secret key and output code using different approaches:

- when the Coding Box is extracted, another random choice will be done to select the right coding to apply
- when the Inserting box is selected, a random number will fix the *type of action* of the introns to be inserted in the output code

At the end of the algorithm the original message is encoded in a new message (the output) that is in the average 8 times longer than the original one (in section 5 this fact will be proved). In a general implementation of the RNA-Crypto System algorithm this number could be different, for example changing the distribution function of the random process devoted to the boxes selections.

As shown below the algorithm is very fast and does not involve the use of complex mathematical operations (like exponentiation or numbers factorization) and its complexity does not depend on the length of the secret keys.

2. The algorithm

As described in section 1, for each step of the algorithm, the method uses one of two different operations (chosen at random and called *boxes*).

In figures 1, 2 is possible to see how these boxes work:

It is useful now to introduce the table 1 (the seek table T). This table is crucial and it is used by both boxes to code and decode the output messages. In the original version the table is $T = \{t_{i,j} = i + j \bmod 4\}$, but,

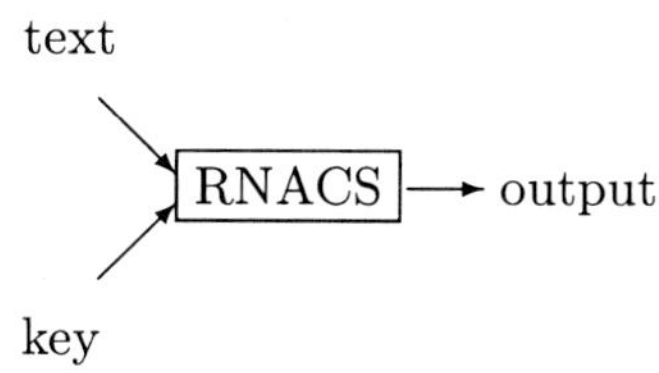

Figure 1. The structure of the RNA-Crypto System Algorithm

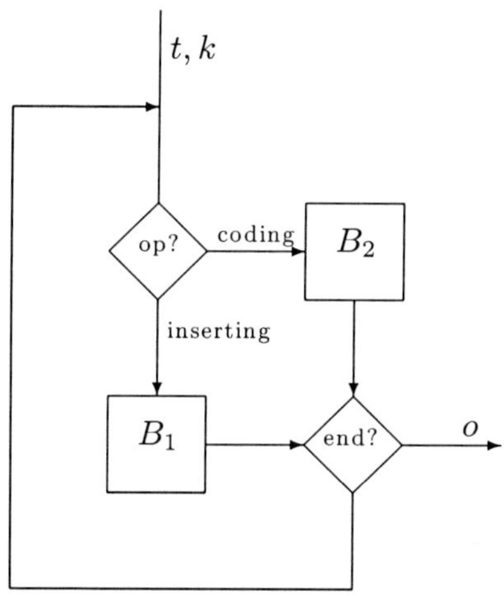

Figure 2. The RNACS in a second level of details

as a further security performance, a secret shared random rows/columns permutations of the table 1 can be used (for example some bits of the shared secret can be used to permute the entries of the original table).

The message is processed (ciphered) in the B_2 box and the algorithm stops only when all the bits in the message have passed through this box.

In the B_1 box there is only insertion of bits in the output code and some manipulation of the secret key: no bits of the message are processed.

This is a crucial point, in fact the stop is not guaranteed by the algorithm due to its statistical behavior and to the fact that box B_1 does not act on message. To guarantee the stop of the algorithm an adjustment to the

	0	1	2	3	4	5	6	7
0	0	1	2	3	0	1	2	3
1	1	2	3	0	1	2	3	0
2	2	3	0	1	2	3	0	1
3	3	0	1	2	3	0	1	2
4	0	1	2	3	0	1	2	3
5	1	2	3	0	1	2	3	0
6	2	3	0	1	2	3	0	1
7	3	0	1	2	3	0	1	2

protocol must be inserted (for example after a fixed $n > 0$ consecutive extractions of the box B_1 a B_2 box must be extracted with probability 1).

3. The boxes

3.1. *The box B_1*

The box B_1 is the place where the *introns* are inserted.

In the RCS algorithm *introns* perform a double role:

- the first is to insert some random, apparently meaningless bits in the output code,
- the second is to move forward or backward the pointer to the next bit of the key to be used (the key–pointer KP).

The latter action introduces some chaotic behavior to the relation between message, secret key and ciphered text.

In each transit in the box B_1, 5 bits $(b_1, b_2, \ldots, b_5)$ will be inserted in the output message o.

Those bits contain information for moving the secret key pointer forward or backwards using the following rule:

- let $0 \le r = b_1 b_2 b_3 \le 7$ the row entry, in binary format, of the table T with certain conditions (see below and table 1)
- let $0 \le d = b_4 b_5 \le 3$ a shift value
- denote $c = k_{KP} k_{KP+1} k_{KP+2}$, where KP is the pointer to the first available bit in the secret key, the column entry of the same table
- and then let $v = \{0|1\}$ a direction selected at random. This selection is crucial and for each passage in the B_1 box a new random choice must be done.

OPERATOR (3 bits)			DATA (2 bits)	
b_1	b_2	b_3	b_4	b_5

Figure 3. The intron packet format

Now if $v = 1$, the r must be set up so that the condition

$$r = \{r \in \{0..7\} | T_{r,c} = 2\} \tag{1}$$

holds. If $v = 0$, then r must be selected keeping the condition

$$r = \{r \in \{0..7\} | T_{r,c} = 3\} \tag{2}$$

Observing table 1 it is possible to see that there are always 2 different rows in which condition 1 or condition 2 holds.

In the first case the KP is moved forward of d bits ($KP \leftarrow KP + d$) otherwise it is moved backward of d bits ($KP \leftarrow KP - d$).

As an example, if the canonical table T (in the version without permutations) is used, it is possible to use the following formula 3 to compute c:

$$c \equiv (6 - k + v) - 4 * \xi \bmod 8 \tag{3}$$

where:

- k is the 3 bit binary value of the key starting from KP (the column index)
- v is 0 or 1 depending on direction
- $\xi = \{0, 1\}$ is a random bit (to select one of the two possibilities in the column k)

otherwise if you use a random row / column permutation of the table T you must seek the right position in column c by hand.

Then, in this version of the algorithm, the role of the *introns* is to modify the access to the shared secret key, but, at present, there are some other interpretations (and behaviors) of the introns for example to add, to the output message, some random bits or to change (permute) the table T.

3.2. *Box* B_2

This box has the role to cipher the message.

If a B_2 box is selected then next bit in the message is processed.

In figure 4 the format of the output packet of box B_2 is displayed

OPERATOR (3 bits)		
b_1	b_2	b_3

Figure 4. The exon packet format

Now let $0 \leq r = b_1 b_2 b_3 \leq 7$ be a specific row entry, in binary format, of the table T with certain conditions (see below and table 1 for details), let $0 \leq c = k_{KP} k_{KP+1} k_{KP+2} \leq 7$ (where KP is the pointer to the first available bit in the secret key) be the column entry of the same table and let $t = m_{MP} = \{0, 1\}$ be the next bit in the message m.

The value of r must be selected so that the condition

$$r = \{r \in \{0, \ldots, 7\} | T_{r,c} = t\} \tag{4}$$

holds.

Also in this case, observing table 1 it is possible to see that there are always 2 different rows in which condition 4 holds.

As an example if the canonical table T (without permutations) is used it is possible to use the following formula 5 to compute c:

$$c \equiv (4 - k + t) - 4 * \xi \bmod 8 \tag{5}$$

where:

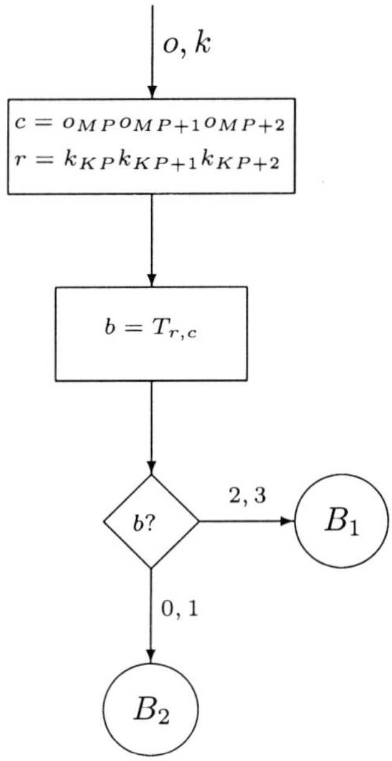

Figure 5. Decoding a message

- k is the 3 bits binary value of the key starting from KP
- t is 0 or 1 depending on message
- $\xi = \{0, 1\}$ is a random bit (to select the first or the second opportunity in the column k)

4. Decoding message

The decoding phase is very fast and easy to explain (see figure 5).

Let $c = o_{MP}o_{MP+1}o_{MP+2}$ and $r = k_{KP}k_{KP+1}k_{KP+2}$ where MP is the pointer to the next available bit in the message and KP is the pointer to the next available bits in the SSK and let $b = T_{r,c}$ where T is the table 1.

If $b \in \{0, 1\}$ the box B_2 will be selected, otherwise if $b \in \{2, 3\}$ the box B_1 will be selected.

The B_2 box add the bit b to the output message and, the KP and MP

counters are update to $KP+3$, $MP+3$ respectively.

The B_1 box extracts the binary value $d = o_{MP+3}o_{MP+4}$ and if $b = 2$ the key pointer KP is moved forward of d bits, otherwise the KP is moved backward of d bits, the counter MP is updated to $MP+5$.

5. Statistical informations about the size of output

Suppose, to fix the idea, to have a message m of l-bit length.

Using a standard RNG, the probability to choose the B_1 box is $\frac{1}{2}$ and then, from the fact that only box B_2 processes the message m, for a l–length message are necessaries $\approx 2l$ iterations ($\approx l$ iterations pass across box B_1 and $\approx l$ iterations pass across box B_2).

Now, for each selection of the B_1 box, 5 bits will be added to the output message, whereas for each selection of the B_2 box, 3 bits will be added to the output message. Then the total length of the output message o is $\approx 5l + 3l = 8l$.

Moreover the following assertion holds:

Assertion 1: Due to the unpredictable contribution of the random generator the output result will be different, in different runs of the protocol, even with the same input message and the same secret key.

As an example: starting from the binary sequence:

$$0100110101000001010110 00$$

and using the small random key:

$$10101010$$

some of the possible outputs are listed below:

```
01110011000110111100111000000010001010011010111
01101110110011001001010011010011010101001011 01
0101110111000011100001111110111101000000101111
1010001110110001101110 10
```

```
00111010000110011011010010111100011011101110 01
00111000111000100001100110010101100001111100 10
01010111100111101110111000100010010100101110 00
011011110111101001110101011111001000111
```

```
11101101000101010101100100000001010011111110001
10100100110000110110001101011111010111010011 00
1110001001100101110000001000010011000101011111 1
0111
```

```
10110111111101010001011001111110001010100010 11
11101111110100111101110011011111001111011010 10
11100100101110001001111000101101011111010110 10
01101001001001110110011101000011000011001001 11
110
```

6. Next step

Actually some peculiarities of the method are under investigations.

At the moment, we are investigating on the complexity of massive attacks to ciphered text, but due to the random behavior of the system it is not easy to calculate this complexity. Of course with a shared secret key of l bits length the number of possible attempts are, as usual, $\approx 2^l$, but with a sufficiently large l this way is impracticable.

But our actual question is: knowing the original message and its ciphered version what about the SSK?

This question is not trivial because of assertion 1. For example, using a shared secret key k, of a dimension l_k bits, to cipher a message m, of a dimension l_m bits, there are roughly speaking l_m passages through the B_1 box and l_m passages through the B_2 box.

In the first case (box B_1) there will be 3 random choices:

- direction (0 or 1)

- shift value (0, 1, 2, 3)
- position of the right value in table T (2 opportunities)

In total there are:

$$2 \times 4 \times 2 = 16 \tag{6}$$

different output codes.

For the B_2 box there will be only 1 random choice about the position in the table T.

So the number of different output messages n_o will be:

$$n_o \approx 16^{l_m} \times 2^{l_m} \approx 32^{l_m} \tag{7}$$

It is important as a next step of this research to:

- analyze the output results of the same input message to understand possible attack to the secret key
- analyze some possible correlations between clear text and all possible output codes
- generalize the algorithm inserting, for example, new operators distinct from the jump operator like:
  - "Inserting random code in output code" box
  - "Inserting the no operation" box (only secret key pointer is modified)
  - and so on

7. Final considerations

Due to the relevant expansion of the ciphered text (approximately 8 times the original one in this standard implementation), this method is applicable in those situations in which the size of the ciphered message does not matter, for example documents of small sizes, email, credit card numbers, passwords and other short information.

However the unpredictability of the output code is a strong tool against the *known text attack* methods in all the cases in which the clear text is approximately known (bank forms, files with a great fixed header like Microsoft Word and so on).

References

1. Lewin, B.,"Gene VI" *Oxford University Press*, 1997.
2. Menezes, A., van Oorschot, P. and Vanstone, S., "Handbook of Applied Cryptography", CRC Press, 1996

SOME ASPECTS OF QUADRATIC GENERALIZED WHITE NOISE FUNCTIONALS

SI SI

Aichi Prefectural Univaersity,
Nagakute, Aichi-ken, 480-1198, Japan

TAKEYUKI HIDA

Professor Emeritus,
Nagoya University and Meijo University Nagoya, Japan

2000 Mathematics Subject Classification 60H40

We shall discuss some particular roles of quadratic generalized white noise functionals. First observation is made from the viewpoint of the so-called "la passage du fini à l'infini". We then come to a dual pairing of spaces formed by quadratic generalized white noise functionals. In this line, we can further discuss quadratic forms of differential operators acting on the space of white noise functionals.

1. Introduction

We review the Fock space:

$$(L^2) = \oplus_0^\infty H_n,$$

where (L^2) is the complex Hilbert space of square integrable functionals of white noise, i.e. $L^2(E^*, \mu)$. The measure μ is the probability distribution of white noise $\dot{B}(t), t \in R$.

The subspace H_n is the collection of homogeneous chaos in the sense of N. Wiener or that of multiple Wiener integrals in the sense of K. Itô, both of degree n.

It is well-known that the space H_n is isomorphic to $\widehat{L}^2(R^n)$, the symmetric space of $L^2(R^n)$ up to the constant $\sqrt{n!}$. There can be a restriction of this isomorphism by introducing a stronger topology in such a way that

$$\widehat{K}^{(n+1)/2}(R^n) \cong H_n^{(n)},$$

where we use the notation $\widehat{K}^m(R^n)$ to denote the symmetric Sobolev space over R^n of degree m. Then we take the dual space of both side of this isomorphism based on symmetric $\widehat{L}^2(R^n)$ and H_n, respectively. We can define $H_n^{(-n)}$ the space of generalized white noise functionals of degree n by the following isomorphism:

$$\widehat{K}^{-(n+1)/2}(R^n) \cong H_n^{(-n)}.$$

Finally, with a suitable choice of a positive increasing sequence c_n, we have the test functional space

$$(L^2)^+ = \oplus c_n H_n^{(n)}$$

and its dual space

$$(L^2)^- = \oplus c_n^{-1} H_n^{(-n)},$$

which is called the space of generalized white noise functionals.

In this note we shall discuss various kind of dualities that exist among subspaces of $(L^2)^-$.

2. Quadratic functionals of white noise

Detailed interpretation of linear functionals of white noise $\dot{B}(t)$ have been given in [5], in particular an identity of $\dot{B}(t)$ has been established. The $\dot{B}(t)$'s are taken to be the variables of white noise functionals.

We are then naturally led to nonlinear functions (actually functionals) of the $\dot{B}(t)$. We claim that among others the subspace $H_2^{(-2)}$ involving quadratic generalized white noise functionals is particularly important. As is mentioned in the last section, we have the isomorphism

$$H_2^{(-2)} \cong \widehat{K}^{-3/2}(R^2).$$

More explicitly, for $\varphi \in H_2^{(-2)}$ we find a function $F(u,v)$ in the space $\widehat{K}^{-3/2}(R^2)$ to have

$$\varphi(\dot{B}) = \int F(u,v) : \dot{B}(u)\dot{B}(v) : dudv.$$

We shall classify those quadratic functionals according to the analytic properties of the kernel. Our method comes from the Lévy's technique called "la passage du fini á l'infini" ([10]).

We therefore start with a quadratic form in the elementary theory of linear algebra, which is expressed in the form:

$$Q(x) = \sum a_{j,k} x_j x_k \ .$$

It is, for convenience, written as a sum of $Q_1(x)$ and $Q_2(x)$:

$$Q(x) = Q_1(x) + Q_2(x),$$

where

$$Q_1(x) = \sum a_j x_j^2, \text{ and } Q_2(x) = \sum a_{j,k} x_j x_k \ .$$

Keeping the idea of passage to infinity, we can consider how to discriminate these two terms. Note that the x_j's are equally weighted variables regardless they are coordinates of finite or infinite dimensional vector. We shall make some quite elementary observations.

i) Suppose x_i's are mutually independent and subject to the standard Gaussian distribution $N(0,1)$. If both are infinite sum, then for $Q_1(x)$ to be convergent the coefficients a_j's should be of trace class, but for $Q_2(x)$ the coefficients $a_{j,k}$ should be square summable. In short, the way of convergence is strictly different.

ii) As for the analytic property, any partial sum of $Q_2(x)$ is harmonic, while each term of $Q_1(x)$ is not.

iii) Start with a Brownian motion $B(t), t \in [0,1]$. Consider an approximation to white noise $\dot{B}(t), t \in [0,1]$ by taking $\frac{\Delta_j B(t)}{\Delta_j}$ for x_j. Let $|\Delta_j|$ tend to 0. Then, each term of Q_1 needs renormalization, but the trick is unnecessary for Q_2.

With the notes mentioned above we now come to the expression of generalized quadratic functionals of white noise, namely representation of quadratic functionals $\varphi(\dot{B}) \in H_2^{(-2)}$. It is expressed in the form

$$\varphi(\dot{B}) = \int\int F(u,v) : \dot{B}(u)\dot{B}(v) : dudv,$$

where $F \in \widehat{K}^{-3/2}(R^2)$.

Applying the S-transform we have the U-functional of the form

$$U(\xi) = \int\int F(u,v)\xi(u)\xi(v)dudv,$$

which is a quadratic form of ξ.

We now remind the entire functionals of the second order due to P. Lévy (see [10] Part I, Chapter 3.). He focuses his attention to the normal form, which is expressible as

$$U(\xi) = \int\int f(u,v)\xi(u)\xi(v)dudv + \int g(t)\xi(t)^2 dt.$$

We tacitly assume suitable conditions posed on f and g.

If we understand in our notation, the generalized function F, which is in the Sobolev space, is chosen with a restriction that a singularity is involved only on the diagonal. Namely, we may understand that $g(u)$ can be considered as $g(\frac{u+v}{2})\delta(u-v)$. Note that singularity appears only on the diagonal.

We are now in a position to remind the observations noted in i), ii) and iii) made just above. If we are permitted to say rather formally, the quadratic form $Q(x)$, which is divided into $Q_1(x)$ and $Q_2(x)$, goes to the Lévy's formula for normal functionals as the dimension of the vector x tends to infinity.

We understand that $Q_1(\xi) = \int g(t)\xi(t)^2 dt$ is in the domain of the Lévy Laplacian and the same for $Q_2(\xi) = \int\int f(u,v)\xi(u)\xi(v)dudv$, in addition, it is harmonic.

A question arises. Why is a $H_2^{(-2)}$-functional having off-diagonal singularities of the kernel $F(u,v)$ not so important ? The answer is just simple; it is not in the domain of the Laplacian.

Remark. It is natural to ask what is the role of quadratic functional that has singularity is off diagonal. For example

$$\int g(u)\dot{B}(u)\dot{B}(u+1)du.$$

It is easy to see that the second order functional derivative does not exist, so that it is not in the domain of the Laplacian.

With some property of the Sobolev space, we take order -3/2, which is important. We can now prove

Theorem 1. If an $H_2^{(-3/2)}$-functional is in the domain of the Lévy Laplacian, then it is a normal functional in the sense of P. Lévy.

3. Duality in the space of quadratic generalized functionals

We can establish an identity of the renormalized square $:\dot{B}(t)^2:$ of white noise, as we did in the case of $\dot{B}(t)$ in $H_1^{(-1)}$ (see § 2.6 in [5]).

Having done this, we can now introduce a subspace L_2^* spanned by quadratic normal functionals of the $\dot{B}(t)$'s. More precisely,

$$L_2^* = \left\{ \int g(u) : \dot{B}(u)^2 : du; g \in K^{-1}(R^1) \right\}.$$

It should be noted that the function g above may be regarded as the restriction of a function f in $\widehat{K}^{-3/2}(R^2)$ down to the diagonal line of R^2. There the trace theorem for Sobolev space is applied.

Obviously the space can be made to be a subspace of $H_2^{(-2)}$ by viewing $g(u)$ to be $g(\frac{u+v}{2})\delta(u-v)$ as the integrand.

Our aim is to prove the following theorem

Theorem 2. There exists a subspace L_2 of $H_2^{(2)}$ such that L_2^* is the dual space of L_2, where the topologies of L_2 comes from that of $H_2^{(2)}$.

Proof. Elementary computations can prove the theorem. But, in reality, there can we see some detailed structure of quadratic generalized white noise functionals. Step by step computations are now in order.

The Fourier transform of $g(\frac{u+v}{2})$ is

$$\frac{1}{2\pi} \int\int e^{i(\lambda_1 u + \lambda_2 v)} g(\frac{u+v}{2})\delta(u-v)dudv = \sqrt{2\pi}\widehat{g}(\lambda_1 + \lambda_2),$$

where $\widehat{g}$ is the Fourier transform of g of one variable. By the definition of the Sobolev space of order 3/2 over R^2

$$\frac{1}{2\pi}\int\int \frac{|\widehat{g}(\lambda_1+\lambda_2)|^2}{(1+\lambda_1^2+\lambda_2^2)^{3/2}}d\lambda_1 d\lambda_2$$

is finite. This fact implies that $2^{-1/2}g(\frac{u}{\sqrt{2}})$ belongs to the Sobolev space $K^1(R^1)$, in addition its norm is equal to the $K^{-3/2}(R^2)$-norm of $g(\frac{u+v}{2})\delta(u-v)$ up to an universal constant.

Numerical values are as follows. Let $\|\cdot\|_{n,m}$ be the Sobolev norm of order m over R^n. Then, actually we have already shown the following equality

$$\|g\|_{2,3/2}^2 = \frac{c}{2\pi}\|g'\|_{1,1}^2,$$

where $c = \int(1+x^2)^{-3/2}dx$ and $g'^{-1/2}g(\frac{u}{\sqrt{2}})$.

Finally, we come to the stage of determinations of the space L_2 and L_2^*. Remind (Ref. [5])

$$H_2^{(2)} = \left\{\varphi(\dot{B}) = \int\int f(u,v) : \dot{B}(u)\dot{B}(v) : dudv, f \in \widehat{K}^{3/2}(R^2)\right\},$$

and introduce an equivalence relation $\sim$ in $H_2^{(2)}$ defined by

$$\int\int f_1(u,v) : \dot{B}(u)\dot{B}(v) : dudv \sim \int\int f_2(u,v) : \dot{B}(u)\dot{B}(v) : dudv$$

if and only if $f_1(u,v) = f_2(u,v)$.

Set

$$H_2^{(2)}/\sim \equiv L_2.$$

Note. Since $f_i, i = 1,2$ is in $K^{3/2}$, the relation to the diagonal $u = v$ is a continuous function. Hence the equivalence relation is defined without any ambiguity.

Then, what we have computed so far can prove that there is the dual pairing between L_2 and L_2^*, which proves the theorem.

This is somewhat a rephrasement, in a formal tone, of the Theorem 2. Suppose that $f \in \widehat{K}^{3/2}(R^2)$ and that $g((u+v)/2)\delta(u-v) \in \widehat{K}^{-3/2}(R^2)$ or $g \in K^1(R^1)$. Then, formal computation shows

$$\left\langle \int g(u) : \dot{B}(u)^2 : du, \int\int f(u,v) : \dot{B}(u)\dot{B}(v) : dudv \right\rangle$$
$$= 2\int g(u)f(u,u)du.$$

This equality is derived from

$$E[(: \dot{B}(t)^2 :)^2] = 2\frac{1}{(dt)^2}.$$

Remark The relationship between $\int : \dot{B}(t)^2 : dt$ and the Lévy Laplacian has been discussed in [8].

4. Quadratic forms of ∂_t's and ∂_t^*'s.

We have discussed algebras generated by differential operators ∂_t's and their adjoint operators ∂_t^*'s in [6], so we do not go into this topic. We shall, however, note the following fact.

We know that there are commutation relations

$$[\partial_t, \partial_s^*] = \delta(t-s)I,$$

where $[\cdot,\cdot]$ is the Lie product. Then, follow the following facts:

Facts The collection of ∂_t's and ∂_s^*s generates a Lie algebra. We can get a finite Lie algebra which is closed under the Lie product, provided $\delta(t-s)I$ can be included.

5. Concluding remarks

It is interesting to note that generalizations of the result reported in this note can be seen for $H_{2n}^{(2n)}$ and $H_{2n}^{(-2n)}$ in-between. Collection of those dual pairings can be one of the characteristics of the space of generalized white noise functionals $(L^2)^-$.

Acknowledgements

The authors are grateful to the organizers of QBIC08 who have given us an opportunity to present our results, although one of the authors missed to attend the conference because of unexpected affair.

References

1. D. Aerts, M. Czachor and T. Durt, eds. Probing the structure of Quantum Mechanics. Nonlinearity, nonlocality, computation, axiomatics. World Scientific Pub. Co. 2002.
2. T. Hida, Analysis of Brownian functionals. Carleton Math. Lecture Notes no. 13, Carleton University, 1975.
3. T. Hida, Brownian motion. Springer-Verlag. 1980.
4. T. Hida and Si Si, An innovation approach to random fields. Application of white noise theory. World Scientific Pub. Co. 2004.
5. T. Hida and Si Si, Lectures on white noise functionals. World. Sci. Pub. Co. 2008.
6. T. Hida, Si Si and T. Shimizu, The $\dot{B}(t)$'s as idealized elemental random variables. Volterra Center Notes N.614. 2008.
7. Si Si, Effective determination of Poisson noise. IDAQP 6 (2003), 609-617.
8. Si Si, An aspect of quadratic Hida distributions in the realization of a duality between Gaussian and Poisson noises. IDAQP 11 (2008) 109-118.
9. P. Lévy, Processus stochastiques et mouvement brownien. Gauthier-Villars. 1948. 2ème ed. with supplement 1965.
10. P. Lévy, Problèmes concrets d'analyse fonctionnelle. Gauthier-Villars. 1951.

ANALYSIS OF SEVERAL SOCIAL MOBILITY DATA USING MEASURE OF DEPARTURE FROM SYMMETRY

KOUJI TAHATA, KOUJI YAMAMOTO, NOBUKO MIYAMOTO, AND SADAO TOMIZAWA*

Department of Information Sciences, Faculty of Science and Technology, Tokyo University of Science
Noda City, Chiba, 278-8510, Japan
**E-mail: tomizawa@is.noda.tus.ac.jp*

For the analysis of square contingency tables with the same row and column classifications, the symmetry model is applied. This paper (1) reviews the measure for representing the degree of departure from the symmetry and (2) using the measure, compares five sets of social-mobility data on the father's and son's occupational status (i) for British in 1954, (ii) for Denmark in 1954, (iii) for Japan in 1955, (iv) for Japan in 1965, and (v) for Japan in 1975.
Keywords: Father-son data, Measure, Occupational status, Social-mobility, Square contingency table, Symmetry.

1. Introduction

Table 1 taken directly from Agresti (1984, p.206) is the data in Glass (1954) which represent father's and his son's occupational status in British. Namely, this is the social-mobility data on the status. These data have been analyzed using various models by several statisticians, including Mosteller (1968), Kullback (1971), Bishop, Fienberg and Holland (1975, p.100), Goodman (1981, 1984), Agresti (1984, p.205), and Tomizawa (1990a, 1990b, 1990c, 1991a).

The data in Table 2, taken directly from Bishop et al. (1975, p.100), describe the cross-classifications of father's and his son's occupational status categories in Denmark. These data have been analyzed by several statisticians, including Mosteller (1968), Kullback (1971), Haberman (1974, p.222), Bishop et al. (1975, p.100), Goodman (1981, 1984), Tomizawa (1990d, 1991b, 1998), and Tomizawa, Miyamoto and Yamamoto (2006).

The data in Tables 3a, 3b and 3c are taken from Tominaga (1979, p.53). These data describe the cross-classification of father's and his son's occupational status categories in Japan which were examined in 1955, 1965 and 1975. These data have been analyzed by Tomizawa and Saitoh (1999), Tomizawa, Miyamoto and Hatanaka (2001), Tomizawa, Miyamoto and Ashihara (2003), Tomizawa, Miyamoto and Yamane (2005), and Tahata, Iwashita and Tomizawa (2006).

For father's and his son's occupational mobility data as Tables 1, 2 and 3, many father-son pairs concentrate on (or near) the main diagonal cells. Therefore the independence between a father's occupational status and his son's occupational status does not hold, namely, a son's occupational status is strongly associated with his father's occupational status. So, instead of independence, we are interested in whether or not a son's occupational status is symmetric to his father's occupational status.

For analyzing square contingency tables with the same row and column classifications as father-son occupational mobility data, Bowker (1948) considered the symmetry model which indicates the structure of symmetry of unknown cell probabilities (see also Bishop et al., 1975, p.282). When the symmetry model does not hold for the given data, we are also interested in seeing what degree the departure from the symmetry is. Tomizawa (1994), and Tomizawa, Seo and Yamamoto (1998) considered the measure to represent the degree of departure from the symmetry.

The purpose of this paper is (1) to review the symmetry model and the measure of the degree of departure from the symmetry and (2) to compare five sets of father-son occupational mobility data in Tables 1, 2 and 3 using the measure.

2. Review of model and measure of symmetry

In this section, we shall review briefly the symmetry model and the measure to represent the degree of departure from symmetry. For an $r \times r$ square contingency table with nominal categories, let p_{ij} denote the probability that an observation will fall in the ith row and jth column of the table $(i = 1, \ldots, r; j = 1, \ldots, r)$. Let X and Y denote the row and column variables, respectively. For the father's and his son's occupational mobility data, X is the father's occupational status and Y is the son's occupational status.

2.1. *Symmetry model*

Bowker (1948) considered the symmetry model defined by

$$p_{ij} = \psi_{ij} \quad (i = 1, \ldots, r; j = 1, \ldots, r),$$

where

$$\psi_{ij} = \psi_{ji};$$

see also, e.g., Bishop et al. (1975, p.282), Agresti (1984, p.202), Tomizawa and Tahata (2007). For the father-son occupational mobility data, this model indicates that the probability that a father's occupational status is i and his son's status is j is equal to the probability that the father's status is j and his son's status is i. Namely, it indicates that the mobility of occupational status of father and his son is symmetric.

2.2. *Measure*

Tomizawa (1994) and Tomizawa et al. (1998) considered the measure to represent the degree of departure from the symmetry below.

Assume that $p_{ij} + p_{ji} > 0$ for $i = 1, \dots, r; j = 1, \dots, r; i \neq j$. Let

$$\delta = \sum\sum_{i \neq j} p_{ij} \quad [= \mathrm{P}(X \neq Y)],$$

$$p_{ij}^{*} = \frac{p_{ij}}{\delta} \quad [= \mathrm{P}(X = i, Y = j \mid X \neq Y)],$$

$$p_{ij}^{S} = \frac{p_{ij}^{*} + p_{ji}^{*}}{2} \quad \left[= \frac{1}{2}\mathrm{P}((X,Y) = (i,j) \text{ or } (j,i) \mid X \neq Y)\right],$$

for $i = 1, \dots, r; j = 1, \dots, r; i \neq j$. The measure, considered by Tomizawa et al. (1998), is given by

$$\Phi^{(\lambda)} = \frac{\lambda(\lambda+1)}{2^{\lambda} - 1} I^{(\lambda)}(\{p_{ij}^{*}\}; \{p_{ij}^{S}\}) \quad \text{for } \lambda > -1,$$

where

$$I^{(\lambda)}(\cdot;\cdot) = \frac{1}{\lambda(\lambda+1)} \sum\sum_{i \neq j} p_{ij}^{*} \left[\left(\frac{p_{ij}^{*}}{p_{ij}^{S}}\right)^{\lambda} - 1\right]$$

and the value at $\lambda = 0$ is taken to be the limit as $\lambda \to 0$, where λ is a real value that is chosen by the user. When $\lambda = 0$, we see

$$\Phi^{(0)} = \lim_{\lambda \to 0} \Phi^{(\lambda)} = \frac{1}{\log 2} I^{(0)}(\{p_{ij}^{*}\}; \{p_{ij}^{S}\}),$$

where

$$I^{(0)}(\cdot;\cdot) = \sum\sum_{i \neq j} p_{ij}^{*} \log\left(\frac{p_{ij}^{*}}{p_{ij}^{S}}\right).$$

Note that $I^{(\lambda)}(\cdot;\cdot)$ is the power divergence between two distributions $\{p_{ij}^*\}$ and $\{p_{ij}^S\}$, and especially $I^{(0)}(\cdot;\cdot)$ is the Kullback-Leibler information between them. For more details of the power divergence $I^{(\lambda)}(\cdot;\cdot)$, see Cressie and Read (1984), and Read and Cressie (1988, p.15). Also, note that the measures $\Phi^{(0)}$ and $\Phi^{(1)}$ are given by earlier Tomizawa (1994).

Let

$$p_{ij}^c = \frac{p_{ij}}{p_{ij}+p_{ji}} \quad [= \mathrm{P}((X,Y)=(i,j) \mid (X,Y)=(i,j) \text{ or } (j,i))]$$

for $i = 1,\ldots,r; j = 1,\ldots,r; i \neq j$. Then the measure may be expressed as

$$\Phi^{(\lambda)} = \sum\sum_{i<j}(p_{ij}^* + p_{ji}^*)\left[1 - \frac{\lambda 2^\lambda}{2^\lambda - 1}H_{ij}^{(\lambda)}(\{p_{ij}^c, p_{ji}^c\})\right] \quad \text{for } \lambda > -1,$$

where

$$H_{ij}^{(\lambda)}(\cdot) = \frac{1}{\lambda}\left[1 - (p_{ij}^c)^{\lambda+1} - (p_{ji}^c)^{\lambda+1}\right];$$

see Tomizawa et al. (1998). Note that $H_{ij}^{(\lambda)}(\cdot)$ is the Patil and Taillie (1982) diversity index of degree λ for the conditional distribution $\{p_{ij}^c, p_{ji}^c\}$, which includes the Shannon entropy, i.e.,

$$H_{ij}^{(0)}(\cdot) = \lim_{\lambda\to 0} H_{ij}^{(\lambda)}(\cdot) = -p_{ij}^c \log p_{ij}^c - p_{ji}^c \log p_{ji}^c.$$

For more details of measure $\Phi^{(\lambda)}$, see Tomizawa (1994) and Tomizawa et al. (1998).

Note that the measure must lie between 0 and 1. For each λ (> -1), (1) the symmetry model holds if and only if $\Phi^{(\lambda)} = 0$, and (2) the degree of departure from symmetry is the largest, in the sense that $p_{ij}^c = 1$ (then $p_{ji}^c = 0$) or $p_{ji}^c = 1$ (then $p_{ij}^c = 0$) for $i \neq j$, if and only if $\Phi^{(\lambda)} = 1$ (see Tomizawa et al., 1998).

3. Approximate confidence interval for measure

Let n_{ij} denote the observed frequency in the ith row and jth column of the square table ($i = 1,\ldots,r; j = 1,\ldots,r$). Assume that the $\{n_{ij}\}$ result from full multinomial sampling. The sample version of measure $\Phi^{(\lambda)}$, i.e., $\hat{\Phi}^{(\lambda)}$, is given by $\Phi^{(\lambda)}$ with $\{p_{ij}\}$ replaced by $\{\hat{p}_{ij}\}$, where $\hat{p}_{ij} = n_{ij}/n$ and $n = \sum\sum n_{ij}$. Using the delta method (e.g., Bishop et al., 1975, Sec. 14.6), $\sqrt{n}(\hat{\Phi}^{(\lambda)} - \Phi^{(\lambda)})$ has asymptotically (as $n \to \infty$) a normal distribution with mean zero and variance $\sigma^2[\Phi^{(\lambda)}]$. For the details of $\sigma^2[\Phi^{(\lambda)}]$, see Tomizawa (1994) and Tomizawa et al. (1998).

Let $\hat{\sigma}^2[\Phi^{(\lambda)}]$ denote $\sigma^2[\Phi^{(\lambda)}]$ with $\{p_{ij}\}$ replaced by $\{\hat{p}_{ij}\}$. Then $\hat{\sigma}[\Phi^{(\lambda)}]/\sqrt{n}$ is an estimated approximate standard error for $\hat{\Phi}^{(\lambda)}$, $\hat{\Phi}^{(\lambda)} \pm z_{p/2}\hat{\sigma}[\Phi^{(\lambda)}]/\sqrt{n}$ is an approximate $100(1-p)$ percent confidence interval for $\Phi^{(\lambda)}$, where $z_{p/2}$ is the percentage point from the standard normal distribution corresponding to a two-tail probability equal to p.

4. Comparison between occupational mobility data

Table 4 gives the values of estimated measure $\hat{\Phi}^{(\lambda)}$ and the approximate 95% confidence interval of measure $\Phi^{(\lambda)}$ for five sets of data of father's and his son's occupational status in Tables 1, 2, 3a, 3b, and 3c. We shall compare these data using the measure and the confidence interval.

The confidence intervals for $\Phi^{(\lambda)}$ applied to the data in each of Table 1, Table 2, and Table 3a, 3b, and 3c, do not include zero for all λ, which indicates that these tables do not have the structure of symmetry.

We shall investigate the degree of departure from symmetry in more details. For instance, when $\lambda = 0$, the estimated measure $\hat{\Phi}^{(0)}$ equals 0.013 for British father-son data (Table 1), 0.013 for Danish data (Table 2), 0.215 for Japanese data in 1955 (Table 3a), 0.409 for Japanese data in 1965 (Table 3b), and 0.356 for Japanese data in 1975 (Table 3c). Thus, (i) for British data, the degree of departure from symmetry is estimated to be 1.3 percent of the maximum degree of departure from symmetry, (ii) for Danish data, it is estimated to be 1.3 percent of the maximum case, and similarly (iii) for Japanese data in 1955, it is estimated to be 21.5 percent of the maximum case, (iv) for Japanese data in 1965, it is estimated to be 40.9 percent of the maximum case, and (v) for Japanese data in 1975, it is estimated to be 35.6 percent of the maximum case.

When the degrees of departure from the symmetry in Tables 1, 2, 3a, 3b, and 3c, are compared using the confidence interval for the measure $\Phi^{(\lambda)}$, (1) it is greater for Japanese data in 1955, 1965, and 1975, than for British data and for Danish data, and (2) it is greater for Japanese data in 1965 and in 1975, than for Japanese data in 1955. However, the comparison between the British data and the Danish data may be impossible because values in the confidence interval for British data are not always greater than values in the confidence interval for Danish data. By the similar reason, the comparison between the Japanese data in 1965 and the Japanese data in 1975 may be impossible.

5. Concluding remarks

The measure $\hat{\Phi}^{(\lambda)}$ always ranges between 0 and 1 independent of the dimension r and sample size n. Therefore, $\hat{\Phi}^{(\lambda)}$ may be useful for *comparing* the degrees of departure from symmetry in several tables (see Tomizawa et al., 1998).

For five sets of father-son occupational mobility data, we have compared the degree of departure from symmetry for the mobility of father's and his son's occupational status, by using the estimated measure $\hat{\Phi}^{(\lambda)}$ and the confidence interval of the measure $\Phi^{(\lambda)}$.

Finally we note that the measure $\Phi^{(\lambda)}$ is invariant under the same arbitrary permutations of row and column categories; so the measure can be applied to square tables with nominal categories and also it may be applied to square tables with ordered categories (as father-son occupational status data) when we may not utilize the information of ordering of categories.

Table 1

Occupational status for British father-son pairs; from Agresti (1984, p.206)

Father's status	Son's status (1)	(2)	(3)	(4)	(5)	Total
(1)	50	45	8	18	8	129
(2)	28	174	84	154	55	495
(3)	11	78	110	223	96	518
(4)	14	150	185	714	447	1510
(5)	3	42	72	320	411	848
Total	106	489	459	1429	1017	3500

Table 2

Occupational status for Danish father-son pairs; from Bishop et al. (1975, p.100)

Father's status	Son's status (1)	(2)	(3)	(4)	(5)	Total
(1)	18	17	16	4	2	57
(2)	24	105	109	59	21	318
(3)	23	84	289	217	95	708
(4)	8	49	175	348	198	778
(5)	6	8	69	201	246	530
Total	79	263	658	829	562	2391

Table 3

Occupational status for Japanese father-son pairs; from Tominaga (1979, p.53)

(a) Examined in 1955

Father's status	Son's status (1)	(2)	(3)	(4)	(5)	(6)	(7)	(8)	Total
(1)	36	4	14	7	8	2	3	8	82
(2)	20	20	27	24	11	11	2	11	126
(3)	9	6	23	12	9	5	3	16	83
(4)	15	14	39	81	17	16	11	15	208
(5)	6	7	22	13	72	20	6	13	159
(6)	3	2	5	12	18	19	9	7	75
(7)	5	3	10	11	21	15	38	25	128
(8)	39	30	76	80	69	52	45	614	1005
Total	133	86	216	240	225	140	117	709	1866

(b) Examined in 1965

Father's status	Son's status (1)	(2)	(3)	(4)	(5)	(6)	(7)	(8)	Total
(1)	27	10	16	3	6	6	1	2	71
(2)	15	38	30	20	8	4	3	7	125
(3)	13	17	32	17	7	16	6	5	113
(4)	12	36	40	132	22	30	13	6	291
(5)	8	22	38	41	91	42	22	9	273
(6)	2	2	7	12	13	16	3	2	57
(7)	3	2	11	11	13	26	30	6	102
(8)	38	44	95	101	132	114	60	309	893
Total	118	171	269	337	292	254	138	346	1925

Table 3 (continued)

(c) Examined in 1975

Father's status	Son's status (1)	(2)	(3)	(4)	(5)	(6)	(7)	(8)	Total
(1)	44	18	28	8	6	8	1	5	118
(2)	15	50	45	20	18	17	4	7	176
(3)	18	25	47	30	24	18	5	7	174
(4)	16	27	53	77	40	29	9	6	257
(5)	18	25	42	31	122	43	17	13	311
(6)	12	15	21	15	36	33	3	8	143
(7)	3	5	8	7	26	21	9	3	82
(8)	44	65	114	92	184	195	58	325	1077
Total	170	230	358	280	456	364	106	374	2338

Table 4

Estimate of $\Phi^{(\lambda)}$, estimated approximate standard error for $\hat{\Phi}^{(\lambda)}$, and approximate 95% confidence interval for $\Phi^{(\lambda)}$, applied to Tables 1, 2, and 3

(a) For Table 1 (British data)

Values of λ	Estimated measure	Standard error	Confidence interval
−0.4	0.009	0.003	(0.003, 0.015)
0	0.013	0.004	(0.005, 0.022)
0.6	0.017	0.005	(0.006, 0.028)
1	0.018	0.006	(0.007, 0.030)
1.4	0.019	0.006	(0.007, 0.030)
1.8	0.019	0.006	(0.007, 0.030)

(b) For Table 2 (Danish data)

Values of λ	Estimated measure	Standard error	Confidence interval
−0.4	0.009	0.004	(0.002, 0.016)
0	0.013	0.005	(0.003, 0.023)
0.6	0.016	0.006	(0.004, 0.029)
1	0.018	0.007	(0.004, 0.031)
1.4	0.018	0.007	(0.004, 0.032)
1.8	0.018	0.007	(0.004, 0.032)

(c) For Table 3a (Japanese data in 1955)

Values of λ	Estimated measure	Standard error	Confidence interval
−0.4	0.157	0.018	(0.123, 0.192)
0	0.215	0.023	(0.171, 0.260)
0.6	0.261	0.026	(0.211, 0.312)
1	0.275	0.027	(0.223, 0.327)
1.4	0.279	0.027	(0.227, 0.332)
1.8	0.277	0.027	(0.225, 0.330)

Table 4 (continued)

(d) For Table 3b (Japanese data in 1965)

Values of λ	Estimated measure	Standard error	Confidence interval
-0.4	0.326	0.021	(0.284, 0.367)
0	0.409	0.023	(0.364, 0.454)
0.6	0.463	0.023	(0.417, 0.508)
1	0.476	0.023	(0.431, 0.521)
1.4	0.480	0.023	(0.435, 0.525)
1.8	0.478	0.023	(0.433, 0.524)

(e) For Table 3c (Japanese data in 1975)

Values of λ	Estimated measure	Standard error	Confidence interval
-0.4	0.282	0.018	(0.247, 0.317)
0	0.356	0.019	(0.318, 0.393)
0.6	0.402	0.019	(0.365, 0.439)
1	0.413	0.019	(0.376, 0.450)
1.4	0.416	0.019	(0.379, 0.453)
1.8	0.415	0.019	(0.378, 0.451)

References

1. Agresti, A. (1984). *Analysis of Ordinal Categorical Data.* Wiley, New York.
2. Bishop, Y. M. M., Fienberg, S. E., and Holland, P. W. (1975). *Discrete Multivariate Analysis: Theory and Practice.* The MIT Press, Cambridge, Massachusetts.
3. Bowker, A. H. (1948). A test for symmetry in contingency tables. *Journal of the American Statistical Association*, **43**, 572-574.
4. Cressie, N. and Read, T. R. C. (1984). Multinomial goodness-of-fit tests. *Journal of the Royal Statistical Society, Series B*, **46**, 440-464.
5. Glass, D. V. (ed.) (1954). *Social Mobility in Britain.* Free Press, Glencoe, Illinois.
6. Goodman, L. A. (1981). Association models and the bivariate normal for contingency tables with ordered categories. *Biometrika*, **68**, 347-355.
7. Goodman, L. A. (1984). *The Analysis of Cross-Classified Data Having Ordered Categories.* Harvard University Press, Cambridge.
8. Haberman, S. J. (1974). *The Analysis of Frequency Data.* University of Chicago Press, Chicago.
9. Kullback, S. (1971). Marginal homogeneity of multidimensional contingency tables. *Annals of Mathematical Statistics*, **42**, 594-606.
10. Mosteller, F. (1968). Association and estimation in contingency tables. *Journal of the American Statistical Association*, **63**, 1-28.
11. Patil, G. P. and Taillie, C. (1982). Diversity as a concept and its measurement. *Journal of the American Statistical Association*, **77**, 548-561.
12. Read, T. R. C. and Cressie, N. (1988). *Goodness-of-Fit Statistics for Discrete Multivariate Data.* Springer-Verlag, New York.
13. Tahata, K., Iwashita, T., and Tomizawa, S. (2006). Measure of departure from symmetry of cumulative marginal probabilities for square contingency tables with ordered categories. *SUT Journal of Mathematics*, **42**, 7-29.
14. Tominaga, K. (1979). *Nippon no Kaisou Kouzou (Japanese Hierarchical Structure).* University of Tokyo Press, Tokyo (in Japanese).
15. Tomizawa, S. (1990a). The decomposition of a quasi-uniform association model and its application to the analysis of British occupational mobility table data. *Behaviormetrika*, **27**, 115-121.
16. Tomizawa, S. (1990b). Polynomial diagonals-parameter symmetry model for square contingency tables with ordered categories. *Statistica*, **50**, 171-178.
17. Tomizawa, S. (1990c). Another linear diagonals-parameter symmetry model for square contingency tables with ordered categories. *South African Statistical Journal*, **24**, 117-125.
18. Tomizawa, S. (1990d). Parsimonious quasi-uniform association models applied to Danish occupational mobility data. *Journal of the Japan Statistical Society*, **20**, 13-22.
19. Tomizawa, S. (1991a). An extended linear diagonals-parameter symmetry model for square contingency tables with ordered categories. *Metron*, **49**, 401-409.
20. Tomizawa, S. (1991b). A model of uniform association plus two-diagonals-

parameter and its application to occupational mobility table data. *Statistical Papers*, **32**, 243-252.
21. Tomizawa, S. (1994). Two kinds of measures of departure from symmetry in square contingency tables having nominal categories. *Statistica Sinica*, **4**, 325-334.
22. Tomizawa, S. (1998). A decomposition of the marginal homogeneity model into three models for square contingency tables with ordered categories. *Sankhyā, Series B*, **60**, 293-300.
23. Tomizawa, S. and Saitoh, K. (1999). Measure of departure from conditional symmetry for square contingency tables with ordered categories. *Journal of the Japan Statistical Society*, **29**, 65-78.
24. Tomizawa, S. and Tahata, K. (2007). The analysis of symmetry and asymmetry: orthogonality of decomposition of symmetry into quasi-symmetry and marginal symmetry for multi-way tables. *Journal de la Société Francaise de Statistique: Journal of the French Statistical Society*, **148**, 3-36.
25. Tomizawa, S., Seo, T., and Yamamoto, H. (1998). Power-divergence-type measure of departure from symmetry for square contingency tables that have nominal categories. *Journal of Applied Statistics*, **25**, 387-398.
26. Tomizawa, S., Miyamoto, N., and Hatanaka, Y. (2001). Measure of asymmetry for square contingency tables having ordered categories. *The Australian and New Zealand Journal of Statistics*, **43**, 335-349.
27. Tomizawa, S., Miyamoto, N., and Ashihara, N. (2003). Measure of departure from marginal homogeneity for square contingency tables having ordered categories. *Behaviormetrika*, **30**, 173-193.
28. Tomizawa, S., Miyamoto, N., and Yamane, S. (2005). Power-divergence-type measure of departure from diagonals-parameter symmetry for square contingency tables with ordered categories. *Statistics*, **39**, 107-115.
29. Tomizawa, S., Miyamoto, N., and Yamamoto, K. (2006). Decomposition for polynomial cumulative symmetry model in square contingency tables with ordered categories. *Metron*, **64**, 303-314.

TIME IN PHYSICS AND LIFE SCIENCE

IGOR.V. VOLOVICH
Steklov Mathematical Institute,
Gubkin St.8, 119991 Moscow, Russia
E-mail: volovich@mi.ras.ru

Some mathematical aspects of the concept of time in physics and life science are discussed. A theoretical model of time machine is a spacetime region with closed timelike curves. Possible production of mini time machines at CERN's Large Hadron Collider (LHC) is considered. It is argued that if the scale of quantum gravity is of the order of few TeVs, proton-proton collisions at the LHC could lead to the formation of traversable wormhole which is a model for the time machine. The wormhole production cross section at the LHC is of the same order as the cross section for the black hole production. We make also some comments on the role of time in life science. It is proposed to describe cells and other life phenomena by using framework of quantum field theory.

1. Introduction

There are many questions related with the notion of time. For instance, is it possible time travel; whether time exists when nothing is changing; irreversibility problem: why *microscopic* equations are reversible in time but the *macroscopic* equations are not; what are the neural mechanisms that account for our experience of time.

In this paper some mathematical aspects of the concept of time in physics and life science will be discussed.

A mathematical definition of the time machine is the following. Time machine is a region in spacetime (spacetime is a manifold with a Lorentz metric) with closed timelike curves. Possible production of mini time machines at accelerators, in particlular at the Large Hadron Collider (LHC) in Geneve is considered. It is argued that if the scale of quantum gravity is of the order of few TeVs, the proton-proton collisions at the LHC could lead to the formation of a traversable wormhole which is a model for the time machine [12]. The wormhole production cross section at the LHC is of the same order as the cross section for the black hole production. The wormholes contain small spacetime regions with closed timelike curves (CTC) which violate the standard causality condition.

Time is a fundamental notion in life science, since life exists only in the process of time development. We use a new paradigm of computability, based on the joint work with M.Ohya on quantum chaotic amplifier, which is more appropriate for life science than the quantum Turing machine, to argue that a living organism is an uncomputable system. Cell can be treated as an "elementary particle" of life. It is proposed to describe cells and other life phenomena by using an extension of the framework of quantum field theory. A living organism can be represented as a non-Markovian, quantum (noncommutative) stochastic process which evolves with (almost) closed timelike curves.

2. Time Travel

Let us discuss whether is it possible time travel. In general relativity a timelike curve in space-time represents a possible path of an object or an observer. Normally such a curve will run from past to future, but in some space-times the curves can intersect themselves, giving a closed timelike curve (CTC) which is interpreted as a time machine. It suggests the possibility of time-travel with its well known paradoxes.

There are many solutions of the Einstein equations with CTCs. A list of such solutions includes Gödel's solution [1], van Stockum and Tipler cylinders , Kerr and Kerr-Newman solutions [2], Gott's time machine [3], Wheeler wormholes (space-time foam) [4], Morris-Thorne traversable wormholes [5], and Ori's dust asymptotically-flat space-time, see [6] for a review.

A wormhole forms a handle-like geometry, whose two mouths join different regions of spacetime. If the wormhole is traversed from mouth to mouth, it acts as a time machine allowing one to travel into the past or into the future.

Violation of normal chronology is so objectionable an occurrence that any such solution could be rejected as unphysical. However, the Einstein equations are local equations and therefore one has to impose additional principles to preserve chronology. There are long debate concerning such principles [6,7,8,9,10]. In particular, in [7] it was shown that acausal CTC in Gott's universe cannot be realized by physical, timelike, sources.

An attempt to save causality and exclude CTCs from general relativity is Hawking's "chronology protection conjecture" which asserts that the law of physics do not allow the appearance of CTC [8]. However, there are not enough convincing arguments for this conjecture. Indeed, it was suggested that divergences in the energy-momentum tensor occur when one has closed causal curves. These divergences may create spacetime singularities which prevent one from traveling through to the region of closed timelike curves. However, it might be that quantum gravitational effects may smear out the divergences. Moreover, if one believes that there exists a full theory of quantum gravity, then chronology protection should be settled by using this theory [9].

Whether the chronology protection conjecture can be derived from the known physics laws or it is an independent postulate is still an open question. We suggest to test it in experiments at the LHC.

Note that the CTC problem probably is related with the irreversibility problem well known in statistical physics. For a discussion of the black hole information paradox see [11] where it is explained that the black hole information paradox is a particular case of the irreversibility problem which is not solved not only for black *hole* but even for the usual black *body*.

Causality is one of fundamental physical principles. We suggest that there is a possibility to test causality in experiments at CERN's Large Hadron Collider. This is related with a possibility of wormhole production in proton-proton collisions at the LHC. The wormholes contain small space-time regions with (CTC) which violate the standard causality condition.

3. Black Hole/Wormhole Production

A possibility of production in ultra-relativistic particle collisions of some objects related with a non-trivial space-time structure is one of long-standing theoretical questions. One of such particular objects is a black hole.

In general relativity there is Thorn's hoop conjecture which says that black holes form when, and only when, a mass M gets compacted into a region whose circumference in every direction is $\mathcal{C} < 4\pi GM$ [13]. The area of the corresponding disk is

$$\pi r_0^2 = 4\pi G^2 M^2 \sim s/M_{\rm Pl}^4 , \tag{1}$$

which gives a rough estimate for the classical geometrical cross-section for black hole production. Here G is the Newton constant, $M_{\rm Pl} = 1/\sqrt{G}$ is the Planck mass and s is the square of the center of mass energy of colliding particles .

A conjecture that in string theory and in quantum gravity at energies much higher than the Planck mass black hole production emerges has been made in [14,15]. It has been proposed to use the Aichelburg-Sexl shock wave metrics to describe ultra-relativistic particles. Under collision of these waves one can expect a production of black holes.

To speak on the production of black holes in quantum theory one should have a notion of a quantum black hole as a state (pure or mixed) in some Hilbert space. We have to compute the transition amplitude from a quantum state describing two particles to a quantum state describing quantum black holes. A quantum gravity approach to this problem is discussed in [16]. One considers the kernel of the transition amplitude

$$\langle h'', \phi'', \Sigma'' | h', \phi', \Sigma' \rangle = \int \exp\{\frac{i}{\hbar} S[g, \Phi]\} \mathcal{D}g \mathcal{D}\Phi \qquad (2)$$

between configuration of the three-metric h'_{ij} and fields ϕ' on an initial spacelike surface Σ' and a configuration h''_{ij} and ϕ'' on a final surface Σ''. In (2) the integral is over all four-geometries $g_{\mu\nu}$, including summation over different topologies, and field configurations Φ, which match given values on the space-like surfaces Σ' and Σ'', i.e. $\Phi|_{\Sigma'} = \phi'$, $g|_{\Sigma'} = h'$ and $\Phi|_{\Sigma''} = \phi''$, $g|_{\Sigma''} = h''$. This formula assumes the Wheeler-de Witt formalism [4], for a recent review see [17].

To get the transition amplitude between two particles and a black hole, or a wormhole one has to integrate the kernel (2) with the wave function $\Psi_{\Sigma'}[h', \phi']$ describing two particles and the wave function $\Psi_{\Sigma''}[h'', \phi'']$ describing black hole or wormhole.

In the case of a semiclassical description of black holes production from particles [16] a leading contribution comes from Σ' being a partial Cauchy surface with asymptotically simple past in a strongly asymptotically predictable space-time and Σ'' being a partial Cauchy surface containing black hole(s), i.e. $\Sigma'' - J^-(\mathcal{T}^+)$ is non empty where $J^-(\mathcal{T}^+)$ is the causal past of future null infinity, see [2].

A possible scenario for creation of black holes by using classical solutions of the Einstein equations has been proposed in [16]. In this scenario it is supposed that ultra-relativistic particles are represented by plane gravitational waves, which interacting collide and produce a black hole. A duality between plane gravitational waves and black holes is used. Trans-Planckian collisions in standard quantum gravity have inaccessible energy scale and cannot be realized in usual conditions. However if the fundamental Planck scale of quantum gravity is of the order of few TeVs [18] then one can argue that there is an exciting possibility of production of black holes, branes, and Kaluza-Klein modes from the extra dimensions in proton-proton collisions at CERN's Large Hadron Collider (LHC) [19,20]. The cross section for creation of a black hole or brane with radius r_0 was postulated to be approximately equal to the geometrical cross section πr_0^2 as in the hoop conjecture (1). The Schwarzschild radius of a $4+n$ dimensional black hole of mass $M = \sqrt{s}$ is approximately,

$$r_0 \sim M_{4+n}^{-1}(s/M_{4+n}^2)^{\frac{1}{2(n+1)}} . \tag{3}$$

Here M_{4+n} is the $4+n$ dimensional Planck mass and the 4 dimensional Planck mass is given by

$$M_{\rm Pl}^2 \sim V_n M_{4+n}^{2+n}, \tag{4}$$

where V_n is the volume of the extra dimensions.

This process can be achieved by scattering of two partons with the center of mass energy $\sqrt{s}$ larger than M and impact parameter smaller than r_0.

D'Eath and Payne have studied the problem of classical collision with zero impact parameter and shown that a closed trapped surface forms. This analysis was extended to a nonzero impact parameter by Eardley and Giddings. The Aichelburg-Sexl solution has the form

$$ds^2 = -dudv + dx^{i2} + \varphi(x^i)\delta(u)du^2, \tag{5}$$

where φ depends only on the transverse coordinates x^i. A marginally trapped surface is constructed in the union of two incoming null hypersurfaces by solving a constraint problem for the Dirichlet Green's function.

The four-dimensional spacetime metric representing a spherically symmetric and static wormhole is given by [5,6]

$$ds^2 = -e^{2\Phi(r)}dt^2 + \frac{dr^2}{1-f(r)/r} + r^2(d\theta^2 + \sin^2\theta d\phi^2) . \tag{6}$$

Here $\Phi(r)$ is designated the redshift function and $f(r)$ is denominated the shape function. The radial coordinate r varies from $r = r_0$ corresponding to the wormhole throat, $f(r_0) = r_0$, to some R. The redshift function supposed to be finite, i.e. the event horizon is absent for $r_o < r < R$ and the shape function should satisfy the following inequality $f'r - f < 0$. For asymptotically flat wormholes $R = \infty$.

As it is well known traversable wormholes exist only for NEC violating stress energy tensors [5]. According to the NEC [2] the stress energy tensor $T_{\mu\nu}$ has to satisfy the requirement $T_{\mu\nu}k^\mu k^\nu \geq 0$, where k^μ is a null vector, $k^\nu k_\nu = 0$. Using the Einstein field equations, $G_{\mu\nu} = M_{\rm Pl}^{-1}T_{\mu\nu}$, one obtains [5] the following expression for the sum of the energy density $\rho(r)$ and the radial pressure $p_r(r)$

$$\rho(r) + p_r(r) = \frac{1}{M_{\rm Pl}}\left(\frac{f'r - f}{r^3} + 2\left(1 - \frac{f}{r}\right)\frac{\Phi'}{r}\right).$$

We see that the embedding condition together with the requirement of finiteness of the redshift function lead to the NEC violation on the wormhole throat.

Several scenario of the NEC violating have been considered in recent years. Generally speaking the NEC violating means unstability. But this is true only under special assumptions. There are examples of stable effective theories where the NEC is violated. In these particular cases the Lorentz invariance is broken and superluminal modes are present. Typical features of NEC violating effective theories is a presence of higher derivative terms and also superluminal modes. Gravitational Lorentz violation and superluminality take place also for wormhole solutions in Euclidean AdS gravity. Note that traversable wormholes may be also supported by the dark energy (see for example [21,22] and refs therein) with the equation of state parameter $w < -1$.

In the brane world scenario, where the Universe is considered as a 3-brane embedded in a D-dimensional bulk, the four-dimensional Einstein field equations contain the effective four-dimensional stress energy tensor,

$$G_{\mu\nu} = M_{\rm Pl}^{-2} T^{\rm eff}_{\mu\nu}. \tag{7}$$

$T^{\rm eff}_{\mu\nu}$ is a sum of the stress energy tensor of a matter confined on the brane, $T_{\mu\nu}$ and correction terms that arise from a projection of the D-dimensional Einstein equation to the four-dimensional space-time. It is possible that $T^{\rm eff}_{\mu\nu}$ supported the four-dimensional whormhole solution violates the NEC meanwhile $T_{\mu\nu}$ does not violate the NEC.

In the simplest brane world scenario where the Universe is considered as a 3-brane embedded in a five-dimensional bulk these correction terms can be written explicitly ,

$$T^{\rm eff}_{\mu\nu} = T_{\mu\nu} + \frac{6}{M_4\lambda}\Pi_{\mu\nu} - E_{\mu\nu}, \tag{8}$$

$$\Pi_{\mu\nu} = \frac{1}{12} T\, T_{\mu\nu} - \frac{1}{4} T_{\mu\alpha}\, T^{\alpha}_{\nu} + \frac{1}{8} g_{\mu\nu}[T_{\alpha\beta}\, T^{\alpha\beta} - \frac{1}{3}T^2], \tag{9}$$

$$E_{\mu\nu} = {}^{(5)}C_{\mu\alpha\nu\beta} n^{\alpha} n^{\beta}, \tag{10}$$

where ${}^{(5)}C_{\mu\alpha\nu\beta}$ is the five-dimensional Weyl tensor, $\alpha, \beta = 0,1,2,3,4$ and n^{α} is the unit normal to the brane. These formulas give the following relation between $\rho^{\rm eff}+p_r^{\rm eff}$ and $\rho+p_r$

$$\rho^{\rm eff}+p_r^{\rm eff}=\rho+p_r-\frac{1}{8\pi}(\epsilon+\sigma_r)+\frac{1}{\lambda}\rho(\rho+p_r)\,. \tag{11}$$

Here $\rho(r)$ and $p_r(r)$ are the energy density and the radial pressure of the matter confined on the brane, ϵ and σ_r are diagonal components of the projected Weyl tensor $diag[\epsilon(r),\sigma_r(r),\,\sigma_t(r),\sigma_t(r)]$. Now to have a wormhole one has to provide the condition

$$8\pi(\rho+p_r)(1+\frac{\rho}{T})<\epsilon+\sigma_r\,. \tag{12}$$

As comparing with four-dimensional wormholes we see a softening of the energy condition. This relaxed condition appears due to corrections from the Weyl tensor in the bulk (compare with the NEC violation from the string field non-local action [21,22]). For some particular examples it is possible to show that the four-dimensional effective stress energy tensor violates the NEC meanwhile the total five-dimensional stress energy tensor does respect the NEC. We do not present here higher-dimensional solutions corresponding to wormholes on 3-brane. It would be interesting to find the wormhole solutions in the context of intersecting D5-branes related with the Standard model. For a general class of solutions one expects the following dependence of the radius of the throat or mouth r_0 from the mass

$$r_0=\gamma_{\rm wh}(D)\frac{1}{M_D}\left(\frac{M_{\rm wh}}{M_D}\right)^{\alpha}\,. \tag{13}$$

This formula is similar to the formula for the Schwarzschild radius for the D-dimensional Schwarzschild solution:

$$ds^2=-(1-(\frac{r_s}{r})^{D-3})dt^2+(1-(\frac{r_s}{r})^{D-3})^{-1}dr^2+r^2d\Omega^2_{D-2}\,, \tag{14}$$

where the Schwarzschild radius r_s is related to the mass of the black hole by the relation

$$r_s=\gamma_{\rm bh}(D)\frac{1}{M_D}\left(\frac{M_{\rm bh}}{M_D}\right)^{\alpha_{\rm bh}}\,,\quad \alpha_{\rm bh}=\frac{1}{D-3}\,, \tag{15}$$

where $\gamma_{\rm bh}(D)=1/\sqrt{\pi}(\frac{8\Gamma(D-1/2)}{D-2})^{1/(D-3)}$. Let us note that for solution (6) the radius of the throat r_0 is larger than the Schwarzschild radius: $r_0>r_s$.

To compute the wormhole production cross section we can follow the approach for computation of the black hole production cross section [19,20]. The wormhole cross section is found from the partonic cross section for partons i and j to form a wormhole:

$$\sigma_{pp\to\text{wh}}(s) \sim \sum_{ij} \int_{\tau_m}^{1} d\tau \int_{\tau}^{1} \frac{dx}{x} f_i(x) f_j(\tau/x) \sigma_{ij\to\text{wh}}(\tau s) \,. \tag{16}$$

Here $\sqrt{s}$ is the center of mass energy, x and τ/x are the parton momentum fractions, and f_i are the parton distribution functions. The parameter $\tau_m = M_{min}^2/s$ where M_{min} corresponds to the minimum mass for a valid wormhole description. f_i are the Parton Distribution Functions (PDFs), (we suppress here transfer momenta). This formula is the same as the formula for black hole production, the difference being only in numerical factors.

The geometrical cross section of the wormhole production is

$$\sigma_{ij\to\text{wh}}(s) = \pi F(\sqrt{s}/M_D) r_0^2(\sqrt{s}, M_D) \,. \tag{17}$$

The form factor $F(\sqrt{s}/M_D)$ incorporates the theoretical uncertainties in description of the process, such as the amount of the initial center mass energy that goes into the wormhole, the distribution of wormhole masses as function of energy. These corrections are similar to corrections in the formula for black hole production.

Possible observable traces of mini-time-machines are discussed in [23].

Causality is the fundamental physical principle. In quantum field theory causality and the spacetime picture of the high energy scattering were considered in many papers. If there are spacetime regions with CTC (time machines) then causality is violated. We suggested [12] to test causality by using experiments at the LHC. We argued that if one can trust the classical geometrical estimate of the cross-section for the black hole production, if there exists an exotic matter similar to the dark energy, and if the scale of quantum gravity is of the order of few TeVs then one can expect the production of time machines/wormholes in the proton-proton collisions at the LHC of the same order as the cross section for the black hole production. This would leads to violation of the standard causality condition. Further studies of the experimental signatures of the wormhole production are required since there are transitions between black holes and wormholes [24].

It would be interesting to explore in some details the formula (2) for the transition amplitude between colliding quantum particles and black holes/wormholes which should be integrated with the wave function of the wormholes.

4. Models of Life

It was stressed by Ohya [25] that a new paradigm for the life science is needed. Here we shall try to consider a possible way to the new paradigm.

It is accepted that life is a characteristic of organisms that exhibit such properties as reproduction, metabolism, homeostasis, adaptation, complexity, self-organization, and growth [26,27]. If we try to make a formal model of life, then, from the point of view of the general modeling theory, see [28], one encounters here the same problems with the notion of a state of the system at a given moment of time which have been discussed in the analysis of financial markets [29].

The smallest unit of living organisms is the cell. It is a biological analogue of the elementary particle in physics. Cells have the following properties: reproduction by cell division; metabolism; cell contents are contained within a cell surface membrane; response to external and internal stimuli; use of enzymes and other proteins coded for by DNA genes and made via messenger RNA intermediates and ribosomes [30].

The living organism is not a machine in the sense of Turing machine, rather it requires a new paradigm similar to the model of computation suggested in [31] which goes beyond the standard model of computability.

Quantum field but not an elementary particle is the fundamental notion in modern physics. Creation and annihilation of particles or collective excitations is described in physics by using the framework of quantum field theory. It seems natural try to model the cell division also by means a kind of quantum field theory. If we accept this view then the basic notion in biology and in life science will be not genes or cells but a more abstract notion of a biological (quantum) field. Moreover this field should be nonlocal in time since a living organism evolves self-consistently. Therefore, mathematical description of biological systems might be similar to the field equations in spacetime with closed timelike curves (time machines), discussed in [12].

Acknowledgements

The work is partially supported by RFFI 08-01-00727-a and NSch-3224.2008.1.

References

1. K. Gödel, Rev. Mod. Phys. 21, 447 (1949).
2. S.W. Hawking and G.R.F. Ellis, *The large scale structure of space-time*, Cambridge University Press, 1973.

3. J.R. Gott, *Closed Timelike Curves Produced By Pairs Of Moving Cosmic Strings: Exact Solutions,* Phys. Rev. Lett. 66, 1126 (1991).
4. J.A. Wheeler, *Geometrodynamics*, (Academic Press, New York, 1962).
5. M.S. Morris and K.S. Thorne, Am.J.Phys. **56**, 395 (1988); M.S. Morris, K.S. Thorne and U. Yurtsever, Phys.Rev.Lett. **61**, 1446 (1989).
6. M. Visser, *Lorentzian Wormholes: From Einstein to Hawking*, AIP Press, USA, 1995.
7. S. Deser, R. Jackiw, G. 't Hooft, *Physical Cosmic Strings Do Not Generate Closed Timelike Curves*, Phys. Rev. Lett. 68, 267 (1992).
8. S.W. Hawking, *Quantum coherence and closed timelike curves*, Phys.Rev. D 52 (1995) 5681.
9. M. Visser, *The quantum physics of chronology protection*, gr-qc/0204022.
10. R.M. Wald, *The History and Present Status of Quantum Field Theory in Curved Spacetime* , gr-qc/0608018.
11. Th.M. Nieuwenhuizen, I.V. Volovich, *Role of Various Entropies in the Black Hole Information Loss Problem* ,in:Beyond the Quantum, World Sci., 2007, 135-145; arXiv:hep-th/0507272.
12. I.Ya. Aref'eva, I.V. Volovich, *Time Machine at the LHC*, Intern. J. of Geom. Methods in Modern Physics,Vol. 5, No. 4 (2008) 641–651; http://arxiv.org/abs/0710.2696.
13. K.S. Thorne, in: "Magic without Magic: John Archibald Wheeler", edited by J.Klauder (Freeman, San Francisco, 1972), p.231.
14. D. Amati, M. Ciafaloni and G. Veneziano, Phys.Lett. B197(1987)81.
15. G. 't Hooft, *Graviton dominance in ultra-high-energy scattering* , Phys.Lett. B198 (1987).
16. I.Ya. Arefeva, K.S. Viswanathan and I.V. Volovich, *Planckian energy scattering, colliding plane gravitational waves and black hole creation,* Nucl.Phys.B452:346-368,1995, *On black hole creation in Planckian energy scattering*, Int.J.Mod.Phys. D5 (1996) 707.
17. D. Giulini, C. Kiefer, *The Canonical Approach to Quantum Gravity: General Ideas and Geometrodynamics*, gr-qc/0611141.
18. N. Arkani-Hamed, S. Dimopoulos and G.R. Dvali, *The Hierarchy Problem and New Dimensions at a Millimeter*, Phys.Lett. B429 (1998) 263.
19. S.B. Giddings and S. Thomas, *High Energy Colliders as Black Hole Factories: The End of Short Distance Physics*, Phys.Rev. D65 (2002) 056010.
20. S. Dimopoulos and G. Landsberg, *Black Holes at the LHC*, Phys.Rev.Lett. 87 (2001) 161602.
21. I.Ya. Aref'eva, *Nonlocal String Tachyon as a Model for Cosmological Dark Energy,* AIP Conf. Proc. 826 (2006) 301.
22. I.Ya. Aref'eva, I.V. Volovich, *On the Null Energy Condition and Cosmology,* Theor.Math.Phys., 155 (2008) 3-12.
23. A. Mironov, A. Morozov, T.N. Tomaras, *If LHC is a Mini-Time-Machines Factory, Can We Notice?*, arXiv:0710.3395.
24. Sean A. Hayward, *Black holes and traversible wormholes: a synthesis*, arXiv:gr-qc/0203051.
25. M. Ohya, Talk at the QBIC 2008 Conference.

26. E. Schrodinger, *What is Life?*, Cambridge University Press, 1944.
27. L. Margulis, D. Sagan, *What is Life?*, University of California Press, 1995.
28. M.N. Khokhlova, I.V. Volovich, *On Modeling Theory and Hypergraph of Classes*, Proceedings of Steklov Mathematical Institute, 2004, vol. 245, p. 281-287.
29. Trelewicz J.Q., Volovich I.V. *Analysis of business connections utilizing theory of topology of random graphs.* In "p-Adic mathematical physics", Ed. A.Yu. Khrennikov, Z.Rakic, I.V. Volovich. AIP, v.826, Melville, New York, 2006, p. 330-344.
30. Maton Anthea et all, *Cells Building Blocks of Life* , New Jersey: Prentice Hall, 1997.
31. M. Ohya, I.V. Volovich, *Quantum computing and chaotic amplifier*, Journal of Optics B:Quantum Semiclass. 2003,v. 5, pp.459-473.

NOTE ON ENTROPIES IN QUANTUM PROCESSES

NOBORU WATANABE
Department of Information Sciences,
Tokyo University of Science
Noda City, Chiba, 278-8510, Japan
E-mail: watanabe@is.noda.tus.ac.jp

We treat the quantum processes of state changes, which can be expressed by quantum channels. In order to investigate the efficiency of information transmission under these quantum processes, three kind of mutual entropy-type measures are used. The quantum mutual entropy, which is called Ohya mutual entropy, for quantum input and output by using the quantum relative entropy was defined by Ohya in 1983. It denotes an amount of quantum information transmitted correctly from the input system to the output system through a quantum process. In this paper, we compare with mutual entropy-type measures for some quantum processes.

1. Introduction

The study of quantum information theory is closely connected with progress of quantum entropy theory. The states of the quantum input system are transmitted under quantum processes. These quantum processes are denoted by quantum channels carried from the quantum input states to the quantum output states, mathematically. In classical information theory, the mutual entropy describes the amount of information correctly sent from the initial system to the final system through the classical channels. It is defined by means of the joint probability distribution between the input and output systems. In quantum system, there does not exist the joint probability distribution in general [33]. Instead of the joint probability distribution, Ohya [17] introduced the quantum (Ohya) compound state, and he formulated the quantum mutual entropy (information), which is called Ohya mutual entropy for quantum input and output systems. Recently Shor [31] and Bennett et al [3,6] took the coherent entropy and defined the mutual entropy type measure to discuss a sort of coding theorem for communication processes.

In this paper, we compare with mutual entropy-type measures for some quantum processes.

2. Quantum Channels

The concept of channel has been fulfilled an important role in the development of the quantum information theory. We briefly explain the definition of the quantum channels.

Let $\mathcal{H}_1, \mathcal{H}_2$ be the complex separable Hilbert spaces of an input and an output systems, respectively, and let $\mathbf{B}(\mathcal{H}_k)$ be the set of all bounded linear operators on $\mathcal{H}_k$. The set of all density operators on $\mathcal{H}_k \quad (k = 1, 2)$ is denoted by

$$\mathfrak{S}(\mathcal{H}_k) \equiv \{\rho \in \mathbf{B}(\mathcal{H}_k); \rho \geq 0, tr\rho = 1\}. \tag{1}$$

A map Λ^* from the quantum input system to the quantum output system is called a **quantum channel**.

1. Λ^* is called a **linear channel** if it satisfies the affine property, i.e.,

$$\Lambda^*\left(\sum_k \lambda_k \rho_k\right) = \sum_k \lambda_k \Lambda^*(\rho_k),$$
$$\forall \rho_k \in \mathfrak{S}(\mathcal{H}_1), \quad \sum_k \lambda_k = 1 \quad (\forall \lambda_k \geq 0).$$

2. $\Lambda^* : \mathfrak{S}(\mathcal{H}_1) \to \mathfrak{S}(\mathcal{H}_2)$ is called a **completely positive (CP) channel** if its dual map Λ satisfies

$$\sum_{j,k=1}^{n} B_j^* \Lambda\left(A_j^* A_k\right) B_k \geq 0 \tag{2}$$

for any $n \in \mathbb{N}$, any $B_j \in \mathbf{B}(\mathcal{H}_1)$ and any $A_k \in \mathbf{B}(\mathcal{H}_2)$, where the map $\Lambda : \mathbf{B}(\mathcal{H}_2) \to \mathbf{B}(\mathcal{H}_1)$ of
$\Lambda^* : \mathfrak{S}(\mathcal{H}_1) \to \mathfrak{S}(\mathcal{H}_2)$ satisfies $tr\rho\Lambda(A) = tr\Lambda^*(\rho)A$ for any $\rho \in \mathfrak{S}(\mathcal{H}_1)$ and any $A \in \mathbf{B}(\mathcal{H}_2)$.

3. Quantum Mutual Entropy Type Measures

3.1. *Ohya Mutual Entropy and Capacity*

The quantum entropy was introduced by von Neumann around 1932 [16], which is defined by

$$S(\rho) \equiv -tr\rho\log\rho$$

for any density operators ρ in $S(\mathcal{H}_1)$. It denotes the amount of information of the quantum state ρ.

In order to define such a quantum mutual entropy, we need the quantum relative entropy and the joint state, which is called a compound state, describing the correlation between an input state ρ and the output state $\Lambda^*\rho$ through a channel Λ^*. For a state $\rho \in \mathfrak{S}(\mathcal{H}_1)$,

$$\rho = \Sigma_k \lambda_k E_k, \tag{3}$$

is called a Schatten decomposition [28] of ρ, where E_k is the one-dimensional orthogonal projection associated with λ_k. The Schatten decomposition is not unique usually depending on a degeneracy of the eigenvalue of ρ.For $\rho \in \mathfrak{S}(\mathcal{H}_1)$ and $\Lambda^* : \mathfrak{S}(\mathcal{H}_1) \to \mathfrak{S}(\mathcal{H}_2)$, the compound states are define by

$$\sigma_E = \sum_n \lambda_n E_n \otimes \Lambda^* E_n, \qquad \sigma_0 = \varphi \otimes \Lambda^* \varphi. \tag{4}$$

The first compound state is called a Ohya compound state associating with the Schatten decomposition $\rho = \Sigma_k \lambda_k E_k$, which generalizes the joint probability in classical dynamical system and it shows a certain correlation between the initial state ρ and the final state $\Lambda^*\rho$.

Ohya mutual entropy with respect to ρ and Λ^* is defined by

$$I(\rho; \Lambda^*) \equiv \sup \{S(\sigma_E, \sigma_0); E = \{E_n\}\}, \tag{5}$$

where $S(\sigma_E, \sigma_0)$ is Umegaki's relative entropy [32]. $I(\rho; \Lambda^*)$ satisfies the Shannon's type inequality :

$$0 \leq I(\rho, \Lambda^*) \leq \min \{S(\rho), S(\Lambda^*\rho)\}.$$

The capacity means the ability of the information transmission of the channel, which is used as a measure for construction of channels. The quantum capacity is formulated by taking the supremum of the Ohya mutual entropy with respect to a certain subset of the initial state space. The capacity of quantum channel was studied in [20,21,22,23].

Let $\mathcal{S}$ be the set of all input states satisfying some physical conditions. Let us consider the ability of information transmission for the quantum channel Λ^*. The answer of this question is the capacity of quantum channel Λ^* for a certain set $\mathcal{S} \subset S(\mathcal{H}_1)$ defined by

$$C_q^{\mathcal{S}}(\Lambda^*) \equiv \sup \{I(\rho; \Lambda^*); \rho \in \mathcal{S}\}. \tag{6}$$

When $\mathcal{S} = S(\mathcal{H}_1)$, the capacity of quantum channel Λ^* is denoted by $C_q(\Lambda^*)$.

3.2. *Coherent Entropy and Lindblad-Nielsen Entropy*

Recently Shor [31] and Bennett et al [3,6] took the coherent entropy and defined the mutual type entropy to discuss a sort of coding theorem for quantum communication. In this section, we compare these mutual types entropy.

Let us discuss the entropy exchange [29]. For a stateρ, a channel Λ^* is denoted by using an operator valued measure $\{A_j\}$ such as

$$\Lambda^* (\cdot) \equiv \sum\nolimits_j A_j^* \cdot A_j, \tag{7}$$

which is called a Stinespring-Sudarshan-Kraus form. Then one can define a matrix $W = (W_{ij})_{i,j}$ with

$$W_{ij} \equiv tr A_i^* \rho A_j, \tag{8}$$

by which the entropy exchange is defined by

$$S_e (\rho, \Lambda^*) = -tr W \log W. \tag{9}$$

By using the entropy exchange, two mutual type entropies are defined as follows:

$$I_C (\rho; \Lambda^*) \equiv S (\Lambda^* \rho) - S_e (\rho, \Lambda^*), \tag{10}$$

$$I_L (\rho; \Lambda^*) \equiv S (\rho) + S (\Lambda^* \rho) - S_e (\rho, \Lambda^*). \tag{11}$$

The first one is called the coherent entropy $I_C (\rho; \Lambda^*)$ [30] and the second one is called the Lindblad entropy $I_L (\rho; \Lambda^*)$ [6].

4. Quantum Measurement Process

One of the simple example of the quantum measurement process is described by using the PVM (projection valued measure) $\{A_j\}$ with $\dim A_j = 1$. The quantum channel Λ^* of the quantum measurement process with the PVM $\{A_j\}$ is defined by

$$\Lambda^* (\rho) \equiv \sum\nolimits_j A_j \rho A_j^*$$

for any on $\rho \in \mathfrak{S}(\mathcal{H})$. By comparing these mutual entropy type measures for this quantum measurement process, we have the following theorem [25,26]:

Theorem 4.1. *Let $\{A_j\}$ be a projection valued measure with $dim A_j = 1$. For arbitrary state ρ in $\mathfrak{S}(\mathcal{H})$ and the quantum channel $\Lambda^* (\cdot) \equiv \sum_j A_j \cdot A_j^*$, one has*

(1) $0 \leq I (\rho; \Lambda^*) \leq \min \{S (\rho), S (\Lambda^* \rho)\}$ *(Ohya mutual entropy),*
(2) $I_C (\rho; \Lambda^*) = 0$ *(coherent entropy),*
(3) $I_L (\rho; \Lambda^*) = S (\rho)$ *(Lindblad entropy).*

5. Quantum Communication Process

Let $\mathcal{K}_1$ and $\mathcal{K}_2$ be two Hilbert spaces expressing noise and loss systems, respectively. Quantum communication process including the influence of noise and loss is described by the quantum channel [17]:

$$\Lambda^* (\rho) \equiv tr_{\mathcal{K}_2}\pi^* (\rho \otimes \xi)$$

for any input state ρ in $\mathfrak{S}(\mathcal{H}_1)$ and a noise state ξ in $\mathfrak{S}(\mathcal{K}_1)$, where the map π^* is a CP channel from $\mathfrak{S}(\mathcal{H}_1 \otimes \mathcal{K}_1)$ to $\mathfrak{S}(\mathcal{H}_2 \otimes \mathcal{K}_2)$ determined by physical properties of the communication device.

5.1. *Attenuation Process*

In particular, an attenuation channel introduced in [17] is one of the most important one for discussing the information transmission in quantum optical communication. One of the example of the quantum communication channels is the attenuation channel Λ_0^* introduced by Ohya [17], which is defined by

$$\Lambda_0^*(\rho) \equiv tr_{\mathcal{K}_2}\pi_0^* (\rho \otimes \xi_0), \quad \xi_0 \equiv |0\rangle \langle 0| \text{ and } \pi_0^* (\cdot) \equiv V_0 (\cdot) V_0^*, \tag{12}$$

where $|0\rangle \langle 0|$ is vacuum state in $\mathcal{H}_1$ and V_0 is a linear mapping from $\mathcal{H}_1 \otimes \mathcal{K}_1$ to $\mathcal{H}_2 \otimes \mathcal{K}_2$ given by

$$V_0 (|n\rangle \otimes |0\rangle) \equiv \sum_{j=0}^{n} \sqrt{\frac{n!}{j!(n-j)!}} \alpha^j \bar{\beta}^{n-j} |j\rangle \otimes |n-j\rangle \tag{13}$$

for any $|n\rangle$ in $\mathcal{H}_1$ and α,β are complex numbers satisfying $|\alpha|^2 + |\beta|^2 = 1$. $\eta = |\alpha|^2$, which is the transmission rate of the channel. π_0^* is called a beam splittings, which means that one beam comes and two beams appear after passing through π_0^*. The attenuation channel is generalized by the noisy optical channel [23,24], which is also reformulated by Accardi and Ohya [1] using the liftings. The noisy optical channel consists of the generalized beam splittings π^*, which was extended on generalized Fock space by Fichtner, Freudenberg and Libsher [9] by means of the concept of compound Hida-Malliavin derivative [12,13] and so on.

For the attenuation channel Λ_0^*, one can obtain the following theorems [25]:

Theorem 5.1. *For any state* $\rho = \sum_n \lambda_n |n\rangle \langle n|$ *and the attenuation channel* Λ_0^* *with* $|\alpha|^2 = |\beta|^2 = \frac{1}{2}$, *one has*

(1) $0 \leq I(\rho; \Lambda_0^*) \leq \min\{S(\rho), S(\Lambda_0^*\rho)\}$ *(Ohya mutual entropy),*

(2) $I_C(\rho; \Lambda_0^*) = 0$ *(coherent entropy),*
(3) $I_L(\rho; \Lambda_0^*) = S(\rho)$ *(Lindblad entropy).*

One can conclude that the Ohya mutual entropy is one of the most suitable measures to discuss the information transmission in quantum communication processes instead of the classical mutual entropy.

For the attenuation channel Λ_0^*, one can obtain the following theorem proved in [25].

Theorem 5.2. *For a subset $\mathcal{S}_n \equiv \{\rho \in S(\mathcal{H}_1); \dim s(\rho) = n\}$, the capacity of the attenuation channel Λ_0^* satisfies*

$$C_q^{\mathcal{S}_n}(\Lambda_0^*) = \log n,$$

where $s(\rho)$ is the support projection of ρ.

When the mean energy of the input state vectors $\{|\tau\theta_k\rangle\}$ can be taken infinite, i.e., $\lim_{\tau\to\infty} |\tau\theta_k|^2 = \infty$, the above theorem tells that the quantum capacity for the attenuation channel Λ_0^* with respect to $\mathcal{S}_n$ becomes $\log n$. It is a natural result, however it is impossible to take the mean energy of input state vector infinite.

6. Quantum Teleportation Process

In teleportation scheme, a particle is not transmitted from Alice's system to Bob's system, but one can reconstruct the initial state by means of the entangled state located at Bob's system. Bennett et al. [4,5,27] proposed a state change, so-called a quantum teleportation process, in terms of EPR entangled state denoted by Bell's base. It is difficult to realize such teleportation scheme because the EPR entangled state dissipates easily. In order to avoid this demerit, Ohya and Fichtner [7,8,27] introduced a new teleportation scheme on boson-Fock space by means of the entangled coherent states and the general beam splitting. The teleportation scheme of general quantum states was discussed in [2,15].

In perfect quantum teleportation scheme, the total process of the quantum teleportation is consist of three systems denoted by the complex Hilbert spaces $\mathcal{H}_1, \mathcal{H}_2, \mathcal{H}_3$. Alice controls $\mathcal{H}_1$,$\mathcal{H}_2$ and Bob treats $\mathcal{H}_3$. Alice has an unknown initial state $\rho^{(1)} \in \mathfrak{S}(\mathcal{H}_1)$ and she teleports it to Bob by using an entangled state $\sigma^{(23)} \in \mathfrak{S}(\mathcal{H}_2 \otimes \mathcal{H}_3)$ belonging to Alice and Bob. At first, Alice measures for the part of $\mathcal{H}_1 \otimes \mathcal{H}_2$ in the state $\rho^{(1)} \otimes \sigma^{(23)}$ by means of an observable $F^{(12)} \equiv \sum_{l,m} z_{lm} F_{lm}^{(12)} \in \mathbb{B}(\mathcal{H}_1 \otimes \mathcal{H}_2)$, where $\left\{F_{lm}^{(12)}\right\}$ is a set of orthogonal projections $F_{lm}^{(12)} \in \mathbb{B}(\mathcal{H}_1 \otimes \mathcal{H}_2)$. If Alice obtains a value z_{lm} after measurement, then the state $\rho^{(1)} \otimes \sigma^{(23)}$ is changed to

$$\rho_{lm}^{(123)} = \frac{\left(F_{lm}^{(12)} \otimes I_3\right)\left(\rho^{(1)} \otimes \sigma^{(23)}\right)\left(F_{lm}^{(12)} \otimes I_3\right)}{tr_{123}\left(F_{lm}^{(12)} \otimes I_3\right)\left(\rho^{(1)} \otimes \sigma^{(23)}\right)\left(F_{lm}^{(12)} \otimes I_3\right)}. \tag{14}$$

Alice send the result z_{lm} of the measurement to Bob through a classical communication channel. Bob reconstructs the unknown initial state $\rho^{(1)}$ from $\Lambda_{lm}^{*}\left(\rho^{(1)}\right) \equiv tr_{12}\rho_{lm}^{(123)} \in \mathfrak{S}(\mathcal{H}_3)$ by applying a unitary key U_{lm} created by the value z_{lm} received from Alice. The total process of the perfect teleportation scheme is described by

$$\rho^{(1)} \equiv U_{lm}\left(\Lambda_{lm}^{*}\left(\rho^{(1)}\right)\right)U_{lm}^{*}. \tag{15}$$

In the perfect teleportation scheme, Bob can reconstruct the unknown initial state $\rho^{(1)}$ by applying the unitary key once. On the contrary, in the non-perfect teleportation scheme, Bob can obtain the unknown initial state $\rho^{(1)}$ by applying the unitary operator V_{nm} more than twice. Applying the unitary operator V_{nm} once to the state $\Lambda_{nm}^{*}\left(\rho^{(1)}\right)$, the state change of the non-perfect teleportation scheme is denoted by

$$\Xi_{nm}^{*}\left(\rho^{(1)}\right) \equiv V_{nm}\Lambda_{nm}^{*}\left(\rho^{(1)}\right)V_{nm}^{*}. \tag{16}$$

For the perfect teleportation, we have the following theorem [14]:

Theorem 6.1. *If the quantum channel Λ_T^{*} describing the perfect teleportation process is linear and the input state $\rho^{(1)}$ is finite rank operator, then one can obtain*

$$I\left(\rho^{(1)}, \Lambda_T^{*}\right) = S\left(\rho^{(1)}\right).$$

7. Quantum Computational Process

Fredkin and Toffoli proposed a conservative gate consisted of three inputs and three outputs, which was developed by Milburn as a quantum gate with quantum input and output. It is called the Fredkin-Toffoli-Milburn (FTM) gate. Recently, FTM gate was reformulated by a quantum channel and the conservation of information for the FTM gate was rigorously studied in [26].

The FTM gate is composed of two input gates I_1, I_2 and one control gate C, which is described by two beam splitters and the optical Kerr medium as follows:

(1) Beam splitters: **(a)** Based on [22], let V_1 be a mapping from $\mathcal{H}_1 \otimes \mathcal{H}_2$ to $\mathcal{H}_1 \otimes \mathcal{H}_2$ with transmission rate η_1 given by

$$V_1\left(|n_1\rangle \otimes |n_2\rangle\right) \equiv \sum_{j=0}^{n_1+n_2} C_j^{n_1,n_2} |j\rangle \otimes |n_1+n_2-j\rangle \tag{17}$$

for any photon number state vectors $|n_1\rangle \otimes |n_2\rangle \in \mathcal{H}_1 \otimes \mathcal{H}_2$. The quantum channel Π_{BS1}^* expressing the first beam splitter (beam splitter 1) is defined by

$$\Pi_{BS1}^*\left(\rho_1 \otimes \rho_2\right) \equiv V_1\left(\rho_1 \otimes \rho_2\right) V_1^* \tag{18}$$

for any states $\rho_1 \otimes \rho_2 \in \mathfrak{S}(\mathcal{H}_1 \otimes \mathcal{H}_2)$. In particular, for an input state in two gates I_1 and I_2 given by the tensor product of two coherent states $\rho_1 \otimes \rho_2 = |\theta_1\rangle\langle\theta_1| \otimes |\theta_2\rangle\langle\theta_2|$, $\Pi_{BS1}^*(\rho_1 \otimes \rho_2)$ is written as

$$\begin{aligned} &\Pi_{BS1}^*(\rho_1 \otimes \rho_2) \\ &= \left|\sqrt{\eta_1}\theta_1 + \sqrt{1-\eta_1}\theta_2\right\rangle \left\langle \sqrt{\eta_1}\theta_1 + \sqrt{1-\eta_1}\theta_2\right| \\ &\quad \otimes \left|-\sqrt{1-\eta_1}\theta_1 + \sqrt{\eta_1}\theta_2\right\rangle \left\langle -\sqrt{1-\eta_1}\theta_1 + \sqrt{\eta_1}\theta_2\right|. \end{aligned} \tag{19}$$

(b) Let V_2 be a mapping from $\mathcal{H}_1 \otimes \mathcal{H}_2$ to $\mathcal{H}_1 \otimes \mathcal{H}_2$ with transmission rate η_2 given by

$$V_2\left(|n_1\rangle \otimes |n_2\rangle\right) \equiv \sum_{j=0}^{n_1+n_2} C_j^{n_2,n_1} |n_1+n_2-j\rangle \otimes |j\rangle \tag{20}$$

for any photon number state vectors $|n_1\rangle \otimes |n_2\rangle \in \mathcal{H}_1 \otimes \mathcal{H}_2$. The quantum channel Π_{BS2}^* expressing the second beam splitter (beam splitter 2) is defined by

$$\Pi_{BS2}^*\left(\rho_1 \otimes \rho_2\right) \equiv V_2\left(\rho_1 \otimes \rho_2\right) V_2^* \tag{21}$$

for any states $\rho_1 \otimes \rho_2 \in \mathfrak{S}(\mathcal{H}_1 \otimes \mathcal{H}_2)$. In particular, for coherent input states $\rho_1 \otimes \rho_2 = |\theta_1\rangle\langle\theta_1| \otimes |\theta_2\rangle\langle\theta_2|$, $\Pi_{BS2}^*(\rho_1 \otimes \rho_2)$ is written as

$$\begin{aligned} \Pi_{BS2}^*(\rho_1 \otimes \rho_2) &= \left|\sqrt{\eta_2}\theta_1 - \sqrt{1-\eta_2}\theta_2\right\rangle \left\langle \sqrt{\eta_2}\theta_1 - \sqrt{1-\eta_2}\theta_2\right| \\ &\quad \otimes \left|\sqrt{1-\eta_2}\theta_1 + \sqrt{\eta_2}\theta_2\right\rangle \left\langle \sqrt{1-\eta_2}\theta_1 + \sqrt{\eta_2}\theta_2\right|. \end{aligned} \tag{22}$$

(2) Optical Kerr medium: The interaction Hamiltonian in the optical Kerr medium is given by the number operators N_1 and N_c for the input system 1 and the Kerr medium, respectively, such as

$$H_{int} = \hbar\chi \left(N_1 \otimes I_2 \otimes N_c\right), \tag{23}$$

where $\hbar$ is the Plank constant divided by 2π, χ is a constant proportional to the susceptibility of the Kerr medium and I_2 is the identity operator on $\mathcal{H}_2$. Let T be the passing time of a beam through the Kerr medium and put $\sqrt{F} = \hbar\chi T$, a parameter exhibiting the power of the Kerr effect. Then the unitary operator U_K describing the evolution for time T in the Kerr medium is given by

$$U_K = \exp\left(-i\sqrt{F}\left(N_1 \otimes I_2 \otimes N_c\right)\right). \tag{24}$$

We assume that an initial (input) state of the control gate is the n photon number state $\xi = |n\rangle\langle n|$, a quantum channel Λ_K^* representing the optical Kerr effect is given by

$$\Lambda_K^*(\rho_1 \otimes \rho_2 \otimes \xi) \equiv U_K(\rho_1 \otimes \rho_2 \otimes \xi)U_K^* \tag{25}$$

for any state $\rho_1 \otimes \rho_2 \otimes \xi \in \mathfrak{S}\left(\mathcal{H}_1 \otimes \mathcal{H}_2 \otimes \mathcal{K}\right)$. In particular, for an initial state $\rho_1 \otimes \rho_2 \otimes \xi = |\theta_1\rangle\langle\theta_1| \otimes |\theta_2\rangle\langle\theta_2| \otimes |n\rangle\langle n|$, $\Lambda_K^*(\rho_1 \otimes \rho_2 \otimes \xi)$ is denoted by

$$\begin{aligned}
&\Lambda_K^*(\rho_1 \otimes \rho_2 \otimes \xi) \\
&= \left| \exp\left(-i\sqrt{F}n\right)\theta_1 \right\rangle \left\langle \exp\left(-i\sqrt{F}n\right)\theta_1 \right| \\
&\quad \otimes |\theta_2\rangle\langle\theta_2| \otimes |n\rangle\langle n|,
\end{aligned} \tag{26}$$

Using the above channels, the quantum channel for the whole FTM gate is constructed as follows: Let both one input and output gates be described by $\mathcal{H}_1$, another input and output gates be described by $\mathcal{H}_2$ and the control gate be done by $\mathcal{K}$, all of which are Fock spaces. For a total state $\rho_1 \otimes \rho_2 \otimes \xi$ of two input states and a control state, the quantum channels $\Lambda_{BS1}^*, \Lambda_{BS2}^*$ from $\mathfrak{S}(\mathcal{H}_1 \otimes \mathcal{H}_2 \otimes \mathcal{K})$ to $\mathfrak{S}(\mathcal{H}_1 \otimes \mathcal{H}_2 \otimes \mathcal{K})$ are written by

$$\Lambda_{BSk}^*(\rho_1 \otimes \rho_2 \otimes \xi) = \Pi_{BSk}^*(\rho_1 \otimes \rho_2) \otimes \xi \quad (k = 1, 2) \tag{27}$$

Therefore, the whole quantum channel Λ_{FTM}^* of the FTM gate is defined by

$$\Lambda_{\mathrm{FTM}}^* \equiv \Lambda_{BS2}^* \circ \Lambda_K^* \circ \Lambda_{BS1}^*. \tag{28}$$

In particular, for an initial state $\rho_1 \otimes \rho_2 \otimes \xi = |\theta_1\rangle\langle\theta_1| \otimes |\theta_2\rangle\langle\theta_2| \otimes |n\rangle\langle n|$, $\Lambda_{\mathrm{FTM}}^*(\rho_1 \otimes \rho_2 \otimes \xi)$ is obtained by

$$
\begin{aligned}
&\Lambda^*_{\mathrm{FTM}}(\rho_1 \otimes \rho_2 \otimes \xi) \\
&= |\mu_n\theta_1 + \nu_n\theta_2\rangle \langle \mu_n\theta_1 + \nu_n\theta_2| \\
&\quad \otimes |\nu_n\theta_1 + \mu_n\theta_2\rangle \langle \nu_n\theta_1 + \mu_n\theta_2| \otimes |n\rangle \langle n| \qquad (29)
\end{aligned}
$$

where

$$\mu_n = \frac{1}{2}\left\{\exp\left(-i\sqrt{F}n\right) + 1\right\}, \qquad (30)$$

$$\nu_n = \frac{1}{2}\left\{\exp\left(-i\sqrt{F}n\right) - 1\right\}, \quad (k = 0, 1, 2, \cdots). \qquad (31)$$

Then we obtain the following theorem:

Theorem 7.1.
If $\sqrt{F}$ *satisfies the conditions* $\sqrt{F}n = 0$ *or* $\sqrt{F}n = (2m+1)\pi$ $(m = 0, 1, 2, \cdots)$, *then one can obtain*

$$I\left(\rho_1 \otimes \rho_2, \widetilde{\Lambda}^*_{FTM}\right) = S(\rho_1) + S(\rho_2)$$

for input states $\rho_1 \otimes \rho_2$ *given by*

$$\rho_i = \lambda_i |0\rangle \langle 0| + (1 - \lambda_i) |x_i\rangle \langle x_i| \in \mathfrak{S}(\mathcal{H}_i), \quad \lambda_i \in [0, 1], \quad (i = 1, 2),$$

where $|x_i\rangle$ *is defined by*

$$|x_i\rangle = \frac{|\theta_i\rangle - |-\theta_i\rangle}{\sqrt{2\left(1 - \exp\left(-2|\theta_i|^2\right)\right)}}.$$

References

1. Accardi, L., and Ohya, M., *Compound channels, transition expectation and liftings*, Appl. Math, Optim., **39**, 33-59 (1999).
2. Accardi, L., and Ohya, M., *Teleportation of general quantum states*, Quantum Information, T. Hida, K. Saito (eds.) World Scientific, 59, (1999).
3. Barnum, H., Nielsen, M.A., and Schumacher, B.W., *Information transmission through a noisy quantum channel*, Physical Review A, **57**, No.6, 4153-4175 (1998).
4. C.H. Bennett, G. Brassard, C. Crepeau, R. Jozsa, A. Peres, and W. K.Wootters, *Teleporting an unknown quantum state via dual classical and EPR channels*, Physical Review Letters, **70**, 1895 (1993).
5. C.H. Bennett, G. Brassard, S. Popescu, B. Schumacher, J.A. Smolin,W.K. Wootters, *Purification of noisy entanglement and faithful teleportation via noisy channels*, Physical Review Letters, **76**, 722 (1996).

6. Bennett, C.H., Shor, P.W., Smolin, J.A., and Thapliyalz, A.V., *Entanglement-assisted capacity of a quantum channel and the reverse Shannon theorem*, quant-ph/0106052.
7. K-H. Fichtner and M. Ohya, *Quantum teleportation with entangled states given by beam splittings*, Communications in Mathematical Physics, **222**, 229 (2001).
8. K-H. Fichtner and M. Ohya, *Quantum teleportation and beam splitting*, Communications in Mathematical Physics, **225**, 67 (2002).
9. Fichtner, K.H., Freudenberg, W., and Liebscher, V., *Beam splittings and time evolutions of Boson systems*, Fakultat fur Mathematik und Informatik, Math/ Inf/96/ **39**, Jena, 105 (1996).
10. Fredkin, E., and Toffoli T., *Conservative logic*, International Journal of Theoretical Physics , **21** , 219-253, (1982).
11. Milburn G.J., *Quantum optical Fredkin gate*, Physical Review Letters , **62**, 2124-2127, (1989).
12. Hida, T., Analysis of Brownian Functionals, Carleton Mathematical Lecture Notes, **13**, (1975).
13. Hida, T., Selected Papers, World Scientific, (2001).
14. K. Inoue, H. Suyari and M. Ohya, *Characterization of quantum teleportation by nonlinear quantum channel and quantum mutual entropy*, Physical D. ,**120**, 117 (1998).
15. Kossakowski, A., and M.Ohya, M., *New scheme of quantum teleportation*, to appear in Infinite Dimensional Analysis Quantum Probability and Related Topics.
16. von Neumann, J., Die Mathematischen Grundlagen der Quantenmechanik, Springer-Berlin, (1932).
17. Ohya, M., *On compound state and mutual information in quantum information theory*, IEEE Trans. Information Theory, **29**, 770-774 (1983).
18. Ohya, M., *Some aspects of quantum information theory and their applications to irreversible processes*, Rep. Math. Phys., **27**, 19-47 (1989).
19. Ohya, M., and Petz, D., Quantum Entropy and its Use, Springer, Berlin, (1993).
20. Ohya, M., Petz, D., and Watanabe, N., *On capacity of quantum channels*, Probability and Mathematical Statistics, **17**, 179-196 (1997).
21. Ohya, M., Petz, D., and Watanabe, N., *Numerical computation of quantum capacity*, International Journal of Theoretical Physics, **37**, No.1, 507-510 (1998).
22. Ohya, M., and Watanabe, N., *Quantum capacity of noisy quantum channel*, Quantum Communication and Measurement, **3**, 213-220 (1997).
23. Ohya, M., and Watanabe, N., Foundation of Quantum Communication Theory (in Japanese), Makino Pub. Co., (1998).
24. Ohya, M., and Watanabe, N., *Construction and analysis of a mathematical model in quantum communication processes*, Electronics and Communications in Japan, Part 1, **68**, No.2, 29-34 (1985).
25. Ohya, M., and Watanabe, N., *Comparison of mutual entropy - type measures*, TUS preprint.

26. Ohya M., and Watanabe N., *On mathematical treatment of optical Fredkin - Toffoli - Milburn gate*, Physica D, **120**, 206-213, (1998).
27. Ohya, M., and Watanabe, N., Quantum Cryptography and Quantum Teleportation(in Japanese), Kyoritsu Pub. Co., (2006).
28. Schatten, R., Norm Ideals of Completely Continuous Operators, Springer-Verlag, (1970).
29. Schumacher, B.W., *Sending entanglement through noisy quantum channels*, Physical Review A, **54**, 2614 (1996).
30. Schumacher, B.W., and Nielsen, M.A., *Quantum data processing and error correction*, Physical Review A, **54**, 2629 (1996).
31. Shor, P., The quantum channel capacity and coherent information, Lecture Notes, MSRI Workshop on Quantum Computation, (2002).
32. Umegaki, H., *Conditional expectations in an operator algebra IV (entropy and information)*, Kodai Math. Sem. Rep., **14**, 59-85 (1962).
33. Urbanik, K., *Joint probability distribution of observables in quantum mechanics*, Studia Mathematica, **21**, 117-133, (1961).

BASICS OF MOLECULAR SIMULATION AND ITS APPLICATION TO BIOMOLECULES

TADASHI ANDO

Department of Biological Science and Technology, Tokyo University of Science 2641 Yamazaki, Noda, Chiba, 278-8510, Japan

ICHIRO YAMATO

Department of Biological Science and Technology, Tokyo University of Science 2641 Yamazaki, Noda, Chiba, 278-8510, Japan

Molecular simulation provides useful information from microscopic view for understanding the properties of molecules. Now molecular dynamics (MD) simulations of large biomolecules can be performed easily on parallel computers. However, accuracy of force field parameters and simulation time scale remain as hard obstacles to biomolecular simulation. In this paper, we will briefly describe basics of MD simulation for biomolecules and discuss the problems of the method. Then, we introduce protein folding problem as a grand challenge of biomolecular simulation. Finally, we will present our approach to overcome the difficulties, a Brownian dynamics, which is able to simulate long-time folding dynamics of some peptides.

1. Introduction

Molecular simulations play an important role for understanding the properties of molecules or groups of molecules. They provide critical information concerning the structures, dynamics, and interaction energies of interesting molecules at atomic resolution, which complements to experimental data. Simulation, sometimes, makes it possible to find out a new fact or law and predict new phenomena. Today, owing to the progressive advance of simulation algorithm and computational facilities, simulating bio-macromolecules, such as proteins, nucleic acids, lipids, and poly-saccharides, are possible. Consequently, molecular simulation is getting to be an indispensable tool for biological research.

Now, there are two main techniques to simulate behaviors of molecules in a computer: Molecular Dynamics (MD) and Monte Carlo (MC). Here, we shall first introduce basics of MD simulation techniques for biomolecules briefly. We will not mention about MC method in this paper. And, there are several good textbooks for molecular simulations of biomolecules [1-3]. Please see these books for details, if necessary. Secondary, the protein folding problem as an example of historic topics in the field of computational biology will be explained. Finally, we will describe our approach to simulate protein folding process, Brownian dynamics simulation.

2. Molecular Dynamics Simulation

2.1. *Governing Equation of Molecular Dynamics*

In MD simulation, we simulate motions of atoms as a function of time according to the Newton's equations of motion. The equations for a system consisting of N atoms can be written as

$$m_i \frac{\mathrm{d}^2 \mathbf{r}_i(t)}{\mathrm{d}t^2} = \mathbf{F}_i(t), \qquad (i = 1, 2, \cdots, N). \tag{1}$$

Here, $\mathbf{r}_i$ and m_i represent the position and mass of atom i and $\mathbf{F}_i(t)$ is the force on atom i at time t. $\mathbf{F}_i(t)$ is given by

$$\mathbf{F}_i = -\nabla_i V(\mathbf{r}_1, \mathbf{r}_2, \cdots, \mathbf{r}_N), \tag{2}$$

where $V(\mathbf{r}^1, \mathbf{r}^2, \cdots, \mathbf{r}^N)$ is the potential energy of the system that depends on the positions of the N atoms in the system.

As described above, MD is based on the classical Newton's equations of motion under the Born-Oppenheimer approximation so that it is impossible to describe the dynamics concomitant with electron motions like bond-breaking or bond-creating reactions.

2.2. *Potential Functions for Biomoleules*

Typically, potential energy, V_{total}, of any given conformation of biomolecules is described as the sum of five interaction terms:

$$\begin{aligned} V_{\text{total}} = & \sum_{\text{bonds}} K_b (r - r_0)^2 + \sum_{\text{angles}} K_\theta (\theta - \theta_0)^2 + \sum_{\text{dihedrals}} K_\phi [1 + \cos(n\varphi - \gamma)] \\ & + \sum_{\substack{\text{van der Waals} \\ i,\, j \text{ pairs}}} \left(\frac{A_{ij}}{r_{ij}^{12}} - \frac{B_{ij}}{r_{ij}^{6}} \right) + \sum_{\substack{\text{electrostatic} \\ i,\, j \text{ pairs}}} \frac{q_i q_j}{\varepsilon r_{ij}}. \end{aligned} \tag{3}$$

Schematic view of each interaction is shown in Fig. 1. The first term is bond-stretching energies. Here K_b, r, and r_0 are the force constant, the bond length, and the equilibrium bond length, respectively. In this term, all covalent bonds of molecules are considered. The second term is angle-bending energies examined for the atom pairs connected with two covalent bonds. Here K_θ, θ, and θ_0 are the force constant, the bond angle, and the equilibrium bond angle, respectively. The third term is torsion or dihedral energies. Here K_φ, φ, n, and γ are the force constant, the dihedral angle, the multiplicity, and phase, respectively. The dihedral angle is determined by four connected-atoms. The fifth term is van der Waals interactions described by Lennard-Jones potential form. Here r_{ij} is the distance between atoms i and j. A_{ij} and B_{ij} are force constants that depend on atom types i and j. The last term is electrostatic interactions expressed by the Coulomb's law. Here q_i and q_j are atomic charges of atoms i and j, respectively, and ε is dielectric constant. Usually the last two terms are called non-bonded interactions since all pairs separated by more than three covalent bonds are considered. On the other hand, first three terms are called bonded interactions. Additionally, although biomolecules adopt various conformations, these parameters are constant throughout the MD simulation.

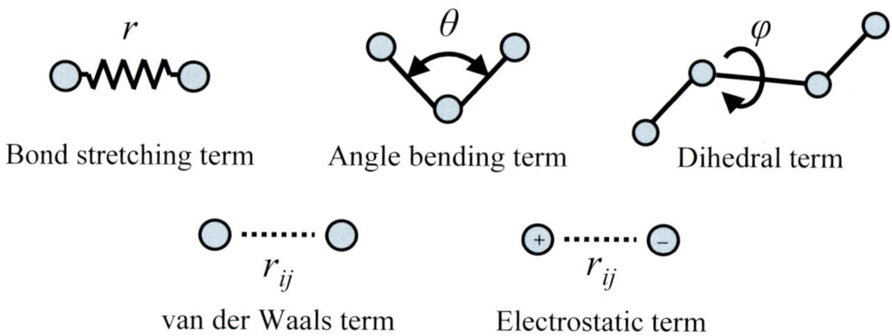

Figure 1. Schematic view of each energy function term used in Eq. 4. Solid and dashed lines represent covalent and non-covalent bonds, respectively. For bond stretching term, covalent bond is shown by spring to emphasize the ability of stretching motion.

In energy calculation, the number of pairs examined in the non-bonded interaction terms grows quadratically with number of atoms, $O(N^2)$, though those in the bonded interaction terms increase only linearly with number of atoms, $O(N)$. For van der Waals interaction, cut-off method is usually used to save computational power. On the other hand, since electrostatic energy decays with the inverse of the distance, truncation of these long-range interactions in MD simulation will strongly affect dynamics of the system. Today, Particle Mesh Ewald (PME) [4] and Fast Multipole Method (FMM) [5] are usually used for calculating electrostatic energy efficiently without truncation, where the complexity of non-bonded energy calculation decreases from $O(N^2)$ to $O(N\log N)$ and $O(N)$ for large system, respectively.

The potential function and force field parameters described above, such as the equilibrium bond length, force constants, atomic charges, and so on, are empirical. Now, we can easily use some lines of force fields, such as AMBER [6], CHARMm [7], GROMOS [8], and OPLS [9], where they have almost the same function form as Eq. 3 and parameters are typically determined by quantum mechanical calculations of small molecules combined with experimental data.

2.3. *Integration Algorithm*

One of the most widely used algorithms for integrating the equations of motion in MD is the velocity Verlet algorithm [10]:

$$\mathbf{r}_i(t+\Delta t)=\mathbf{r}_i(t)+\mathbf{v}_i(t)\Delta t+\frac{\Delta t^2}{2m_i}\mathbf{F}_i(t) \tag{4a}$$

$$\mathbf{v}_i(t+\Delta t)=\mathbf{v}_i(t)+\frac{\Delta t}{2m_i}\{\mathbf{F}_i(t)+\mathbf{F}_i(t+\Delta t)\}. \tag{4b}$$

The method is symplectic and time-reversible, and can be derived from a general Trotter factorization associated with the Liouville operator [11]. Furthermore, the method is easily applied to the multiple time step algorithm where fast varying bonding forces are calculated more frequently that the slow varying non-bonding forces to reduce computation time [11].

In MD simulation for biomolecules, time step is limited up to ~1 femtoseconds. This restriction comes from the fact that this short time step is necessary to simulate the fastest motion in the system such as O-H vibration whose period is about 10 femtoseconds. Now, bonds constraining method, SHAKE[12], RATTLE[13], LINCS[14], are usually used. However time step of 2 femtoseconds is upper limit for stable simulation of biomoleules even using these constraining methods.

2.4. *Simulation Conditions*

Biomolecules work in aqueous solution. Therefore, treating a lot of water molecules around a solute is essential for biomolecular simulations. Typically, biomolecules are soaked in a solvent box and simulation is performed in periodic boundary condition to minimize effects arising from the atoms at surface (Fig. 2).

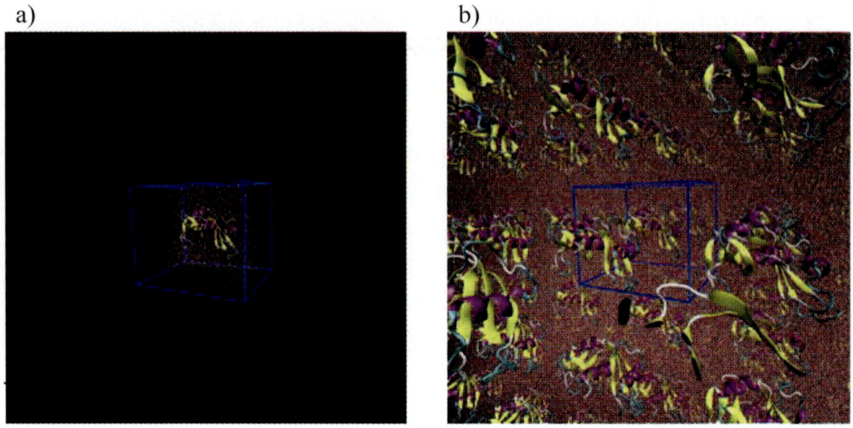

Figure 2. Periodic boundary condition. A protein molecule is solvated with solvent molecules in a box (a). In periodic boundary condition, the box is replicated toward infinity in all the three Cartesian coordinates. The figures are generated with VMD [37].

In addition to the necessity of huge number of solvent molecules, biomolecular simulation is typically performed in constant temperature, T, and pressure, P. The isothermal-isobaric, or NPT ensemble can be generate by Nose-Andersen method [15-17], where additional degrees of freedom to represent the environment of the molecular system are introduced. Sometimes other ensembles such as canonical, NVT (V is volume), ensemble are also used for biomolecular simulation with respective methods [15-17].

2.5. *Current Limitations of Molecular Dynamics*

In 1977, the first MD simulation of a protein, basic pancreatic trypsin inhibitor, was conducted by McCammon, Gelin, and Karplus [18]. At that time number of atoms was 885 and simulation length was only 9.2 ps in vacuum condition. Today, thanks to the continuous developments of simulation algorithm and dazzling progress of computational power, even simulating bio-macromolecules with surrounding solvent molecules consisting of hundred thousand of atoms for tens of ns is well performed on parallel computer. MD simulation has contributed toward understanding dynamics and functions of biomolecules at atomic resolution.

However, as we have been able to simulate much larger systems for much longer time, it has become clear that there are two big obstacles of the current MD simulation technique: simulation time scale and force field. Fig. 3 shows typical examples of protein motions and its time scale. Protein dynamics covers a wide range of amplitudes (0.01 to 100 Å) and time scales (10^{-15} to 10^{3} s) (Fig.

3). For example, torsion angle rotation takes place in 100 picoseconds order. Protein folding and conformational change accompanying with the expression of its own function (*i.e.* protein-protein or protein-ligand binding) occur from sub-millisecond to second. On the other hand, even by using up-to-date super computers and algorithms, simulation time that we can follow is practically just tens of nanoseconds. Furthermore, at present, there is no "universal force field". Therefore, results of a MD simulation would depend on the force field used for the simulation.

In order to overcome these difficulties, various approaches are introduced. Replica exchange is one of the most famous methods to enhance the sampling of broad energies and conformations, in which simulations of replica systems at various temperature are carried out on parallel, and the replicas are periodically exchanged based on the Metropolis criterion [19]. Multicanonical method also makes it possible to search wide range of conformational space by using non-Boltzmann weight function for sampling [20]. To reduce the computational time greatly, continuum solvent or implicit solvent models are introduced [21-23]. Additionally, force fields for biomolecules are being developed extensively [24-26]. However, we are still on the way to a solution.

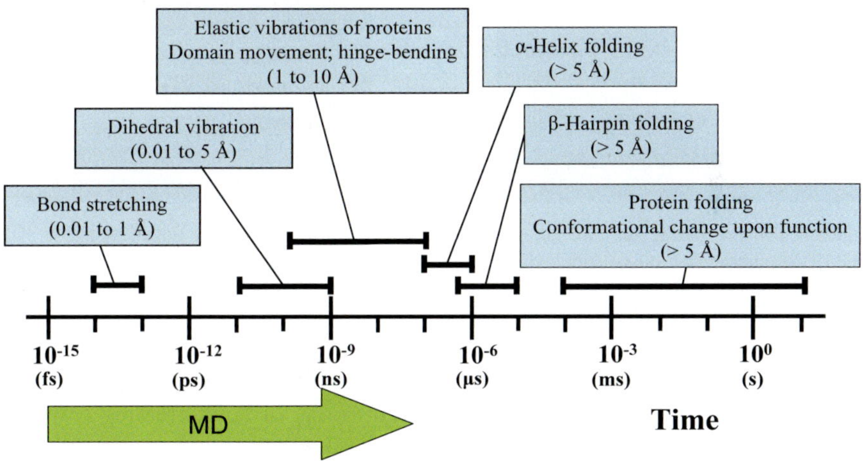

Figure 3. Protein motions. Ranges of spatial amplitudes are shown in parentheses. At the bottom, time scale that current MD simulation can follow is shown for comparison.

3. Protein Folding Problem

Until now, we mentioned about basic techniques for MD and its problems. Here, we would like to introduce "protein folding problem" as an example of grand challenges of computational biology to clarify the importance of solving the problems in current MD.

Proteins are synthesized as unbranched long-chain polymers of just twenty kinds of amino acids in a cell. Unlike most of polymers, each chain can fold spontaneously into a well-ordered three-dimensional structure depending on its amino acid sequence encoded by the gene (Fig. 4) [27]. Remarkable precision and fidelity of protein folding form the basis of all living systems. However, we do not have any complete answer to "Why does a protein can fold into a specific conformation within a finite time?" and "Can we predict a protein structure from its amino acid sequence?" Protein folding is a central problem in not only molecular biology but also physics and chemistry. In order to solve the problem, protein folding has been studied by experimental, theoretical, and computational approaches [28].

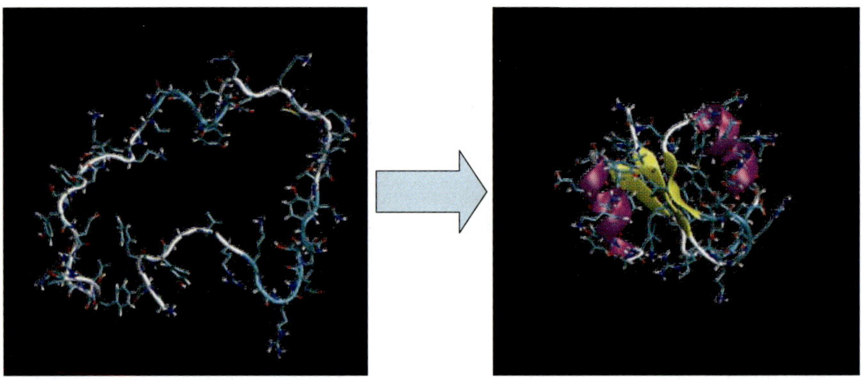

Figure 4. Protein folding from an unfolded or random-coil state (left) to its folded or native state (right). An unbranched poly-peptide chain synthesized on a ribosome in a specific order based on information encoded within DNA sequence folds into its native structure spontaneously. The figures are generated with VMD [37].

Molecular simulation technique is essential for understanding of protein folding from microscopic view. Since bond breaking or bond-creating reactions do not occur upon the protein folding, simulation of folding dynamics of a protein is a good target of MD. However, when we try simulating whole protein folding reaction by MD, we face to the same two difficult problems as we mentioned in previous section. Protein folding typically takes place from sub-microseconds to seconds (See Fig. 3). On the other hand, ten nanoseconds simulation of a protein-solvent system necessitates about a week of computation time. Therefore, simulation time scale we can follow is not good enough to follow the whole process of folding. Additionally, although protein folding is very dynamic and protein adopts huge number of conformations until peptides reach at their native structures, we do not know whether current force fields can describe precisely both protein's intramolecular interactions and intermolecular interactions with solvent for these arbitrary conformations. In 1998, Duan and Kollman carried out "brute force" MD simulation for one microsecond with explicit solvent molecules to watch a folding reaction of 36-residue helical protein [29]. However, the protein did not reach its stable native conformation but folded into a helical structure that resembles the native conformation.

In these respects, sometimes simulating whole folding process of any protein molecules is called "the Holy Grail" because it is extremely desirable to possess but extremely elusive to obtain [30].

4. Brownian Dynamics Simulation

In our group, an atomistic Brownian dynamics (BD) simulation with multiple time step algorithm and a new implicit solvent model to describe the protein folding process at atomic resolution have been developed [31-34]. In this approach, solvent molecules around a protein are treated in implicit manner so that computational time is greatly reduced compared to the conventional molecular dynamics simulation with explicit solvent model.

4.1. *Brownian Dynamics Algorithm*

By treating the effects of solvent as a dissipative random force, the Langevin equation can be expressed as

$$m_i \frac{\mathrm{d}^2 \mathbf{r}_i}{\mathrm{d}t^2} = -\zeta_i \frac{\mathrm{d}\mathbf{r}_i}{\mathrm{d}t} + \mathbf{F}_i + \mathbf{R}_i. \tag{5}$$

Here, m_i and $\mathbf{r}_i$ represent the mass and position of atom i, respectively. ζ_i is a frictional coefficient and is determined by the Stokes' law, that is, $\zeta_i = 6\pi a_i^{\mathrm{Stokes}}\eta$ in which a_i^{Stokes} is a Stokes radius of atom i and η is the viscosity of water. $\mathbf{F}_i$ is the systematic force on atom i. $\mathbf{R}_i$ is a random force on atom i having a zero mean $\langle\mathbf{R}_i(t)\rangle = 0$ and a variance $\langle\mathbf{R}_i(t)\mathbf{R}_j(0)\rangle = 2\zeta_i k_{\mathrm{B}} T\delta_{ij}\delta(\mathrm{t})\mathbf{I}$ where k_{B} is the Boltzmann's constant, T is absolute temperature, δ_{ij} is the Kronecker delta, $\delta(t)$ is the Dirac delta, and $\mathbf{I}$ is 3×3 unit tensor; this derives from the effects of solvent.

For the overdamped limit (the solvent damping is large and the inertial memory is lost in a very short time), we set the left side of Eq. 5 to zero,

$$\zeta_i \frac{\mathrm{d}\mathbf{r}_i}{\mathrm{d}t} = \mathbf{F}_i + \mathbf{R}_i. \tag{6}$$

Integrated equation of Eq. 6 is called Brownian dynamics;

$$\mathbf{r}_i(t+h) = \mathbf{r}_i(t) + \frac{\mathbf{F}_i(t)}{\zeta_i}h + \sqrt{\frac{2k_B T}{\zeta_i}}h\boldsymbol{\omega}_i, \tag{7}$$

where h is a time step and $\boldsymbol{\omega}_i$ is a random noise vector obtained from Gaussian distribution. This BD algorithm was introduced by Ermak and McCammon [35]

For multiple time step algorithm, short time step, $\Delta\tau$, of 5 fs and long time step, Δt, of 40 fs were used [32]. Cut-off method was not used. All bond lengths were constrained with LINCS algorithm [14]. Stokes radius of each atom was its van der Waals radius plus 1.4 Å. Coordinates and energies were recorded every 100 ps during the simulation. For analysis, the structures collected for first 10 ns were removed.

4.2. *Force Field*

We used the AMBER91 united-atom force field for amino acids [36] with angle-dependent, 12-10 hydrogen-bond potential [31].

To reproduce the solvation effects, three implicit solvent models were used: distance-dependent dielectric model (DD), solvent-accessible surface area model (SA), and effective charge model (EC) [33]. In the DD model, $\varepsilon = 2r_{ij}$ was used. The atomic solvation parameters used in the SA model were $\sigma(\mathrm{C}) = 12$ cal/mol/Å^2, $\sigma(\mathrm{O, N}) = -116$ cal/mol/Å^2, $\sigma(\mathrm{S}) = -18$ cal/mol/Å^2, and $\sigma(\mathrm{O^-/N^+}) = -280$ cal/mol/Å^2 [34]. The EC model was introduced by us to represent the shielding effect of oriented water molecules around a point charge [33], in which atomic charge of atom i, q_i, is neutralized as a function of solvent-accessible surface area of the atom, $SA_i(\mathbf{r}^N)$, in a given atomic coordinate $\mathbf{r}^N$ ($\mathbf{r}^N$ is a position vector of Nth atom):

$$q_i' = q_i\left[\frac{1 - SA_i(\mathbf{r}^N)/S_i}{\alpha_{\mathrm{int}}} + \frac{SA_i(\mathbf{r}^N)/S_i}{\alpha_{\mathrm{ext}}}\right] \tag{8}$$

Here q_i' is the effective charge of atom i, S_i is the total solvent-accessible surface area of isolated atom i, α_{int} is a shielding parameter against interior of the solute (wherein α_{int} is set at unity), and α_{ext} is a shielding parameter for exterior water. In this study, $\alpha_{\mathrm{ext}} = 5$ was used.

4.3. *Folding Simulations of α-Helical and β-Hairpin Peptides*

To study the folding mechanism of the key structural elements of proteins, we performed long time simulations of α-helical, peptide III, and β-hairpin,

BH8, peptides at 298 K using our BD with MTS algorithm and the implicit solvent DD/SA/EC models from the fully extended conformations. Folding simulations of the peptides were performed five times using different random seeds for each peptide.

4.3.1. *Folding trajectories*

Figure 5 shows the fraction of native contacts (Q) of the two peptides as a function of simulation time. For peptide III, although there were few states having $Q > 0.8$ due to lack of hydrogen bonds at C-terminus, the peptide reached the folded states from the extended states within 400 ns in all simulations. Because the formation of perfect helix ($Q = 1.0$) accompanies with large entropic cost of conformation, this state is not expected to exist in a significant amount. In the simulations of BH8, the peptide also folded from the extended structures in all trajectories.

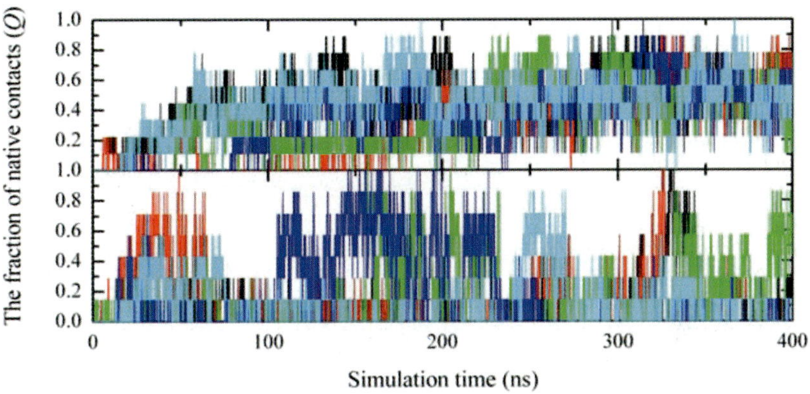

Figure 5. Time evolutions of the fraction of native contacts during BD simulations of peptide III (upper) and BH8 (lower). Five trajectories (black, red, green, blue and sky blue) obtained by the BD simulations using different random number seeds are shown.

4.3.2. *Energy components*

The average effective energy (effective energy is the intra-protein energy plus solvation free energy) and its components (van der Waals term, E_{vdW}, electrostatic term including the effects of DD/SA/EC implicit solvent models, E_{elec}') of the two peptides as a function of Q are shown in Figure 6. The total effective energy showed downhill profile for both peptides. The negative gradient of the total effective energy of peptide III was much larger than that of BH8. However, since variances of the total energies were too large, there were many non-native structures having lower effective energies than the energies of the native structures in both systems. This result indicates that it is impossible to predict the native states of the peptides based on the energy alone. For peptide III, the average values of E_{vdW} and E_{elec}' decreased with Q. For BH8, although the average value of E_{vdW} decreased with Q, the slope of E_{elec}' is quite flat. These results indicate that the effective driving energy contributions to the folding of the peptides are concluded to be derived from both van der Waals and electrostatic terms for the α-helical peptide, peptide III, and from van der Waals term for β-hairpin peptide, BH8.

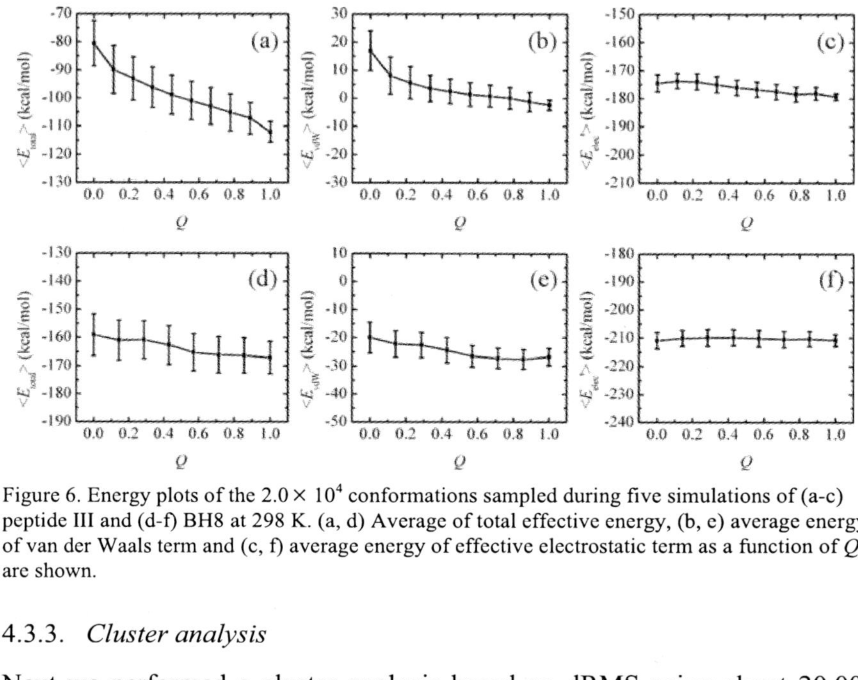

Figure 6. Energy plots of the 2.0×10^4 conformations sampled during five simulations of (a-c) peptide III and (d-f) BH8 at 298 K. (a, d) Average of total effective energy, (b, e) average energy of van der Waals term and (c, f) average energy of effective electrostatic term as a function of Q are shown.

4.3.3. *Cluster analysis*

Next we performed a cluster analysis based on dRMS using about 20,000 structures obtained by the simulations of each peptide. The structures of the centers of the three most populated clusters for both peptides are shown in Figure 7. The most populated clusters of peptide III and BH8 contain 7% and 8% of all the conformations, respectively. Interestingly, the most populated cluster had higher average value of Q than that of other clusters and the folded structures belonged to these most populated clusters for both peptides. The central structure of cluster 1 of peptide III had a helical conformation throughout the peptide. For BH8, the central structure of the most populated cluster was a β-hairpin conformation that had side-chains of Ile3, Val5, Lys8, and Tyr10 protruding on the same side of plane of the strands, which is consistent with the NMR data. An important point is that the cluster analysis makes it possible to predict the native folded states from the structures obtained by the BD simulations.

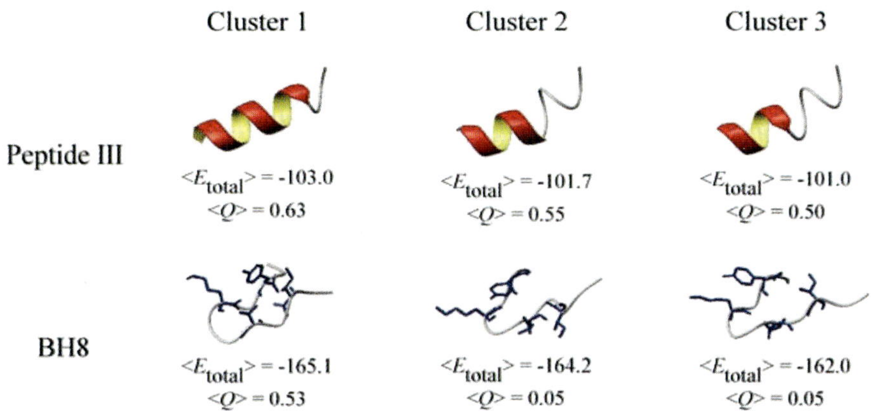

Figure 7. Ribbon representations of the central structures of the three most populated clusters for peptide III (upper) and for BH8 (lower). From left to right, cluster 1, cluster 2 and cluster 3 are shown. The values of total effective energy (E_{total} in kcal/mol) and the fraction of native contacts, Q, averaged over the cluster are listed under each central structure. For BH8, residues of Ile3, Val5, Lys8 and Tyr10 are shown in sticks. The figures are generated with MOLMOL [38].

5. Conclusions

Molecular simulation is often referred to "the third way" of doing research in the field of physics and chemistry, bridging between theory and experiment. In the field of biological science, this is becoming unexceptional. Solving the problems in force field and simulation time scale would be a key step, we believe, to shift today's molecular biology from descriptive science to predictive one. Many researchers have tried to overcome these difficulties. And we believe that our BD method and implicit solvent models would be promising approaches to the problems. We hope this material will help not only biologist but also physicist, chemist, and mathematician to become interested in this field.

Acknowledgments

T. A. acknowledges support from the Academic Frontier Project from MEXT (Ministry of Education, Culture, Sports, Science and Technology of Japan). This work was supported by the High-Tech Research Center Project from MEXT to T.A. and I. Y.

References

1. T. Schlick, *Molecular Modeling and Simulation*. Springer-Verlag: New York, 2002.
2. C. L. Brooks, M. Karplus and B. M. Pettitt, *Proteins: a theoretical perspective of dynamics, structure, and thermodynamics*. John Wiley & Sons: New York, 1988.
3. *Computer simulation of biomolecular systems, Vol.2,* ed. by W. F. van Gunsteren, P. K. Weiner and A. J. Wilkinson. ESCOM: Netherlands, 1993.
4. T. Darden, D. York and L. Pederson, *J. Chem. Phys.* **98**, 10089 (1993).
5. L. Greengard and V. Rokhlin, *J. Comput. Phys.* **73**, 325 (1987).
6. D. A. Case, T. E. Cheatham, III, T. Darden, H. Gohlke, R. Luo, K. M. Merz, Jr., A. Onufriev, C. Simmerling, B. Wang and R. Woods. *J. Computat. Chem.* **26**, 1668 (2005).
7. A. D. MacKerell, Jr., B. Brooks,C. L. Brooks, III, L. Nilsson, B. Roux, Y. Won and M. Karplus, The Encyclopedia of Computational Chemistry, 1, 271-277, P. v. R. Schleyer et al., editors (John Wiley & Sons: Chichester, 1998)
8. M. Christen, P. H. Hünenberger, D. Bakowies, R. Baron, R. Bürgi, D. P. Geerke, T. N. Heinz, M. A. Kastenholz, V. Kräutler, C. Oostenbrink, C. Peter, D. Trzesniak and W.F. van Gunsteren, *J. Comput. Chem.* **26**, 1719 (2005).
9. W. L. Jorgensen, D. S. Maxwell and J. Tirado-Rives, *J. Am. Chem. Soc.*, **118**, 11225 (1996).
10. W. C. Swope, H. C. Andersen, P. H. Berens and K. R. Wilson, *J. Chem. Phys.* **76**, 637 (1982).
11. M. Tuckerman, B. J. Berne and G. J. Martyna. *J. Chem. Phys.* **97**, 1990 (1992).
12. J. Ryckaert, G. Ciccotti and H. J. C. Berendsen, *J. Comp. Phys.* **23**, 327 (1977).
13. H. C. Andersen, *J. Comp. Phys.* **52**, 24 (1983).
14. Hess, B., H. Bekker, H. J. C. Berendsen and J. G. E. M. Fraaije, *J. Comput. Chem.* **18**, 1463 (1997).
15. S. Nose, *Mol. Phys.* **52**, 255 (1984).
16. H. C. Andersen, *J. Chem. Phys.* **72**, 2384 (1980).
17. W. G. Hoover, *Phys. Rev. A.* **31**.1695 (1985).
18. J.A. McCammon, B.R. Gelin and M. Karplus, *Nature* **267**, 585 (1977).
19. Y. Sugita and Y. Okamoto, *Chem. Phys. Lett.* **314**, 141 (1999).
20. U. H. E. Hansmann and Y. Okamoto, *J. Comput. Chem.* **14**, 1333 (1993).
21. W. C. Still, A. Tempczyk, R. C. Hawley and T. Hendrickson, *J. Am. Chem. Soc.* **112** 6127 (1990).
22. D. Bashford and D. A. Case, *Annu. Rev. Phys. Chem.***51**, 129 (2000).
23. M. Feig and C. L. Brooks. *Curr. Opin. Struct. Biol.* **14**, 217 (2004).

24. Y. Duan, C. Wu, S. Chowdhury, M. C. Lee, G. Xiong, W. Zhang, R. Yang, P. Cieplak, R. Luo, T. Lee, J. Caldwell, J. Wang and P. Kollman. *J. Comp. Chem.* **24**,1999 (2003).
25. A. D. MacKerell, M. Feig and C. L. Brooks, *J. Comp. Chem.* **25**, 1400 (2004).
26. A. D. MacKerell, *J. Comput. Chem.* **25**,1584 (2004).
27. C. B. Anfinsen, *Science* **181**, 223 (1973).
28. C. N. Dobson, A. Sali and M. Karplus, *Angew. Chem. Int. Ed.* **37**, 868 (1998).
29. Y. Duan and P. A. Kollman, *Science* **282**, 740 (1998).
30. H. J. C. Berendsen, *Science* **282**, 642 (1998).
31. T. Ando, T. Meguro and I. Yamato, *J. Comput. Chem., Jpn.* **1**, 103 (2002).
32. T. Ando, T. Meguro and I. Yamato, *Mol. Simul.* **29**, 471 (2003).
33. T. Ando, T. Meguro and I. Yamato, *J. Comput. Chem., Jpn.* **3**, 129 (2004).
34. T. Ando and I. Yamato, *Mol. Simul.* **31**, 683 (2005).
35. D. L. Ermak and J. A. McCammon, *J. Chem. Phys.* **69**, 1352 (1978).
36. S. J. Weiner, P. A. Kollman, D. A. Case, U. C. Singh, C. Ghio, G. Alagona, S. Profeta and P. Weiner, *J. Am. Chem. Soc.* **106**, 765 (1984).
37. W. Humphrey, A. Dalke and K. Schulten, *J. Molec. Graphics* **14**, 33 (1996).
38. R. Koradi, M. Billeter and K. Wuthrich, *J. Mol. Graphics* **14**, 51 (1996).

THEORY OF PROTON-INDUCED SUPERIONIC CONDUCTION IN HYDROGEN-BONDED SYSTEMS

HIROSHI KAMIMURA

Research Institute for Science and Technology,
Tokyo University of Science
1-3 Kagurazaka, Shinjuku-ku, Tokyo 162-8601, Japan
E-mail: kamimura@rs.kagu.tus.ac.jp

This paper consists of three sections. After the Introduction of Section 1, in Section 2 we describe a quantum mechanical mechanism of proton-induced ionic conduction in the superionic phase in zero-dimensional hydrogen-bonded $M_3H(XO_4)_2$ [M=K,Rb,Cs, X=S,Se] materials, by giving a review on the theory developed by Ito and Kamimura. Then we discuss the characteristic difference between quantum mechanical and classical mechanisms in the case of proton-induced superionic conducction, in paticular, by comparing characteristic time scales in quantum mechanical and classical diffusions in hydrogen-bonded systems. In Section 3 a theory is proposed with regard to the origin of the divergent behavior of ionic conductivity near the phase-transition temperature in the ferroelatic phase of $M_3H(XO_4)_2$.

Keywords: Superionic conduction Successive proton tunneling First-principles theory Ferroelastic to paraelastic phase transition Hydrogen-bonded $M_3H(XO_4)_2$.

1. Introduction

Among superionic conduction phenomena in various ionic materials, the conduction phenomenon associated with the motion of protons in hydrogen-bonded systems has aroused considerable interests with regard to a problem of whether the proton motion should be treated quantum mechanically or classically. In this context the zero-dimensional hydrogen-bonded crystals of $M_3H(XO_4)_2$ [M=K,Rb,Cs, X=S,Se] type dielectric crystals have been extensively studied. This type of crystals exhibits a ferroelastic to paraelastic transition at high temperatures such as 400 K. For examples, the values of T_c for $K_3H(SeO_4)_2$ and $Rb_3H(SeO_4)_2$ are 390 K and 449 K, respectively [1].

Figure 1 shows the crystal structure of rubidium hydroselenate $Rb_3H(SeO_4)_2$ in the ferroelastic phase, where Fig. 1(a) shows its projection on the $a-c$ plane while Fig 1. (b) shows its projection on the $a-b$ plane at room temperature, where Rb atoms are not shown for simplicity [1].

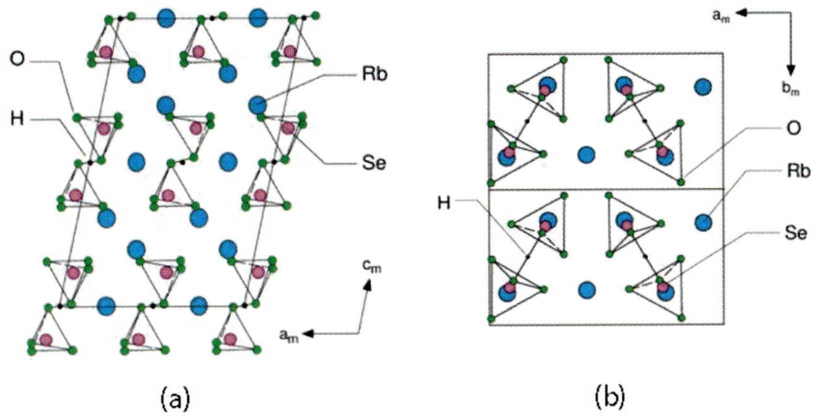

Figure 1. (a) The crystal structure of the $M_3H(XO_4)_2$ [M=K, Rb,Cs, X=S,Se] in the room temperature. (b) Projection on the *ab* plane of the $M_3H(XO_4)_2$, where M atoms are not shown for simplicity.

From Fig. 1(a) we note that the top and bottom oxygen of the neighboring tetrahedrons lie nearly at the same height along the *c*-axis and that a hydrogen bond is formed in between these top and bottom oxygen whose distance is about 2.5 Å. These hydrogen bonds are isolated and thus do not form a network. Such isolation can be seen clearly in Fig. 1(b). Because of this localized nature of the hydrogen bonds, $M_3H(XO_4)_2$ type crystals are called "zero-dimensional hydrogen-bonded crystals". In the paraelastic phase ($T > T_c$), each tetrahedron is tilted so as for Rb, Se and O to stand in line perpendicular to the $a-b$ plane (i.e. c^*-axis). For top or bottom oxygen of each tetrahedron there are three equivalent positions for tilting. Since each tetrahedron thermally rotates very fast at high temperatures above T_c, the hydrogen bonds become weak and the three equivalent positions for tilting appear equally. As a result a crystal structure in the paraelastic phase has·the three-fold symmetry along the c^*-axis and the length of the hydrogen bonds becomes equal, where the length of a hydrogen bond is about 2.9 Å. This length is much longer than that in the ferroelastic phase. Reflecting the three-fold symmetry as to the average geometrical arrangement of hydrogen bonds, the space group of the paraelastic phase is $R\bar{3}m$, When the temperature decreases, an disorder-order phase transition occurs as to the geometrical arrangement of hydrogen-bonds, and the space group changes from $R\bar{3}m$ of a three-fold symmetry to a two-fold symmetry of a monoclinic system in the $a - b$ plane.

In these crystals the anomalous increase of electrical conductivity near the phase transition temperature T_c has been observed. Further the conductivity in the paraelastic phase ($T > T_c$) is very high, and it is of the order of 10^{-4} to 10^{-2} S/cm. Because of this fact we call the paraelastic phase as *superionic phase.*

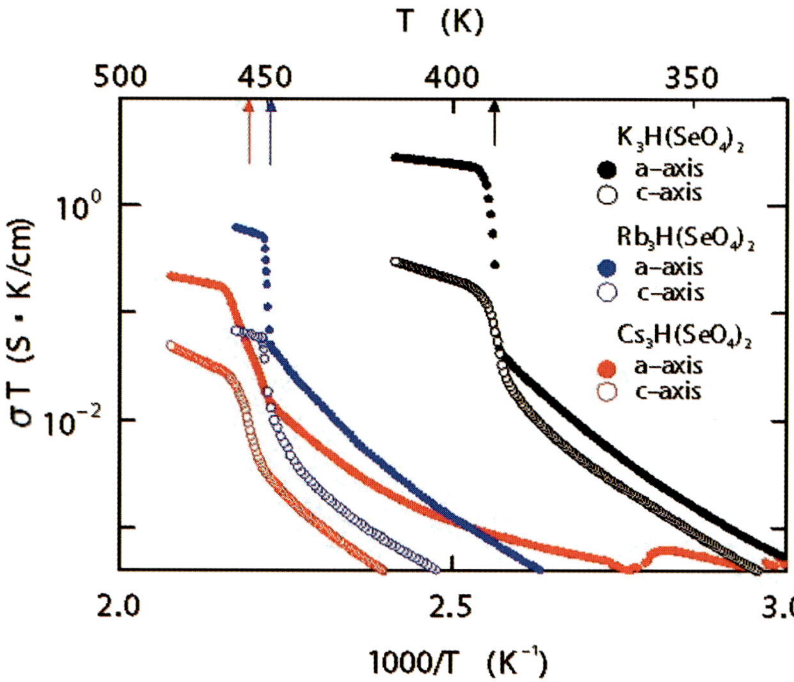

Figure 2. Arrhenius plot of the ionic conductivity in $M_3H(XO_4)_2$ as a function of temperature T.

In Fig. 2 we present the experimental results on the temperature dependence of the electrical conductivity in $M_3H(SeO_4)_2$ crystals (M: K, Rb, Cs) [1]. This static conductivity was obtained by extrapolating the observed frequency-dependent conductivity in an applied alternating electric field to the zero frequency. In the superionic phase it is evident that all crystals show the high electrical conductivity of 10^{-4} to 10^{-2} S/cm, although the phase transition temperature depends on a kind of alkaline metal and on the direction of the electrical conduction. The electrical conductivities along the a-axis are about 10^2 times larger than those along the c^*-axis. It is also noted that in the superionic phase, the conductivity obeys the Arrhenius law in its temperature dependence. The activation energy of the $M_3H(SeO_4)_2$ crystals along the a-axis becomes about 0.2 eV. This value is 2 times smaller than that along the c^*-axis, indicating that the $M_3H(SeO_4)_2$ crystals display the quasi-two dimensional conductivity. Therefore, it is deduced that the electrical conductivity is closely related to the itinerant motion of protons in the $a-b$ plane accompanied by the breaking of the hydrogen bond between SeO_4 tetrahedrons. In addition to the behavior of the electrical conductivity in the superionic phase, a considerable increase of the electrical conductivity appears with increasing temperature even in the ferroelastic phase. That is also one of the characteristic features in ferroelastic $M_3H(SeO_4)_2$ crystals. The conductivity just below T_c seems to increase following the $(T_c - T + b)^{-1/2}$ power law when the temperature approaches T_c from the lower temperature.

As regards a mechanism of superionic conduction, Plakida and Salejda [2] developed a theory of the superionic conduction in $Rb_3H(SeO_4)_2$, based on the phenomenological Ginzburg-Landau theory for the ferroelastic transition [3]. In 1998 Ito and Kamimura [4] developed the quantum mechanical theory of superionic conduction using the theory of Matsubara and Toyozawa for impurity conduction in doped semiconductors [5] and by the Kubo formula [6]. In 2001 Kamimura and Watanabe proposed a new approach to the mechanism of ionic conductivity below and at the ferroelastic phase transition and predicted in their preliminary report that the temperature dependence of ionic conductivity follows the $(T_c - T)^{-1/2}$ power law just below T_c [7].

The present paper consists of three sections including this first section. In section 2 we first describe the mechanism of proton-induced superionic conduction in the superionic phase ($T > T_c$) developed by Ito and Kamimura, and we try to clarify the important features of the quantum mechanical mechanism for superionic conduction. Then we discuss the key factors in the difference between the quantum mechanical mechanism by Ito and Kamimura and a classical mechanism in the proton-induced superionic conduction in hydrogen-bonded systems. In the case where heavier ions than a proton are involved in the superionic conduction, a classical hopping mechanism holds, so that, in order to clarify whether a quantum mechanical mechanism plays an important role or not in the superionic conduction, it is very important to investigate a behavior of a proton in the proton-induced superionic conduction. Thus in this paper, we pay attention to hydrogen-bonded materials and choose $M_3H(XO_4)_2$ [M=Rb,Cs, X=S,Se] materials as an object of study.

Section 3 is devoted to trying to clarify the origin of the divergent behavior of the ionic conductivity at and just below the phase transition temperature ($T < T_c$) . First the Ginzburg-Landau (G-L) theory for the ferroelastic phase transition is developed following the theory by Plakida and Saledja. It will be shown that the G-L theory cannot explain the divergent behavior of the ionic conduction just below T_c. In this context a new theory of ionic conduction just below T_c is proposed, taking account of the precursor effect of the phase transition. Then the formula of calculating the ionic conductivity at and near T_c is developed. It is shown that a divergent behavior of ionic conductivity at T_c can be explained by the precursor effect of the phase transition, and in particular, it is shown that the successive tunneling of protons between the expanded hydrogen-bonds plays an essential role in producing the divergent behavior.

2. Theory of superionic conduction in the superionic phase

2.1. *Mechanism of proton-induced superionic conduction*

In this section we describe a mechanism of superionic conduction in $M_3H(XO_4)_2$ crystals proposed by Ito and Kamimura [4]. In the superionic phase the tetrahedrons rotate as mentioned in a previous section. Thus the strength of hydrogen-bonds becomes weak and the distance between neighboring XO_4 tetrahedrons becomes longer than that in a dimer. As a result all the distances between XO_4 tetrahedrons become equal and hence it becomes possible for a proton in an elongated hydrogen bond to tunnel to an empty site in another elongated hydrogen-bond-site quantum mechanically.

In order to describe the correlated behavior of proton-successive tunneling following the rotation of tetrahedrons quantum mechanically, let us consider a certain arrangement of hydrogen-bonds shown in Fig. 3(a), where $XO_4 - H - XO_4$ dimers are represented by the thick solid lines.

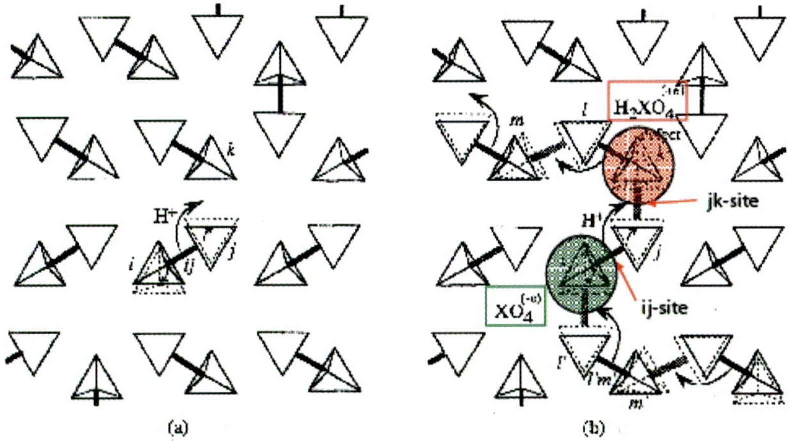

Figure 3. (a)Schematic view for the breaking of a hydrogen-bond by the rotational displacement of two neighboring tetrahedrons. (b) Schematic view for the formation of two kinds of excited states, $H_2XO_4^{(+e)}$ and $XO_4^{(-e)}$, and for the coherent motion of these excited states due to successive proton tunneling.

As Suwa *et al.* showed theoretically in the case of alkali-hydrosulfates or alkali-hydroselenates, there exists no double-well potential but a single-well type potential for a hydrogen-bond in $XO_4 - H - XO_4$ dimers, while in the case of deuterium-bonded materials there exists a double-well potential [8]. Thus, taking account of anharmonicity, we can show that the strength of the hydrogen-bonds becomes weak with increasing temperature. In this circumstance we consider that a hydrogen-bond between two tetrahedrons marked by i and j is broken thermally by the rotational displacement of top and bottom oxygen of neighboring tetrahedrons as shown by the short arrows in Fig. 3(a). Then, when the tetrahedron j is tilted and displaced toward a neighboring tetrahedron, say k, as shown by the dotted lines in this figure, a proton in the elongated hydrogen-bond denoted by ij hops to an interstitial position between the tetrahedrons j and k as shown by the long arrow. Then the tetrahedron k is tilted to the tetrahedron j so as for the energies before and after a hop to be equal with each other. Simultaneously an electron separated from the proton is accommodated in the tetrahedron i. As a result two kinds of defect states, $H_2XO_4^{(+e)}$ and $XO_4^{(-e)}$, are formed as excited states to carry an ionic current in the superionic phase, as indicated in the dotted circles as shown in Fig. 3(b).

The formation process of these ionic excited states is considered to be of the thermal activation type, because the breaking of a weak hydrogen-bond into a proton and an electron is caused by the thermal rotational motion of the tetrahedrons. In the case of a static dielectric constant of 8 like that of present materials, the activation energy is estimated as 0.2 eV, which is consistent with the experimental result mentioned in a previous section. In the $H_2XO_4^{(+e)}$ excited state, each of the two elongated hydrogen bonds denoted by the thick solid and gray lines in Fig. 3(b) is attached to the top and bottom oxygen, while in the $XO_4^{(-e)}$ excited state there are no hydrogen-bonds. The (+e) and (−e) represent the extra charges yielded in the respective one of the two tetrahedrons when a hydrogen-bond is broken, and these ionic excited states carry an ionic current.

First we shall pay attention to the $H_2XO_4^{(+e)}$ excited state in Fig. 3(b). When a new weak hydrogen-bond is formed between the tetrahedrons j and k, this brings about the breaking of the hydrogen-bond between the tetrahedrons k and l. Then, when the tetrahedron l is tilted to a neighboring tetrahedron, say m, an extra proton in the hydrogen-bond kl tunnels to an interstitial position between the tetrahedrons l and m, because the energies before and after the tunneling of a proton is equal. As a result the $H_2XO_4^{(+e)}$ excited state which was originally located at k moves to the tetrahedron m without scattering. In the same way, the $XO_4^{(-e)}$ excited state moves from i to a neighboring tetrahedron, say m', when a proton in the hydrogen-bond $l'm'$ tunnels to an interstitial position between the tetrahedrons i and l' as shown by the arrow in Fig. 3(b).

In this way, the successive breaking and formation of weak hydrogen-bonds at different positions occur. Since the thermal rotation of the tetrahedrons is very fast at higher temperatures above T_c of about 400 K, compared with the time of the proton hopping, the coherent proton-induced transport phenomenon takes place, and a band-like state is expected to be formed for each of $H_2XO_4^{(+e)}$ and $XO_4^{(-e)}$ excited states, which itinerate toward opposite directions in the presence of an applied alternating electric field. Then a question arises why a band-like conduction has finite resistivity. As seen in Figs. 3(a) and (b), there are four directions for the hopping of a proton at each tetrahedron site. Random selection of a hopping direction causes a finite conductivity. Ito and Kamimura proved that this remarkable feature of the itinerant motion of two excited states shown in Figs. 3(a) and (b) causes the quasi-one-dimensional characteristics of the Bethe lattice(Cayley tree) in the energy spectra of these excited states, although the ionic conduction takes place in the two-dimensional $a-b$ plane.

We consider that the transfer interactions of the $H_2XO_4^{(+e)}$ and $XO_4^{(-e)}$ excited states due to the proton tunneling, Γ and Γ', are of the same order of magnitudes as the rotational energy of an XO_4 type molecule, because a proton interacts with strain induced by the rotation of XO_4. Thus it may be estimated that the absolute magnitudes of Γ and Γ' are of the order of 10^{-3} to 10^{-4} eV. Since the energy for the thermal reorientation of the tetrahedrons is of the order of k_BT with k_B and T being the Boltzmann constant and the absolute temperature, the time of the transfer of a proton between the neighboring sites, $\hbar/\Gamma$ (or $\hbar/\Gamma'$), is of the same order as the time constant of the reorientation of tetrahedrons, $\hbar/k_BT$, for $T = 400$ K, where $\hbar = h/2\pi$, and h is the Planck constant. As a result the coherent motions of both the $H_2XO_4^{(+e)}$ and $XO_4^{(-e)}$ excited states take place.

In this context, choosing the site energy of the excited state as the origin of energy, the model Hamiltonian to describe the coherent motion of the $H_2XO_4^{(+e)}$ and $XO_4^{(-e)}$ excited states are expressed, respectively, by

$$H = \sum_k \sum_j \sum_i \Gamma a_{jk}^{\dagger} a_{ij} \left(\sum_{j'=1}^{3} a_{ij'}^{\dagger} a_{ij'} - 1 \right) \tag{1}$$

and

$$H = \sum_k \sum_j \sum_i \Gamma' a_{ij}^{\dagger} a_{jk} \left(1 - \sum_{j'=1}^{3} a_{ij'}^{\dagger} a_{ij'} \right), \tag{2}$$

where Γ and Γ' represent the transfer integrals of the $H_2XO_4^{(+e)}$ and $XO_4^{(-e)}$ excited states, respectively: In other words, the above Hamiltonians are considered to represent the transfer interaction of a proton between adjacent available sites. The symbols $a_{ij}^{\dagger}$ and a_{ij} are the creation and annihilation operators of a proton in the site ij, respectively. Here the summation over j' in the parentheses is taken over the nearest-neighbor sites around the site i.

It should be noted that the numbers of the nearest neighboring sites are 3 both for $H_2XO_4^{(+e)}$ and $XO_4^{(-e)}$. The factor $(\sum_{j'=1}^{3} a_{ij'}^{\dagger} a_{ij'} - 1)$ in Eq. (1) is 1 when $H_2XO_4^{(+e)}$ excited state lies at the $i-$position while otherwise zero, because of two hydrogen-bonds attached to $H_2XO_4^{(+e)}$. On the other hand, $(1 - \sum_{j'=1}^{3} a^{\dagger}_{ij'} a_{ij'})$ in Eq. (2) is 1 when $XO_4^{(-e)}$ excited state lies at the $i-$position while otherwise zero, because an $XO_4^{(-e)}$ state does not have hydrogen bonds.

Based on this model, Ito and Kamimura calculated the density of states for the two kinds of excited states [4]. Considering that the excited states have the features of the Bethe lattice, they calculated the density of states (DOS) by using the recursion formula by Haydock et al [4,9,10]. We show their calculated results for DOS in Figs. 4(a) and (b). These DOS curves show the feature of a twin-peak structure, reflecting the characteristics of the Bethe lattice(Cayley tree). This means that the coherent motion of $H_2XO_4^{(+e)}$ and $XO_4^{(-e)}$ exited states exhibit a one-dimensional feature although the itineration of these excited states occurs over a two-dimensional layer in the $a - b$ plane. The finite height of the twin peaks is due to the fact that the coherent motion of the excited states are disturbed by the random selection of hopping direction at each tetrahedron site. They showed that the excited states have the feature of the quasi-one-dimensional Bethe lattice.

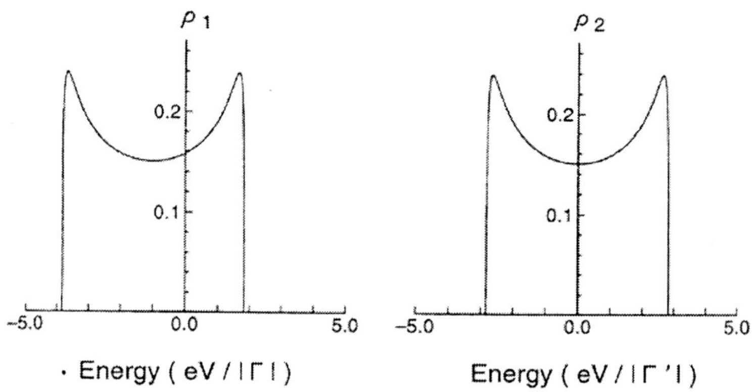

Figure 4. (a) Density of states for the $H_2XO_4^{(+e)}$ excited state. (b) Density of states for the $XO_4^{(-e)}$ excited state.

2.2. *Formula for proton-induced ionic mobility in the superionic phase*

Ito and Kamimura [4] derived a formula calculating the mobility of the $H_2XO_4^{(+e)}$ and $XO_4^{(-e)}$ excited states by using the Kubo formula [6]. According to them, we characterize an excited state around each tetrahedron as α, β, *etc.* by their positions and orientations of the two attached hydrogen bonds, as shown in Fig. 5.

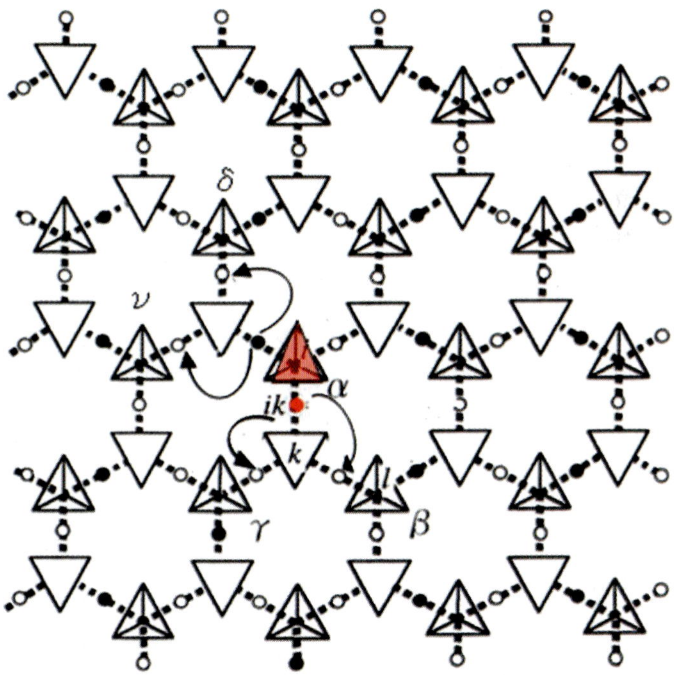

Figure 5. The $H_2XO_4^{(+e)}$ excited state situated at the i-th tetrahedron. Different states of this excited state by its positions and orientations of the two attached hydrogen bonds are shown as α, β, γ, δ, and ν. The possible paths for the transfer motion of this excited state is represented by the solid arrows.

Then the mobility μ is given by

$$\mu = \left(\frac{\pi\hbar}{2qk_BT}\right) \int e^{-E/k_BT}\Xi(E)dE \Big/ \int e^{-E/k_BT}\rho(E)dE, \qquad (3)$$

where

$$\Xi(E) = -\frac{1}{N_{ex}} \sum_{\alpha} \sum_{\beta} \sum_{\gamma} \sum_{\delta} G_{\alpha\delta}(E) H_{\delta\gamma} R_{\delta\gamma} G_{\gamma\beta}(E) H_{\beta\alpha} R_{\beta\alpha}. \quad (4)$$

Here N_{ex} is the total number of the excited states, q an effective charge of an excited state, and $R_{\beta\alpha}$ represents the displacement vector of a proton associated with the change of an excited state from α to β, which is equal to the thermally elongated distance between the tetrahedrons i and j, R_{ij}, in Fig. 3. Further $H_{\beta\alpha}$ is the matrix element of Hamiltonian (1) or (2) between the states β and α in the excited states.

As seen above, the calculations of μ are ascribed to calculating the Green's function $G_{\beta\alpha}(E)$, which is defined by

$$G^{(+)}_{\beta\alpha}(E) = \left\langle \beta \left| \frac{1}{E + i\epsilon - H} \right| \alpha \right\rangle$$

$$G_{\beta\alpha}(E) = -\frac{1}{\pi} \mathrm{Im} G^{(+)}_{\beta\alpha}(E), \quad (5)$$

where E is the energy of a system, and ϵ a positive infinitesimal. Ito and Kamimura calculated $\Xi(E)$ in Eq. (4) by using the method of Matsubara and Toyozawa [5] which was developed to calculate the conductivity for impurity conduction in doped semiconductors, because the random choice of a proton hopping among four directions at each XO_4 tetrahedron site can be treated like the hopping between impurities distributed randomly in doped semiconductors. Then they obtained the ionic mobility for the coherent transfer motion of $H_2XO_4^{(+e)}$ and $XO_4^{(-e)}$ excited states as follows;

$$\mu = \frac{q\Gamma R_0^2}{\hbar k_B T} \times 1.070 \qquad \text{for } H_2XO_4^{(+e)} \text{ excited state}, \quad (6)$$

$$\mu' = \frac{|q'||\Gamma'|R_0^2}{\hbar k_B T} \times 1.066 \qquad \text{for } XO_4^{(-e)} \text{ excited state}, \quad (7)$$

where R_0 is the average hopping distance of the $H_2XO_4^{(+e)}$ or $XO_4^{(-e)}$ excited states, and q and q' are the effective charges of an $H_2XO_4^{(+e)}$ and $XO_4^{(-e)}$ states, respectively. Although the proton motion is coherent in the quantum-mechanical sense, the obtained formula for the ionic mobility show the same form as the Einstein relation with regard to the temperature dependence as seen in Eqs. (6) and (7). However, it should be emphasized that the conduction mechanism is quantum-mechanical, but not classical.

Further we must make another important remark. That is for the numerical factors of 1.070 and 1.066 in Eqs. (6) and (7). Using the method of Matsubara and Toyozawa, Ito and Kamimura expanded the off-diagonal Green's function in Eq. (4) in power series of z/H, where $z = E + i\epsilon$. The result is

$$zG_{\beta\alpha}^{(+)}(E) = \delta_{\beta\alpha} + \frac{1}{z}H_{\beta\alpha} + \sum_{\nu=1}^{\infty} \frac{1}{z^{\nu+1}} \sum_{\gamma_1} \sum_{\gamma_2} \cdots \sum_{\gamma_\nu} H_{\beta\gamma_\nu} \cdots H_{\gamma_1\gamma_2} H_{\gamma_1\alpha}. \tag{8}$$

In calculating the off-diagonal Green's function, we have to sum up all the paths which start from a state α and end at a state β via intermediate states. We can say that among a considerable number of paths, the paths in which intermediates states compose a linear route contribute to the mobility mainly. By accumulating contributions from the paths of linear routes, we obtain the numerical values of 1.070 and 1.066 in Eqs.(6) and (7), respectively.

On the other hand, in the case of classical hopping mechanism which means the diffusive atomic motion described in terms of random (over-barrier) hopping from site to site according to Boettger and Bryksin [11], a proton in the superionic phase hops from each hydrogen bond to a neighboring empty hydrogen-bond-site by the interaction with the vibration-rotation motion of surrounding XO_4 tetrahedrons. If one adopts this classical hopping mechanism, one might treat the elementary jump process of a proton as a classical over-barrier hopping with an activation energy given by the barrier height at sufficiently high temperature like the present case of $T > T_c$. In that case, following Boettger and Bryksin the hopping rate is given by $\nu \exp(-E_a/k_B T)$, where ν is a frequency factor and E_a is the activation energy. This exponential temperature dependence in the over-barrier hopping is an origin for the observed Arrhenius type temperature dependence in the case of a classical diffusion mechanism. Since in the present quantum mechanical diffusion the Boltzmann distribution of the excited state densities is an origin of the Arrhenius type temperature dependence, the origin of the observed temperature dependence is quite different in the classical and quantum-mechanical diffusion mechanisms. As regards the magnitude of mobility, the mobility by the classical over-barrier hopping is two orders of magnitude smaller than those given by Eqs.(6) and (7), as will be seen in the next section. In summary we can say that the mobility of the proton-induced superionic conduction is remarkably high, compared with those of superionic conduction due to heavier ions than a proton.

2.3. *Comparison between quantum mechanical and classical diffusion times*

In this section we first define the quantum mechanical diffusion constant, for example, for $H_2XO_4^{(+e)}$, as follows;

$$D_{qm} = \frac{|\Gamma| R_0^2}{\hbar} \times 1.070. \tag{9}$$

For the diffusion constant of $XO_4^{(-e)}$, a numerical factor of 1.070 in Eq.(9) should be replaced by 1.066. From this diffusion constant we can define the quantum mechanical diffusion time, $\tau_{qm} = \hbar/\Gamma$. On the other hand, the classical diffusion time to move over a distance R with a frequency factor ν, τ_{cl}, is defined as

$$\tau_{cl} = \frac{1}{\nu \exp(-E_a/k_B T)}. \tag{10}$$

By inserting an observed value of $E_a = 0.26$ eV, $T = 300$ K and $\nu = 10^{13}\text{sec}^{-1}$, τ_{cl} is estimated as 1ns, while τ_{qm} is estimated as 10 ps for a value of $|\Gamma| = 10^{-4}$ eV. Thus τ_{cl} is much longer than τ_{qm}. Accordingly the quantum mechanical mobility defined by Eqs.(6) and (7) which we denote μ_{qm} is 10 to 100 times faster than the classical mobility which we denote μ_{cl}.

2.4. *Estimation of the quantum mechanical mobility and ionic conductivity for $Rb_3H(SeO_4)_2$*

Now let us estimate μ_{qm} and the ionic conductivity. In doing so we recall that two kinds of excited states are formed thermally by breaking a hydrogen-bond within a layer. Thus the concentration of excited states, n, is expressed in the form of the Boltzmann distribution,

$$n = N\mathrm{e}^{-E_a/k_B T}, \tag{11}$$

where E_a is the formation energy for creating simultaneously an $H_2XO_4^{(+e)}$ and an $XO_4^{(-e)}$ excited state thermally, and N the total number of the hydrogen bonds. Thus, using Eqs.(6) and (7) the static ionic conductivity is obtained as follows;

$$\begin{aligned} \sigma &= qn(\mu + \mu') \\ &= \frac{q^2 R_0^2 N (1.070|\Gamma| + 1.066|\Gamma'|)}{\hbar k_B T} \mathrm{e}^{-E_a/k_B T}, \end{aligned} \tag{12}$$

where we have taken $q = |q'|$. We estimate the magnitude of σ for the case of $Rb_3H(SeO_4)_2$. From experimental results, we obtain $E_a = 0.26$ eV, $R_0 = 5.8 \times 10^{-8}$ cm, and $T_c = 449$ K. Then, choosing that $q = |q'| = e$, $N = 10^{22}$ cm^{-3}, and $|\Gamma| = |\Gamma'| = 10^{-4}$ eV, we estimate the mobility as

$$\mu_{total} = \mu + \mu'^{-3}\text{cm}^2/\text{Vsec}. \tag{13}$$

Thus the conductivity σ has the order of a magnitude of 10^{-2} S/cm, consistent with experimental results.

2.5. *Summary of Section 2 and concluding remarks*

In this paper we have described a quantum mechanical mechanism of proton-induced ionic conduction in the superionic phase in zero-dimensional hydrogen-bonded $M_3H(XO_4)_2$ [M=K,Rb,Cs, X=S,Se] materials. We have showed that ionic conduction occurs by the coherent motion of two kinds of excited states, $H_2(XO_4)^{(+e)}$ and $(XO_4)^{(-e)}$, induced by the successive tunneling of protons in the elongated hydrogen-bonds. Therefore, though the same excited states carry ionic currents, the protons which induced the coherent conductivity do not itinerate in a crystal. Different protons in respective hydrogen-bonded areas are involved in successive tunneling from an elongated hydrogen-bond site to a different site. In this context we should emphasize that the key feature of the conduction mechanism mentioned in this paper is quantum-mechanical so that the present mechanism is applicable only to the materials of proton-induced ionic conduction. As regards the ordinary superionic conduction by heavier ions than a proton, Ishii and Kamishima have developed a microscopic mechanism extensively by taking account of microscopic interactions of carries with strain [12].

Finally a prediction is made on an interesting phenomena in which the suppression of the superionic conduction might occur in the presence of an alternating electric field, when a frequency of the alternating field increases up to the terahertz region. When we calculate a frequency-dependent conductivity $\sigma(\omega)$, we must remember that the quantum mechanical mechanism proposed by Ito and Kamimura holds in the condition that, during the coherent motion of the $H_2XO_4^{(+e)}$ and $XO_4^{(-e)}$ excited states, a direction of the alternating electric field of an electromagnetic wave must not be changed. Thus, when the direction of an alternating electric field $E = E_0 \exp(i\omega t)$ is reversed, the present mechanism of the superionic conduction is suppressed.

Here we derive a condition with regard to a time scale in which the superionic conduction holds. For the alternating electric field $E(\omega)$, let us introduce a time τ defined by $\tau = 2\pi/\omega$. Then, when τ is longer than a time τ_{qm} defined by $\tau_{qm} = \hbar/|\Gamma|$, the superionic conduction continues to occur. However, when τ is shorter than τ_{qm}, the present mechanism of superionic conduction is suppressed. Thus the frequency dependent conductivity $\sigma(\omega)$ is expected to decrease sharply at $\omega = \omega_{qm}$, when ω increases from zero, where $\omega_{qm} = 2\pi/\tau_{qm}$. Since the values of $|\Gamma|$ are in the range of the order of 10^{-3} to 10^{-4} eV, the value of ω_{qm} is estimated to be in the boundary between gigahertz and terahertz. Thus, when ω will exceed ω_{qm}, we may expect that $\sigma(\omega)$ exhibits the characteristic features of coupled rotation and vibration spectra of a single XO_4 tetrahedron and/or coupled XO_4 tetrahedrons, instead of those of superionic conduction. In fact, Kamimura, Ikehata, Matsuo and Yoshida succeeded recently in observing such an anomalous spectra for $Rb_3H(SeO_4)_2$ in the terahertz region for the first time [13].

3. Theory of superionic conduction below and at the ferroelastic phase transition

3.1. *The Ginzburg-Landau theory for the ferroelastic phase transition in $M_3H(XO_4)_2$*

In this section we study the temperature dependence of an order parameter to express a change of the ordering of the hydrogen bonds in the ferroelastic phase following the G-L theory developed by Plakida and Saljda (1988) for the first-order phase-transition from the ferroelastic to paraelastic phase in $M_3H(XO_4)_2$ [2]. In the framework of the G-L theory Plakida and Salejda introduced an order parameter ϕ in order to represent the magnitude and the direction of a spontaneous strain in the ferroelastic phase which occurs when each tetrahedron forms a hydrogen-bonded pair. They took one component of ϕ, say ϕ_1 along the direction of a spontaneous strain in the $a - b$ plane as an order parameter and the other two components to vanish. Following them, in this paper we write an order parameter as r instead of ϕ_1. Thus, when $r = 0$, the hydrogen-bonded pairs disappear and the distance between all the tetrahedron becomes equal to R_0, which is the distance between neighboring (XO_4) tetrahedrons in the paraelastic phase above T_c. Following Plakida and Salejda, the free energy near the first order phase transition is expressed in terms of r as follows;

$$F = F_0 + \frac{\alpha}{2}r^2 + \frac{\beta}{4}r^4 + \frac{\beta_0}{6}r^6, \tag{14}$$

where $\alpha = \alpha_0(T - T_c)$ with α being a positive sign of α_0. The coefficients α and β depend on the elastic constants of materials and the coupling between the order parameter and elastic strains.

By minimizing F with regard to r, $\partial F/\partial r = 0$, we obtain the following form for the order parameter $r(T)$ as to its temperature dependence,

$$r(T) = \frac{1}{2}\sqrt{\beta + \sqrt{\beta^2 + \frac{\alpha_0}{\beta_0}(T_c - T)}} \ . \tag{15}$$

Near T_c, we can approximate Eq. (15) as

$$r(T) = u_0\sqrt{(T_c - T) + b} \ . \tag{16}$$

Thus, when we adopt the mean-field approximation for the ferroelastic phase transition in the zero-dimensional hydrogen-bonded dielectric materials, the order parameter to express the geometrical arrangement of hydrogen-bonds in the ferroelastic phase changes following $\sqrt{(T_c - T) + b}$ in its temperature dependence. We call this phase as the G-L phase.

On the basis of Eq. (16) we can calculate the ionic conductivity due to the hopping of a proton from a hydrogen-bonded $(XO_4)_2$ pair to a neighboring site between $(XO_4)_2$ in an insulating ferroelastic phase. Since the number of the hydrogen-bonded $(XO_4)_2$ pairs is constant and the distance between $(XO_4)_2$ pair without a hydrogen is much longer than the distance between $(XO_4)_2$ pair bonded by a hydrogen, the temperature dependence of the ionic conductivity may be expressed as an activation type such as $\sigma_0 = \exp(-E_a/k_B T)$ reflecting a semiconducting behavior in the ferroelastic phase, where the activation energy E_a depends on the geometrical arrangement of hydrogen-bonds, i.e., the order parameter $r(T)$. Thus E_a changes with temperature through the order parameter r and becomes a constant at $T = T_c$. Therefore, it is clear that the G-L phase can not explain the observed power-law divergent behavior of the conductivity at $T = T_c$. In this context we would like to propose a new idea to explain the divergent behavior of ionic conduction for $T \leq T_c$ in the next section.

3.2. *A new idea for explaining the divergent behavior of ionic conduction just below and at* $\mathbf{T = T_c}$ *in the ferroelastic phase*

In this section we propose a new idea to explain the divergent behavior of the ionic conduction in the ferroelastic phase of $M_3H(XO_4)_2$ for $T \leq T_c$. Just below T_c we consider that a hydrogen-bonded system in $M_3H(XO_4)_2$

consists of a mixed phase in the presence of an applied electric field by precursor effects of a phase transition. In this mixed phase a ferroelastic phase near T_c consists of a mixture of the following two regions; the quasi-paraelastic region of the superionic phase and the G-L regions by precursor effects. In the quasi-paraelastic regions all the distances between (XO_4) tetrahedrons are the same while in the G-L regions (XO_4)'s form still the hydrogen-bonded dimmers. A schematic picture of this "mixed phase" by precursor effects near T_c is shown in Fig.6.

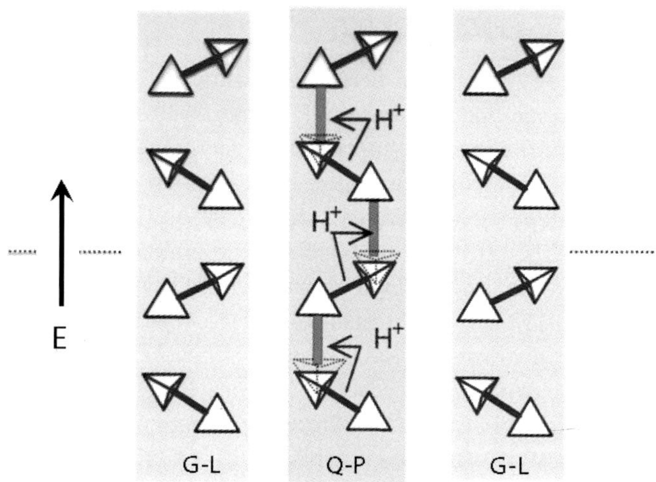

Figure 6. Schematic picture of a mixed phase just below the phase transition, where G-L and Q-P in the figure represent a Ginzburg-Landau region and a quasi-paraelastic region, respectively.

In order to realize such mixed phase, the energy of the mixed phase must be lower than that of the G-L phase for the same temperature. In the following section, by using a simple theoretical treatment we will show that the energy of the mixed phase becomes lower than that of the G-L phase when the successive transfer motion of the protons along the direction of an applied electric field is possible. Meantime, in this section we describe an underlying physical concept of why a mixed phase is necessary for explaining the divergent behavior of ionic conductivity just below T_c and of why the Ginzburg-Landau theory must be modified. When the quasi-paraelastic regions are introduced into the G-L regions which are dominant in the ferroelastic phase, the elastic energy of (XO_4) tetrahedron-pairs in the quasi-paraelastic regions increase, because the distances between (XO_4) tetrahedrons in the quasi-paraelastic regions become longer. Now let us apply an alternating electric field $E = E_0 e^{i\omega t}$ to the mixed phase. Then the hydrogen in certain hydrogen-bonds in each quasi-paraelastic region are ionized to protons and electrons by the effect of the rotational displacement of tetrahedrons within quasi-paraelastic regions at high temperature near T_c. Since all the distances between (XO_4) tetrahedrons are of the same in the quasi-paraelastic regions, a proton at the ij site in a quasi-paraelastic region can tunnel to a neighboring empty site, say a jk site between tetrahedrons j and k, and so on, as seen in Fig. 7.

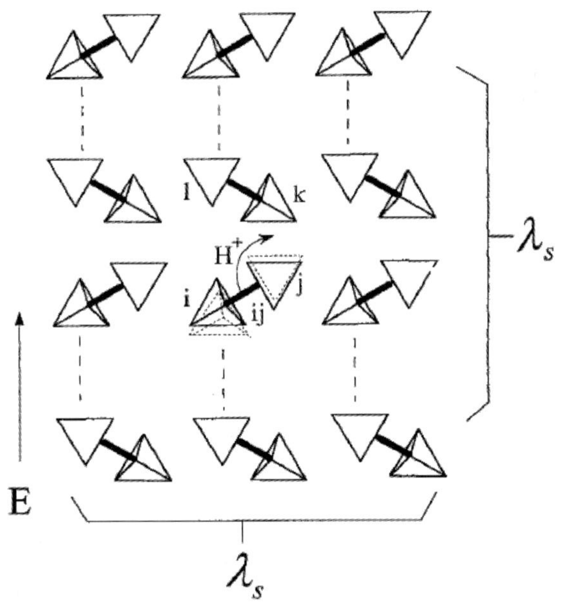

Figure 7. Microscopic view of the quasi-paraelastic (Q-P) region.

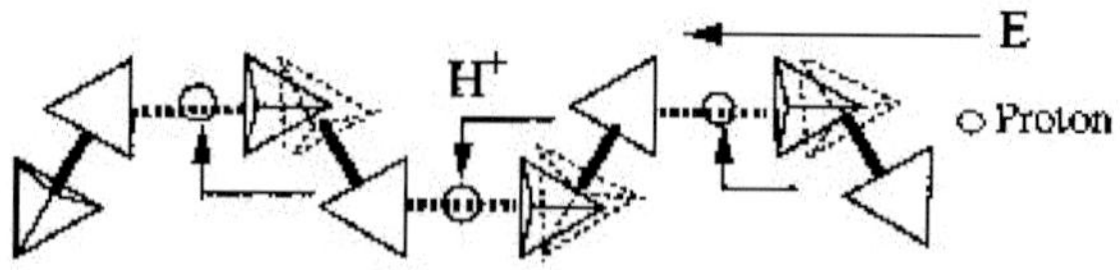

Figure 8. Schematic view of the coherent motion of the $H_2(XO_4)^{(+e)}$ excited state along the *zigzag paths* induced by the successive transfer of protons in the elongated hydrogen bonds along the alternating electric field E.

As a result the *zigzag paths* for the successive motion of protons are created along the electric field as seen in Fig. 8. Thus the kinetic energy of a proton along the *zigzag path* is reduced by the transfer interaction Γ. When the reduction of the kinetic energy of protons overcomes the increase of the elastic energy, a mixed phase is realized.

3.3. *Derivation of an energy criterion for the realization of the mixed phase by a simple consideration*

Based on the physical concept described in a previous section, in this section we will derive an energy criterion in order for a mixed phase to be realized. Since a general theoretical treatment must take into account the elastic energy for a whole hydrogen-bonded system in a crystal expressed by the strain, the interaction energy among strain, the rotation of XO_4 tetrahedrons and the transferring motions of protons, and etc., it is very complicated to derive a complete formula for the energy criterion. In order to avoid such a detailed theoretical treatment, here we use a simple argument for deriving the energy criterions, as shown below.

Let us denote the number of the *zigzag paths* along the electric field within a quasi-paraelastic (Q-P) region λ_s, as indicated in Fig. 7. As explained in a previous section, in the Q-P region the elastic energy is increased, because the distance between (XO_4) tetrahedrons is forced to become longer. Suppose that this distance is the same as that in the paraelastic phase above T_c, R_0. Then we may express simply an elastic energy per hydrogen-bond in a Q-P region in the form of $k{R_0}^2/2$. As to the G-L region we introduce R for expressing the distance between (XO_4) tetrahedrons in each dimer, which is bonded by hydrogen. Since R corresponds to the length of a hydrogen-bond in the ferroelastic phase, R may be expressed as $R = R_0 - r$ with use of the order parameter r. At the phase transition from the ferroelastic to the paraelasitc phase in the G-L phase, r vanishes. Thus the energy difference between the mixed phase and the G-L phase per unit area, $\Delta\epsilon = F_{mix} - F_{G-L}$, is expressed as

$$\Delta\epsilon(T) = \frac{k}{2}({R_0}^2 - R^2) \quad , \tag{17}$$

where F_{mix} and F_{G-L} represent the energies of the mixed phase and the G-L phase, respectively. Since $\Delta\epsilon$ depends on temperature T through the order parameter r, we have indicated the temperature dependence explicitly in Eq. (17). By inserting $R = R_0 - r$ into Eq.(17), we obtain

$$\Delta\epsilon(T) = krR_0 - \frac{k}{2}r^2 \quad . \tag{18}$$

In the present model the area of a Q-P region in the *ab* plane is taken to be a square whose area is λ_s^2, as shown in Fig. 7, where the distance between the neighboring *zigzag paths* is taken to be unity. Thus in the presence of an electric field $E = E_0 e^{i\omega t}$, a static part of the change of the total energy, ΔE, is given by

$$\Delta E(T) = \Delta\epsilon\lambda_s^2 - a\Gamma\lambda_s E_0 \quad , \tag{19}$$

where a is a constant with a dimension of volt^{-1}. The second term in the right hand side of Eq. (19) represents the energy decrease due to the transfer motion of a proton along the direction of an applied electric field. Inserting Eq. (18) into the first term of the right-handed side of Eq. (19), we obtain

$$\Delta E = (krR_0 - \frac{k}{2}r^2)\lambda_s^2 - a\Gamma\lambda_s E_0 \ . \tag{20}$$

By minimizing ΔE with regard to λ_s, that is from $\partial(\Delta E)/\partial\lambda_s = 0$, we obtain an optimized λ_s ($= \lambda_{sm}$) in the following form;

$$\lambda_{sm}(T) = \frac{a\Gamma E_0}{2kR_0 r - kr^2} \ . \tag{21}$$

Thus, when the number of the hydrogen bonds along a *zigzag path* is given by Eq. (21) for temperature T, the mixed phase begins to form. Thus the energy gain of the mixed phase compared with that of the G-L phase is given by

$$\Delta E(\lambda_{sm}) = -\frac{1}{2}a\Gamma\lambda_{sm}E_0 \ . \tag{22}$$

This means that the mixed phase becomes stable for a certain temperature T just below T_c, compared with the G-L phase. Since the order parameter

r vanishes at the ferroelastic transition temperature T_c, we can neglect the r^2 term in the denominator of Eq. (21) near the ferroelastic to paraelastic phase transition. Then Eq. (21) is approximately expressed as

$$\lambda_{sm} = \frac{a\Gamma E_0}{2R_0 kr} \ . \tag{23}$$

When we insert Eq. (16) into r in Eq. (23), λ_{sm} is expressed as

$$\lambda_{sm}(T) = \frac{a\Gamma E_0}{2R_0 k u_0 \sqrt{(T_c - T) + b}} \ . \tag{24}$$

This expression is consistent with the observed temperature dependence of the conductivity, as will be seen in the following section.

3.4. *Expression of the ionic conductivity for* $T \leq T_c$

Now let us calculate the conductivity below and at the phase transition. As shown in Fig. 8, an $H_2(XO_4)^{(+e)}$ excited state moves along the direction of an applied electric field within a quasi-paraelastic region. Thus the ionic current flows on a *zigzag path* along which an $H_2(XO_4)^{(+e)}$ excited state moves. Since the number of the *zigzag paths* along the electric field is λ_s, as showed in Fig. 7, the ionic current I is expressed in the following form,

$$I = q\lambda_{sm}\mu = \frac{1.070 q^2 \tilde{R}^2 \Gamma}{\hbar k_B T} \frac{a\Gamma}{2kR_0 u_0 \sqrt{(T_c - T) + b}} E \ . \tag{25}$$

where we have inserted Eq. (24) for λ_{sm} and Eq. (6) for μ.

Thus the conductivity below the transition temperature T_c is given by

$$\sigma = \frac{1.070 a}{\sqrt{(T_c - T) + b}} \frac{q^2 \tilde{R}^2 \Gamma^2}{2k u_0 R_0 \hbar k_B T} \ . \tag{26}$$

Thus the conductivity exhibits a divergence behavior expressed as the $(T_c - T + b)^{-1/2}$ power-law at $T = T_c$. This is consistent with experiment result shown in Fig. 2.

3.5. *Summary of Section 3*

In section 3 we have proposed a mechanism of ionic conduction near and just below the phase transition temperature in $M_3H(XO_4)_2$ [M=K,Rb,Cs, X=S,Se]. We have showed that ionic conduction occurs by the coherent motion of two kinds of excited states, $H_2(XO_4)^{(+e)}$ and $(XO_4)^{(-e)}$, induced by the successive tunneling of different protons in the expanded hydrogen-bonds in the quasi-paraelastic regions. In this context we emphasize that the key feature of the present mechanism is quantum-mechanical despite the ionic conduction. The clarified features of ionic conduction at the phase transitions are the followings: (1) When the temperature is increased toward T_c, ferroelastic $M_3H(XO_4)_2$ materials with the zero-dimensional hydrogen-bond network are governed by a mixed phase consisting of the G-L regions and of the quasi-paraelastic (Q-P) regions in the presence of an electric field ; (2) below T_c the energy of a mixed phase is lower than that of the G-L phase in the presence of an applied electric field; (3) since in each Q-P region all the distances between the neighboring tetrahedrons are elongated and equal, protons in the Q-P regions can tunnel from an elongated hydrogen-bonded position between (XO_4) tetrahedrons to a neighboring intervening position between (XO_4) tetrahedrons along the *zigzag paths* in the direction of an alternating electric field; and (4) this results in the coherent transfers of the $H_2(XO_4)^{(+e)}$ and $(XO_4)^{(-e)}$ excited states along the electric field; and thus (5) the temperature dependence of conductivity is described by the power-law of $(T_c - T + b)^{-1/2}$ near the phase transition temperature.

To our knowledge, this work is the first attempt for the detailed theoretical study of transport phenomenon in $M_3H(XO_4)_2$ materials in their ferroelastic phase, although Kamimura and Watanabe proposed an idea in their preliminary work [7].

Acknowledgments

It is a pleasure to acknowledge the collaboration of Takuo Ito and Satoru Watanabe on the theoretical work mentioned in this paper. I would like also to acknowledge many helpful and stimulating conversations on the experimental results mentioned in this paper with Professor Yasumitsu Matsuo of Setsunan University, Professor Seiichiro Ikehata, Dr. Junko Hatori and Dr. Yukihiko Yoshida of Tokyo University of Science.

References

1. H. Kamimura, Y. Matuso, S. Ikehata, T. Ito, M. Komukae, and T. Osaka, *Phys. Stat. Solidi* (b) **241** (2004) 61.
2. N.W. Plakida and W. Salejda, *Phys. Stat. Solidi* (b) **148**(!988) 473.
3. L.D. Landau and E. M. Lifshitz, *Statistical Physics* (London, Pergamon, 1958).
4. Y. Ito and H. Kamimura, *J. Phy. Soc. Jpn.* **67** (1998) 1999.
5. T. Matsubara and Y. Toyozawa, *Prog. Theor. Phys.* **26** (1961) 739.
6. R. Kubo, *J. Phys. Sco. Jpn.* **12** (1957) 570.

7. H. Kamimura and S. Watanabe, 2001, *Philos. Mag.* **81** (2001) 1011.
8. Y. Suwa, J. Yamauchi, H. Kageshima and S. Tsuneyuki, *Material Science and Engineering B* **79** (2001) 31, and 98.
9. R. Haydoc, V. Heine, and M.J. Kelly, J. Phys. C **5** (1972) 2846.
10. R. Haydoc, V. Heine, and M.J. Kelly, J. Phys. C **8** (1975) 2591.
11. H. Boettger and V.V. Bryksim, *Hopping Conduction in Solids*, Chapter 4 (Akdemie-Verlag Berlin, 1985)
12. T. Ishii and O. Kamishima, *J. Phys. Soc. Jpn.* **70** (2001) 159, related references therein.
13. H. Kamimura, S. Ikehata, Y. Matsuo and Y. Yoshida, *Physics of Semiconductors, 28th International Conference* **CP893**, edited by W. Jantsch and F. Schaeffler, 2007, 1485.

MASSIVE COLLECTION OF FULL-LENGTH COMPLEMENTARY DNA CLONES AND MICROARRAY ANALYSES: KEYS TO RICE TRANSCRIPTOME ANALYSIS*

SHOSHI KIKUCHI†
Plant Genome Research Unit
Division of Genome and Biodiversity Research
National Institute of Agrobiological Sciences (NIAS)
Kannondai 2-1-2
Tsukuba, Ibaraki 305-8602, Japan

Completion of the high-precision genome sequence analysis of rice led to the collection of about 35,000 full-length cDNA clones and the determination of their complete sequences. Mapping of these full-length cDNA sequences has given us information on (1) the number of genes expressed in the rice genome; (2) the start and end positions and exon–intron structures of rice genes; (3) alternative transcripts; (4) possible encoded proteins; (5) non-protein-coding (np) RNAs; (6) the density of gene localization on the chromosome; (7) setting the parameters of gene prediction programs; and (8) the construction of a microarray system that monitors global gene expression. Manual curation for rice gene annotation by using mapping information on full-length cDNA and EST assemblies has revealed about 32,000 expressed genes in the rice genome. Analysis of major gene families, such as those encoding membrane transport proteins (pumps, ion channels, and secondary transporters), along with the evolution from bacteria to higher animals and plants, reveals how gene numbers have increased through adaptation to circumstances. Family-based gene annotation also gives us a new way of comparing organisms. Massive amounts of data on gene expression under many kinds of physiological conditions are being accumulated in rice oligoarrays (22K and 44K) based on full-length cDNA sequences. Cluster analyses of genes that have the same promoter *cis*-elements, that have similar expression profiles, or that encode enzymes in the same metabolic pathways or signal transduction cascades give us clues to understanding the networks of gene expression in rice. As a tool for that purpose, we recently developed "RiCES", a tool for searching for *cis*-elements in the promoter regions of clustered genes.

*This work was partly supported by a grant from the Ministry of Agriculture, Forestry, and Fisheries of Japan (Integrated Research Project on Plants, Insects, and Animals Using Genome Technology, no. QT1003) and by a grant from the budget of the Generation Challenge Program SP4 project.

1. Introduction

1.1. *Massive collection of full-length cDNA clones and their sequence information: starting materials for transcriptome analysis*

Highly accurate genome sequences of rice are available (1–4). Genome sequences of rice (*O. sativa* ssp. *japonica* 'Nipponbare') have been assembled independently by The Institute for Genome Research (TIGR), the International Rice Genome Sequencing Project (IRGSP), and the Beijing Genomics Institute (BGI). In addition, full-length complementary DNA (FL-cDNA) sequences (5, 6) and expressed sequence tags (ESTs) (7–9) from rice have served as valuable resources for genomic and genetic studies.

FL-cDNA collections have been established for several organisms (10–12). The cDNA information has been used extensively to determine gene annotation, structures, and the start and end sites of transcription. The cDNA information is also indispensable for validation of gene function by reverse genetics. The rice FL-cDNA collection has also been used to complement the information obtained by genome sequencing. The first report of a rice FL-cDNA collection, published in 2003 (5), described the characteristics and annotation of 28K FL-cDNA sequences. The collection has now expanded to 578K FL-cDNA clones, among which 35K cDNA clones were completely sequenced and annotated (available at Knowledge-based Oryza Molecular biological Encyclopedia (KOME): http://cdna01.dna.affrc.go.jp/cDNA/).

Annotation of the complete genome of Nipponbare was performed by the Rice Annotation Project (RAP). All functional annotations for proteins and non-protein-coding RNA (npRNA) candidates were manually curated. Functions were identified or inferred for 19,969 (70%) of the proteins, and 131 possible npRNAs (including 58 antisense transcripts) were found. Almost 5000 annotated protein-coding genes were found to be disrupted in insertional mutant lines. The rice loci were determined by using cDNA sequences obtained from rice and other representative cereals. Our conservative estimate based on these loci and an extrapolation suggested that there are about 32,000 rice genes—fewer than previous estimates (13).

Here we show the comparison of the transcription units (TUs) mapped by the full-length cDNA sequences to the rice genome sequence and the loci predicted by the gene annotation computer program (TIGR-CDS). About 5400 TUs are generated only by the full-length cDNA mapping and are not predicted by the gene annotation program. Those are so-called non-annotated expressed (NAE) genes. Detailed structural, gene expression, and homology analyses have revealed that many of the rice NAE genes are similar to the npRNA genes in mouse.

1.2. *Gene-family–based functional annotation*

Family-based gene annotation was launched after global gene annotation. We will first focus here on the genes encoding membrane-associated proteins. Cells maintain their biological activities by importing and exporting various substances. Provision of energy and nutrients and efflux of salts, biochemicals, and ions are necessary to maintain biological activity in prokaryotic and eukaryotic cells. Environmental situations within cells differ among organisms: unicellular organisms cannot control the ion concentrations outside cells, but multicellular eukaryotes (especially animals) can precisely regulate the ion concentrations of their cellular environments within micromolar ranges. Therefore, we can expect organisms to differ in gene number, structure, and function according to their biological abilities and environmental situations.

We searched for orthologs of known membrane transport genes among the 35,180 full-length rice cDNA sequences (5, 6) and genomic sequence data from Arabidopsis (14) and *japonica* rice (1,4), and among global functional gene annotations in Arabidopsis and rice (Munich Information Center for Protein Sequences [MIPS] data service, http://mips.gsf.de/proj/plant/jsf/ (15, 16); Rice Annotation Project Data Base [RAP-DB], http://rapdb.lab.nig.ac.jp/ (13, 17); TIGR Rice Genome Annotation, http://www.tigr.org/tdb/e2k1/osa1/index.shtml (18)). Transmembrane proteins have a hydrophobic structure, a pore-forming sequence, and molecule-binding sites. Because of these specific structural features, the identification of membrane transport orthologs is clear from computer calculations. Previous reports have characterized individual transporter protein families but have not extended to whole transport systems in general (19–22). In a more general analysis of various organisms, the features of prokaryotes were contrasted with those of eukaryotes (23). However, differences among eukaryotes—especially animals and plants—were not a focus of that analysis. We also searched for orthologs of membrane transport genes in various organism databases (Human Gene Nomenclature Database Search Engine, http://www.genenames.org/cgi-bin/hgnc_search.pl (24); Genomic Comparison of Membrane Transport Systems [TransportDB], http://www.membranetransport.org/index.html (25); Functional Genomics of Plant Transporters [PlantsT], http://plantst.genomics.purdue.edu/ (26); ARAMEMNON, http://aramemnon.botanik.uni-koeln.de/ (27)). We compare total membrane transport systems from diverse organisms and conclude that membrane transport genes exemplify the evolutionary diversity of homeostatic systems. The evolutionary changes in gene families indicate the dynamics of alterations in biological systems and gene networks. Therefore, analysis of large categories of gene families may reveal many basic concepts of biological systems.

1.3. *Establishment of rice microarray systems and development of a tool for searching for* cis*-elements in the promoter regions of clustered genes*

On the basis of the results of our large-scale FL-cDNA analysis (5), we have constructed a monitoring system that uses an oligonucleotide array to monitor gene transcriptional levels and to develop genome-wide functional analysis of rice. The array (22K rice oligoarray) was composed of 21,938 probes with 60-mer oligonucleotides synthesized at gene-specific regions (28–30) from 32,127 FL-cDNAs. Mapping of these cDNA clones to genomic DNA revealed that there were about 20,500 TUs, and clustering of the clones revealed a unique clone set. Two 22K array platforms are registered in NCBI-GEO (National Center for Biotechnology Information – Gene Expression Omnibus). Platform GPL477 is a prototype version of the 22K array. One study, which was the comparison of gene expression profiles of rice callus by the Gibberellic acid and Abscisic acid treatments has used this array system (31). Platform GPL892 is the commercial version of the 22K array (Agilent Technologies catalog number G4138A) and has been used to accumulate many environmental stress data. At the Rice Genome Resource Center at our institute we have used these arrays and a new 4 × 44K format from Agilent Technologies to develop a new rice oligoarray (GPL6864) covering all the genes expressed in the rice genome.

If good-quality RNA samples are prepared, the oligomicroarray system provides us with highly reproducible gene expression data. It takes just a few days to obtain the gene expression data, but the subsequent data mining process can take many months. For systematic data analyses, functional annotation data on each probed gene must be well facilitated and good tools must be available. One such useful tool would search for *cis*-elements in the promoter regions of clustered gene sets after microarray analysis. The existence of common *cis*-elements in the promoter regions of clustered gene sets may suggest that those genes are controlled by the same transcription factor.

Cis-elements in the promoter regions of genes and *trans*-acting transcription factors are major biological features to be characterized if we are to achieve an understanding of the systems that regulate gene expression. Identification of candidate *cis*-elements corresponding to genes is now practicable through the use of available sequence and genome mapping information, combined with information about the responses of genes to specific experimental conditions; such responses have been elucidated by using the gene expression profiles now publicly available.

Exhaustive sequence analysis using available public databases can identify *cis*-element candidate motifs for further examination, but such approaches are not efficient. One confounding factor is that public databases are independently constructed and not generally optimized to facilitate the integration of information from many sources with local experimental data. A more perplexing issue for experimental researchers who are not familiar with bioinformatics techniques is the challenge of finding unknown but biologically notable relationships among genes, *cis*-elements, and experimental conditions from the huge number of possible combinations generated by large experimental datasets.

To resolve some of these issues, we developed a novel data mining tool to identify *cis*-elements in the rice genome. It performs the complex bioinformatics analysis mentioned above, and then lists *cis*-element candidates for genes. The genes can be grouped by similarity of expression profiles and other criteria for assessment by researchers, and the tool then annotates them with related public database information.

Similar tools have been developed previously. Helden released RSAT, which includes a program that can detect motifs over-represented in the upstream regions of co-regulated genes (32). Holt et al. established CoReg, which links the hierarchical clustering of co-expressed gene sets with frequency tables of promoter elements (33). Zhao et al. established TRED, which integrates a database and a system for predicting *cis*- and *trans*-elements in mammals (34).

Our novel tool searches for *cis*-element candidates in the upstream, downstream, or coding regions of differentially regulated genes. The tool first lists *cis*-element candidates by motif searching based on the supposition that if there are *cis*-elements playing important roles in the regulation of a given set of genes then they will be statistically over-represented and will be conserved. Then it evaluates the likelihood scores of the listed candidate motifs by association rule analysis. This strategy depends on the idea that motifs over-represented in the promoter region could play specific roles in the regulation of expression of these genes. The tool is designed so that any biological researchers can use it easily at the publicly accessible Internet site http://hpc.irri.cgiar.org/tool/nias/ces. We evaluated the accuracy and utility of the tool by using a dataset of auxin-inducible genes that have well-studied *cis*-

elements. The test showed the effectiveness of the tool in identifying significant relationships between *cis*-element candidates and related sets of genes.

2. Results and Discussion

2.1 *Comparison of gene models predicted by gene prediction programs and transcription units identified by the mapping of full-length cDNA sequences*

We mapped 578,000 rice (ssp. *japonica*, 'Nipponbare') FL-cDNA clones (DDBJ accession numbers: completely sequenced FL-cDNA: AK058203–068528, AK068530–068912, AK068914–70720, AK070722–074028, AK098843–112119, AK119160–122186, AK240633–243692; one-pass sequences of FL-cDNA, 5′ end: CI285358–311811 and CI563340–778739, 3′ end: CI000001–285357 and CI311812–563339) to five genome assemblies: TIGR4, IRGSP3, IRGSP4, the Nipponbare genome determined by Syngenta, and the 93-11 genome by BGI (Table 1). The mapping criteria were 95% identity and 90% coverage. For proper comparison between the assemblies, the results of mapped cDNAs in the respective assemblies were compared with those of TIGR4. The numbers of mapped FL-cDNA clones differed among assemblies, with the highest in TIGR4. The orientation in which some cDNA clones were mapped onto a chromosome and the chromosome on which some clones were mapped were not consistent among the assemblies. Of the 32,775 completely sequenced FL-cDNA clones mapped in TIGR4, 29,925 were also mapped in all of the other assemblies; however, the number of clones commonly mapped in both TIGR4 and a given assembly differed (Table 1). The maximum and minimum numbers of common clones were 32,730 in IRGSP4 and 30,162 in 93-11, respectively. The number of mapped clones was greater in the *japonica* rice genomes than in the *indica* genomes; this might reflect differences in the genome sequences between subspecies. The number of clones common to TIGR4 and IRGSP4 was close to that common to TIGR4 and IRGSP3, and both numbers were greater than the number of clones common to TIGR4 and the Syngenta sequence. This suggests that the differences in numbers of common clones may have resulted from differences in the sequencing methods adopted in the assemblies (TIGR4 and IRGSP by the map-based method; Syngenta sequence by the whole-genome shotgun method). Mapping of 578K FL-cDNA clones identified about 28,500 loci in the *japonica* genome and 27,800 loci in the *indica* genome. A total of 29,925 completely sequenced FL-cDNAs were mapped in all the genome assemblies, and more than 90% of the FL-cDNAs were mapped in all five assemblies (Table 1). We therefore decided to use only the mapping results of TIGR4 for further analyses, not those from the other assemblies. The number of predicted loci was about 56K, which was sufficient for our data analysis but probably not sufficient to reach complete accuracy of gene prediction and annotation of TIGR4.

Table 1. FL-cDNA clones mapped to five rice genome assemblies.
(Source: Satoh et al. PLoS One 2, e1235(6))

	origin sequencing All	*japonica* genome				*indica* genome
		Map-base cloning			whole shotgun	
		TIGR	IRGSP4	IRGSP3	Syngenta	93-11
FL-cDNA	35,187	32,775	32,745	32,640	31,928	30,354
5endFLEST	241,854	212,598	212,539	211,564	208,606	199,001
3endFLEST	536,885	483,657	484,358	482,909	482,665	465,775
FL-cDNA locus	Chr1	4,026	4,021	4,039	4,050	3,940
	Chr2	3,196	3,198	3,215	3,186	3,153
	Chr3	3,569	3,567	3,566	3,597	3,607
	Chr4	2,531	2,530	2,534	2,477	2,493
	Chr5	2,313	2,305	2,310	2,338	2,329
	Chr6	2,292	2,293	2,290	2,262	2,266
	Chr7	2,183	2,185	2,193	2,165	2,021
	Chr8	1,933	1,934	1,939	1,912	1,827
	Chr9	1,605	1,605	1,574	1,545	1,515
	Chr10	1,538	1,528	1,536	1,502	1,416
	Chr11	1,685	1,683	1,675	1,486	1,333
	Chr12	1,693	1,692	1,705	1,523	1,435
	Chr0 [(a)]				434	497
	Total	28,564	28,541	28,576	28,477	27,832
Comparison of FL-cDNA mapping with TIGR4		Both mapped	32730	32623	31741	30162
		Same Chr-Same Strand	32646	32611	30422	28760
		Same Chr-Reverse Strand	80	10	317	335
		Differential Chr.	4	2	1002	1067
		Mapped on only TIGR	45	152	1034	2613
		Unmapped on only TIGR	15	17	187	192
		Both unmapped	2397	2395	2225	2220
		mapped on all assemblies			29925	
		unmapped on all assemblies			2186	

*: sequence-assembled contigs that were not localized to one of the 12 chromosomes.
doi:10.1371/journal.pone.0001235.t001

A total of 55,890 gene loci were predicted in the rice genome according to TIGR OSA1 release 4. Mapping of FL-cDNA clones on TIGR4 revealed that 533,667 FL-cDNA clones were derived from 28,564 FL-cDNA loci (Table 2). FL-cDNA loci were cross-referenced with TIGR4 loci to examine the overlaps between the two groups. According to the sources of mapped loci and the occurrence of overlap, the loci were classified as follows: (1) when a FL-cDNA locus overlapped with a TIGR4 locus, the FL-cDNA locus was defined as FL-AE (annotated expressed) and the TIGR4 locus was defined as coding-sequence–AE (CDS-AE); (2) a FL-cDNA locus that did not overlap with any TIGR4 locus was defined as FL-NAE (non-annotated expressed); and (3) a

TIGR4 locus that did not overlap with any FL-cDNA locus was defined as CDS-ANE (annotated non-expressed). On the basis of these definitions, the loci were classified into 23,117 FL-AE, 23,193 CDS-AE, 5447 FL-NAE, and 32,697 CDS-ANE (Table 2).

Table 2. Comparisons of FL-cDNA loci and TIGR4 loci
(Source: Satoh et al. PLoS One 2, e1235)

		Class		
		AE	NAE	ANE
TIGR CDS		23193	0	32697
FL-locus		23117	5447	0
mapping information	FL-cDNA	29808	2967	0
	5endFLEST	201343	11255	0
	3endFLEST	465816	17481	0
	FL-clones	511817	21850	0

doi:10.1371/journal.pone.0001235.t002

The classification of loci as defined above raised questions about whether any characteristic distinctions existed between FL-AE and FL-NAE, and why FL-NAE loci were not predicted in TIGR4. To answer these questions, we analyzed the structures of genes belonging to the respective groups.

Open reading frames of FL-cDNA clones mapped on the genome

We mapped 32,775 FL-cDNAs at 22,943 FL-cDNA loci (FL-AE, 20,324; FL-NAE, 2619). The numbers of FL-cDNAs mapped to FL-AE and FL-NAE were 29,808 and 2967, respectively (Table 2). The median lengths of the FL-cDNA mapped to FL-AE and FL-NAE were 1540 and 1173 bp, respectively (Figure 1a, Table 3).

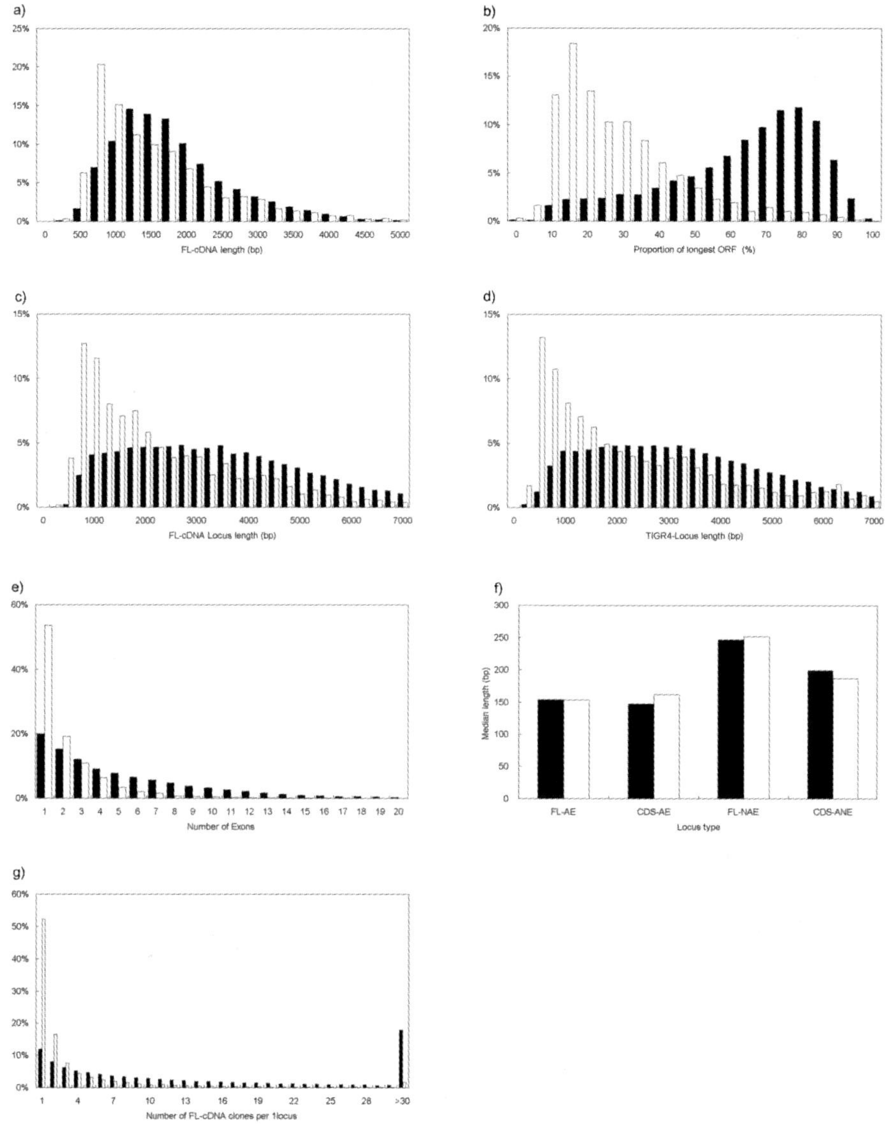

Figure 1 Gene structure analysis in rice. (Source: Satoh et al. PLoS One 2, e1235)
(a) Length distribution of FL-cDNA for FL-AE (black) and FL-NAE (white). (b) Distribution of open reading frame (ORF) proportions for FL-AE (black) and FL-NAE (white). (c) Distribution of FL-cDNA locus lengths for FL-AE (black) and FL-NAE (white). (d) Distribution of locus lengths for CDS-AE (black) and CDS-ANE (white) in TIGR4. (e) Distribution of number of exons for FL-AE (black) and FL-NAE (white). (f) Distribution of exon (black) and intron (white) lengths for the respective locus types. (g) Distribution of number of FL-cDNA clones mapped per single FL-AE (black) and FL-NAE (white) locus.

The proportions of the length of the longest open reading frames (ORFs) to that of FL-cDNA were also considerably different between FL-cDNAs mapped to FL-AE and those mapped to FL-NAE. The median proportion of the longest ORF in FL-AE was 66%, versus only 21% in FL-NAE (Figure 1b, Table 3). The results indicate that FL-NAE clones generally encode shorter peptides, and that the ORF lengths differ considerably between FL-AE and FL-NAE (based on the clone length × ORF proportion).

Table 3 Structural characteristics of locus types (Source: Satoh et al. PLoS One 2, e1235)

	FL-cDNA length (median)	ORF ratio (median)	Locus length (median) [a]	Variation of locus length	Number of exons (average) [b]	Exon length (median) [b]	Intron length (median) [c]	Ave.nunber of mapped FL-cDNA clones
FL-AE	1540	66%	3354	Rich	5.3	154	153	22.3
FL-NAE	1173	21%	1727	Poor (short)	2.4	247	251	4.1
CDS-AE	-	-	3173	Rich	5.8	147	162	-
CDS-ANE	-	-	1643	middle	3.9	199	186	-

[a]: For the calculation of locus lengths, we used the maximum lengths of individual loci.
[b]: Exons shorter than 10 bp were excluded from the analysis. Thus, the definition of an exon in FL-cDNA loci differs from that in TIGR OSA1.
[c]: Introns shorter than 10 bp were excluded from the analysis. Thus, the definition of an intron in FL-cDNA loci differs from that in TIGR OSA1.
doi:10.1371/journal.pone.0001235.t003

Locus length

The start and end sites for transcription were determined in 24,164 loci (FL-AE, 21,263; FL-NAE, 2901) out of 28,564 FL-cDNA loci. The distance between the start and the end sites (i.e., locus length) was calculated by using TIGR4. The median locus lengths differed considerably between FL-AE and FL-NAE, with lengths of 3354 and 1727 bp, respectively (Figure 1c, Table 3). The average ratio of locus length to FL-cDNA clone length was greater than 2 for FL-AE, but less than 1.5 for FL-NAE. The median locus lengths in CDS-AE and CDS-ANE were 3173 and 1643 bp, respectively (Figure 1d). The locus lengths in CDS-ANE appeared to be more variable than those in CDS-AE. The patterns of locus length variation in CDS-AE and CDS-ANE were similar to those in FL-AE and FL-NAE, respectively (Figure 1c, d, Table 3). We could not compare the lengths of FL-cDNA loci with those of the TIGR4 loci, because FL-cDNAs were constructed from coding and 5′- and 3′-end untranslated regions, whereas many gene structures in TIGR4 were predicted only from coding regions. However, from the results above, we expect that FL-AE and FL-NAE would differ from each other in this characteristic, as would CDS-AE and CDS-ANE.

Exon-intron structure

The average numbers of exons per FL-AE and FL-NAE were 5.3 and 2.3, respectively (Figure 1e, Table 3). The frequency of loci with a single exon was highest in both FL-AE and FL-NAE when the loci were distributed according to the numbers of exons. However, the proportions of loci with a single exon were significantly different between FL-AE and FL-NAE, with more than 50% in FL-NAE and about 20% in FL-AE (Figure 1e). When FL-cDNA loci with one exon were excluded, the ratio between the lengths of exons and introns in individual loci was approximately 1 irrespective of the locus type, but exon and intron lengths were significantly different ($P < 0.01$; Student's *t*-test) between FL-AE and FL-NAE, with median lengths of about 150 and 250 bp, respectively (Figure 1f, Table 3). We also analyzed the exon–intron structures of CDS-AE and CDS-ANE from the information at TIGR OSA1. The number of exons was higher in CDS-AE (5.8) than in CDS-ANE (3.9) (Table 3). The exon and intron lengths also differed significantly ($P < 0.01$, as calculated by Student's *t*-test) between CDS-AE and CDS-ANE, with median exon lengths of 147 and 199 bp and median intron lengths of 162 and 186 bp, respectively (Figure 1f, Table 3).

Number of mapped FL-cDNA clones at a locus

FL-NAE accounted for about 19% of the entire FL-cDNA loci, whereas the proportion of FL-cDNA clones mapped as FL-NAE was only 5% (21,850) of all mapped FL-cDNA clones (533,667) (Table 2). The average numbers of FL-cDNA clones mapped per locus (collection efficiency) differed significantly between FL-AE (22.3 clones) and FL-NAE (4.1) (Figure 1g, Table 3). The collection efficiency was 1 for more than half of the FL-NAE loci (Figure 1g, Table 3), suggesting that FL-cDNA clones derived from FL-NAE are more difficult to collect than those from FL-AE.

Gene annotation

We analyzed the homology of FL-cDNA mapped on TIGR4 with Arabidopsis CDSs in The Arabidopsis Information Resource (TAIR6, http://www.arabidopsis.org/) (35) using BlastX software. On the basis of the significance of similarity, FL-cDNA clones were classified into highly homologous (E-value $< 10^{-50}$), weakly homologous ($10^{-50} <$ E-value $< 10^{-10}$), and non-homologous (E-value $> 10^{-10}$). Under these criteria, the numbers of FL-cDNA clones classified as highly homologous, weakly homologous, and non-homologous were 17,759, 7103, and 7913, respectively. Of these clones, 99.5% of highly homologous FL-cDNAs were mapped to 59% of FL-AE, and 92% of

FL-NAE coded non-homologous genes (Table 4). Thus, the results indicate that nearly all highly homologous genes were derived from FL-AE, and that most FL-NAE loci encoded genes likely to be specific to rice or other monocots. This is consistent with the findings of a previous report that some rice genes with no homologs in Arabidopsis are similar to genes in the sorghum genome (3).

Causes of inconsistency between gene prediction and FL-cDNA mapping

Table 4. Frequency of occurrence of Arabidopsis homologous genes at each FL-locus (Source: Satoh et al. PLoS One 2, e1235(6))

	FL-AE		FL-NAE		Total	
homology [(a)]	Locus	FLcDNA	Locus	FLcDNA	Locus	FL-cDNA
HH	11898	17669	75	90	11973	17759
LH	4763	6941	140	162	4903	7103
NH	3663	5198	2404	2715	6067	7913
Total	20324	29808	2619	2967	22943	32775

[a]: HH, LH, NH: highly-, low- or non-homologous FL-cDNA with Arabidopsis CDS
doi:10.1371/journal.pone.0001235.t004

Cross-examination between FL-cDNA and TIGR4 loci revealed the existence of FL-NAE clones. The results of our analyses of the locus structures of FL-AE and FL-NAE suggest explanations of why some expressed genes were not annotated. One possible reason is the characteristic lengths of ORFs in the different classes of loci. The proportion of the longest ORF in FL-NAE (median ratio 21%) is significantly lower than that in FL-AE (median ratio 66%) indicating that the transcripts from FL-NAE are more likely to encode either small peptides or no peptide. In the TIGR OSA1, 687 CDSs encoding fewer than 50 amino acids were excluded from the predicted gene model. Thus, even though the number of excluded CDSs was less than the number of FL-NAE sequences, FL-cDNA sequences overlapping with the excluded CDSs in TIGR4 might have been mapped as FL-AE. Another possible reason may be the difference in exon–intron structure between FL-NAE and FL-AE. The locus lengths of FL-NAE were generally shorter than those of FL-AE, and more than half of the FL-NAE loci contained only one exon (Figure 1e). Meanwhile, the lengths of both exon and intron in FL-NAE were generally greater than those in FL-AE, CDS-AE, and CDS-ANE (Figure 1f). If we consider the structures of FL-AE, CDS-AEs, and CDS-ANE as standard for rice genes, then the structure of many FL-NAE loci may be recognized as an irregular form. In light of the unique features of ORFs and the exon–intron structures in FL-NAE, it may have been difficult to assign proper annotations to the genes in FL-NAE through the use of gene prediction software.

The structural difference between FL-AE and NAE is also similar to that between protein-coding mRNA and mRNA-like non-coding RNA (npRNA) in the mouse (36). In the mouse, the total length of npRNA is shorter and the exon is longer than in mRNA. Moreover, more than 70% of npRNA is constructed from one exon. In our classification, FL-AE overlaps with predicted CDSs that encode >50 amino acid sequences, so cDNAs mapped on FL-AE originate from protein-encoding mRNAs. In addition, the diversity of FL-cDNA lengths between FL-AE and FL-NAE is not large and the proportions of ORFs between FL-AE and FL-NAE are reversed (Figure 1a, b, Table 3). So it seems that the proportion of protein-coding FL-NAE loci is not large, and many FL-NAE loci encode mRNA-like npRNA. Therefore, the structural diversity between FL-AE and FL-NAE may correspond to the difference between protein-coding mRNA and mRNA-like npRNA in rice. In addition, these results suggest that the structural differences between protein-coding mRNA and mRNA-like npRNA are also conserved between plants and mammals.

We categorized the gene loci in TIGR4 into three types by cross-examination between FL-cDNA loci and TIGR4 loci. Collection efficiency varied considerably depending on locus type (FL-AE = 22.3; FL-NAE = 4.1; CDS-ANE = 0). These differences may be associated with the levels of mRNA or the transcription activity of each locus type. Moreover, the general features of locus structure and the average levels of homology with Arabidopsis genes were distinctively different among the locus types. Thus, these findings may indicate an interrelationship among locus structure, transcription activity, and the assignability of gene annotation. An association of transcription activity with locus structure has been reported in plants (37): highly expressed genes have longer primary transcripts, ORFs, and exon and intron sequences and have more exons than weakly expressed genes. The results from a previous report (37) are consistent with our hypothesis that locus structure affects transcription activity, except that the results for intron length differed from our results. The locus length of FL-AE was greater than that of FL-NAE, and the collection efficiency of FL-AE was also greater than that of FL-NAE. Moreover, the cloning efficiency of mRNA-like npRNA in the mouse is lower than that of protein-coding mRNA, and half of the npRNA in the mouse has an efficiency of 1 (36). Cloning efficiency features in the mouse are also similar to those of FL-NAE; this may imply that many FL-NAE loci encode mRNA-like npRNA.

We analyzed the diversity between FL-AE and FL-NAE identified from FL-cDNA mapping and found some differences between FL-AE and FL-NAE. The difference between FL-AE and FL-NAE is similar to that between protein-coding mRNA and mRNA-like npRNA. In our classification, FL-AE implies a protein-coding locus and FL-NAE is an npRNA-coding locus, which might explain why FL-NAE loci are not predicted in TIGR4. Although gene prediction software can identify loci that encode proteins, it does not detect loci that are transcribed into ncRNA. Therefore, prediction software cannot find an FL-NAE locus that encodes npRNA.

2.2 *Gene-family–based annotation of rice genes, such as membrane-transport-protein–coding genes, contributes to study of comparative biology from bacteria to higher plants*

Membrane transport proteins carry various materials for homeostasis. The transport proteins have many clear domain features (e.g., transmembrane, pore-forming, ATP-binding, molecular capture) and functional features. In accordance with their structures and functional systems, membrane transport proteins have been divided into three categories: pump, channel, and secondary transporter. The pump system is the slowest system (1–103 molecules/s) but is environmentally independent and consumes energy (mainly ATP) for transport. The channel system is the most rapid system (107–108 molecules/s) and is non-energy-consuming, but it needs concentration gradients previously (transport directions are only according to the gradients). The secondary transport system adapts the movement energy of co-transport molecules to carry molecules. Therefore, it needs co-transport molecules, and the transport direction depends on the environmental conditions; the speed of this system (102–104 molecules/s) is midway between those of the pump and channel systems. We summarized all three categories (pump, channel, and secondary transporter) of genes and compared the total numbers of membrane transport genes in *Escherichia coli*, *Arabidopsis thaliana*, *Oryza sativa*, *Caenorhabditis elegans*, *Drosophila melanogaster*, *Homo sapiens*, *Neurospora crassa*, and *Saccharomyces cerevisiae*. The genome sizes among these organisms were diverse (4.6–3150 megabases), and the numbers of transmembrane genes ranged from 300 to 350 in *E. coli*, fungi, and yeast to about 1000–1200 in Arabidopsis and rice. This suggests that a minimum number of about 300 gene species is required to retain cell homeostasis. The greater numbers of transmembrane transport genes in Arabidopsis and rice suggest additional redundancy as well as the modification of genes for new roles (e.g., addition of new substances, adaptation of systems for regulating transport, divergence of stage- and tissue-specific material transport), specialization for the various tissues and cells of multicellular organisms, and the increased complexity of cells, which in eukaryotes have many additional organelles. The greater relative increase (plant versus bacterium, fungus, and yeast) in the numbers of membrane transport genes was less than has been reported for other gene categories, such as transcription factor genes and metabolic enzyme genes, in higher eukaryotes (38). This suggests that adaptations in membrane transport are critical for the survival of organisms during evolution. The total numbers of membrane transport genes in higher plants (Arabidopsis, about 1000; rice, 1200) are 1.2–2.0 times those in animals (fly, 600; nematode, 650; human, 750). These differences in numbers of transporter genes may be related to differences in the need for efflux and influx systems in restricted habitation environments. Because of their immobility and the simplicity of their uptake systems, plant cells have more opportunity than animals to absorb inappropriate substances; they can also absorb greater amounts of substrates and synthesize larger amounts of secondary products.

We compared the composition ratios of the three classes of protein (pump, channel, and secondary transporter; Fig. 2). The numbers of pump genes in animals (72–82) were almost the same as in bacteria (70). The numbers of secondary transporter (animals, 350; bacteria, 230) and channel (animals, 160–320; bacteria, 15) genes were increased in animals. In particular, vertebrates (humans) had more (322) channel gene species than plants (130–180). We considered that this gene diversity in the development of channel systems was caused by the acquisition of a nervous system. The electrical transmission systems in the nervous systems supplying organs (e.g. muscles, kidneys) need precisely controlled ion concentrations and the ability to make immediate changes in gradients. The development of active transport systems in animals allowed the regulation of rapid movements of the body and organs. Therefore, animals presumably acquired genes for the fastest transport-system channels. Plants also had more channel gene species than bacteria, although fewer than animals. Because plants do not transmit signals for quick movement of their organs, they do not need to regulate membrane voltages as precisely as animals. Additionally, signal-transmitting systems with ligand molecules (e.g. neurotransmitters) are not specific, unlike in animals. Therefore, the numbers of voltage-gated ion channels (VICs) and ligand-dependent channels were smaller in plants than in higher animals (Table 5). On the other hand, higher plants had increased numbers of genes for pumps (170–250) and secondary active transporters (660–760). Plant cells have chloroplasts, which synthesize carbohydrates for many biological activities, including protein synthesis and functioning of ATP-dependent pumps. Plants presumably use ATP-consuming systems more easily than animals, and the pumps transport the molecules that act as the driving forces of the secondary active transporters. Additionally, plant-specific organelles and vacuoles provide pools of ions and catabolite molecules. Co-transport molecules for secondary transport are also safely and stably stored in the vacuoles. Therefore, plants are presumably able to constantly supply co-transport molecules for secondary active transporters, independently of environmental conditions. The existence of vacuoles gives plant cells more self-sufficiency than animal cells and explains the evolution of membrane transport genes for individual cell homeostasis in plants. Therefore, pump and secondary transporter systems in plants are more divergent than in animals.

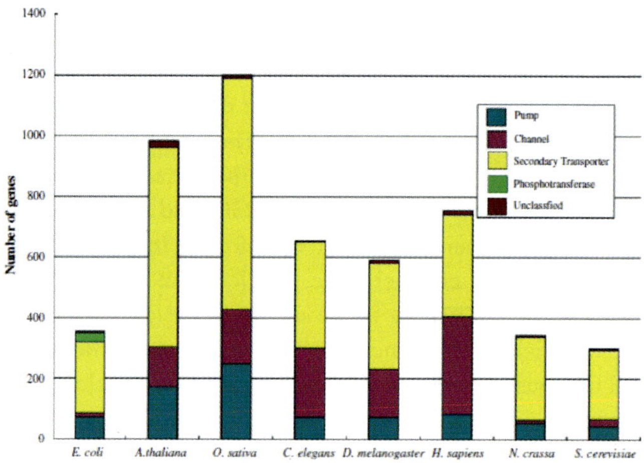

Fig. 2 **Numbers of membrane transporter proteins of each class**. Membrane transporter proteins were categorized into three classes (ATP-dependent [pump], channel, and secondary transporter) and compared among *Escherichia coli* K12-MG1655, *Arabidopsis thaliana*, *Oryza sativa*, *Caenorhabditis elegans*, *Drosophila melanogaster*, *Homo sapiens* NCBI, *Neurospora crassa* 74-OR23-IVA, and *Saccharomyces cerevisiae* S228C (Source: Nagata et al. Plant Mol Biol. 2008 Apr;66(6):565–85. Epub 2008 Feb 22.(58))

Table 5 Comparison of genome size and total and membrane transport gene numbers in various organisms

	E. coli K12	*A. thaliana*	*O. sativa*	*C. elegans*	*D. melanogast*	*H. sapiens*	*N. crassa 74*	*S. cerevisiae*
Genome Size (Mb)	4.6	125	430	97	120	3150	40	13
Total gene number	4,290	26,000	32,000	20,621	13,489	30,000	10,082	5,804
Total Transporter Proteins	354	984	1200	654	590	754	344	300
Transporters per Mb genome	76.96	7.87	2.79	6.74	4.92	0.24	8.60	23.08
Transporters per whole gene	0.08	0.04	0.04	0.03	0.04	0.03	0.03	0.05
ATP-dependent	72	173	245	73	91	99	63	70
	20.3%	17.6%	20.4%	11.2%	15.4%	13.1%	18.3%	23.3%
Ion Channels	15	160	144	230	180	353	14	22
	4.2%	16.3%	12.0%	35.2%	30.5%	46.8%	4.1%	7.3%
Phosphotransferase	29	0	0	0	0	0	0	0
	8.2%	0.0%	0.0%	0.0%	0.0%	0.0%	0.0%	0.0%
Secondary Transporters	233	635	721	349	359	337	271	223
	65.8%	64.5%	60.1%	53.4%	60.8%	44.7%	78.8%	74.3%
Unclassified	4	20	11	4	9	14	6	6
	1.1%	2.0%	0.9%	0.6%	1.5%	1.9%	1.7%	2.0%

Comparative analysis of membrane transporters among these eight diverse organisms reveals differences in the types of cell homeostasis, as evidenced by the patterns of gene conservation and diversification. Evolutionary changes in gene families, in general, indicate the dynamics of alterations in biological systems and gene networks. Therefore, analysis of large categories of gene families may reveal many basic concepts of biological systems. In practice, analyses of the membrane transporter mechanism are useful in revealing changes in the absorption of molecules by, or their efflux from, cells and tissues. This information also is useful for examining changes in soil adaptability, nutritional demand, and stress tolerance in plants. It may also help to improve the harvest of crop cultivars or extend the areas habitable by plant species. Gene networks are intricately related, and analysis of the whole genetic structure is needed if we are to gain a full understanding of biological phenomena and systems of gene regulation. We are continuing to analyze whole categories of genes in an effort to develop an overview of total gene networks.

2.3 *RiCES: a tool for* cis*-element searches in the promoter regions of clustered genes after microarray analysis*

The tool, called Rice *Cis*-Element Searcher (RiCES), consists of a *cis*-element–searching pipeline controlled via a Web-based user interface. Figure 3 summarizes the procedure. The pipeline first reads a list of gene identifiers from the user and retrieves the promoter sequences corresponding to the listed genes. Then a preliminary list of *cis*-element candidates is built by aligning information from the built-in list of plausible motifs, or by *ab initio* motif searching of the sequence data. Association rule analysis is carried out and reported to support the candidacy of the resulting *cis*-element list.

Gene list

RiCES assumes that a user has already identified genes of interest from experimental analysis (e.g., clusters of coordinately regulated genes). The list of identifiers is input into a Web-based data entry form. RiCES recognizes GenBank accession numbers, identifiers of TUs as defined in the TIGR pseudomolecular assemblies (18), and several other major gene identification systems. Using the list, it retrieves the set of associated upstream, downstream, or coding region sequences flanking the specified genes from available genomic sequence data.

Preliminary *cis*-element candidate list

The second step of the analysis is the compilation of a list of motifs as candidate *cis*-elements. RiCES supports two methods of achieving this. The first method depends on *ab initio* motif searching based on the supposition that if there are *cis*-elements playing important roles in the regulation of a given set of genes, they will be statistically over-represented in the associated promoter sequences as conserved motifs that can be identified by using a suitable motif search program. There are several programs implementing several algorithms. We have chosen to use MEME, which is a publicly available motif discovery program (39) supporting an expectation maximization algorithm. In our analysis algorithm, MEME is invoked to identify motifs 6 to 8 bp long that look highly conserved among promoter sequences of the selected genes. Users can modify some of the search parameters of the MEME program via the Web form. The second method relies on the hypothesis that common, known *cis*-elements play important roles under the experimental conditions that gave rise to the list of genes specified by the user. Therefore, RiCES searches for matches to a pre-compiled list of known *cis*-elements. Several databases of plant *cis*-elements are publicly available. PLACE (40) is one of the most popular databases of known *cis*-elements in plant genomes. AtcisDB, a part of AGRIS (41), includes information on *cis*-elements involved in gene regulation in *Arabidopsis thaliana*. Although these databases are extremely useful resources, it is not straightforward to cross-link information from them directly to the researcher's own data. Current databases are not exhaustive enough to distinguish 'core' motifs, which decide the function of *cis*-elements, from co-existing sequences in neighboring regions. As a result, many *cis*-element sequence data in these databases include superficial core motifs for which no evidence of functionality has been obtained. The use of such data prohibits effective information analysis. The *cis*-elements are collected from reports of experiments such as gel shift assays and footprint analyses, categorized by transcription factor, and documented with respect to known activity in the plant genome. Some *cis*-elements known only in organisms other than plants are also listed, in consideration of their possible, albeit unknown, roles in plants. The database includes four types of *cis*-elements: (1) G-box and E-box, which bind to common sequences such as bHLH or bZIP in many organisms; (2) A-box, T-box, and GGTTTAG repeats, which bind to common sequences in many organisms, such as homeodomain and Myb; (3) CArG boxes and GCC-box, which bind to plant MADS, zinc finger, and AP2/EREBP elements; and (4) other *cis*-elements binding only in animals, such as HSF, PcG, and HMG.

Association rule analysis

The third step of the analysis is the likelihood evaluation of the *cis*-element candidates by association rule analysis, which is a data mining method designed to discover significant relationships between pairs of characteristics observed in datasets. Candidates showing the highest likelihood (specificity) are retained in the final *cis*-element candidate list. Association rule analysis has been applied to mechanisms that regulate gene expression (e.g. 42, 43). We used it to find relationships between identified *cis*-elements and gene expression profiles. The strategy depends on the idea that motifs over-represented in the promoter region of the genes of interest could play specific roles in regulation of the expression of those genes. Implied cause-and-effect relationships documented as 'rules' are evaluated by using several well-known indices of likelihood, including support, confidence, and lift (42). On the basis of sample datasets, the lift index appeared to best discriminate significant relationships between experimental conditions and *cis*-element candidates. If the presence of motif X in a gene implies that the gene is a member of group Y, then lift is the ratio of the posterior probability (the probability that the gene is in group Y if it possess motif X) to the prior probability (the probability of X possession, irrespective of the membership of Y). When lift > 1.0, coexistence of X and Y is not a random occurrence but suggests some causal relationship between them. If lift < 1.0, it is not considered probabilistically significant. Consequently, we set the default threshold of lift to 1.0, and the *cis*-element candidates are included in the final candidate list only if their lift values are higher than this threshold. RiCES also evaluates pairwise combinations of motifs in the preliminary candidate list (upper right-hand box in Fig. 3), in consideration of possible protein–protein interactions of multiple transcription elements binding *cis*-elements, as illustrated by experimental evidence (44, 45).

Output

The final *cis*-element candidate list is presented as an association table with the identifier of the submitted genes (TU identifiers based on TIGR gene model annotation are used in the current version) annotated with any available corresponding information from RiceCyc (http://www.gramene.org/pathway/) and Gene Ontology (46). RiCES also provides information on candidate motifs, including the positions of the element in the promoter regions of corresponding TUs, the sequence, and related information from AtcisDB (41). The position of the *cis*-element candidates is also presented in both text and graphics.

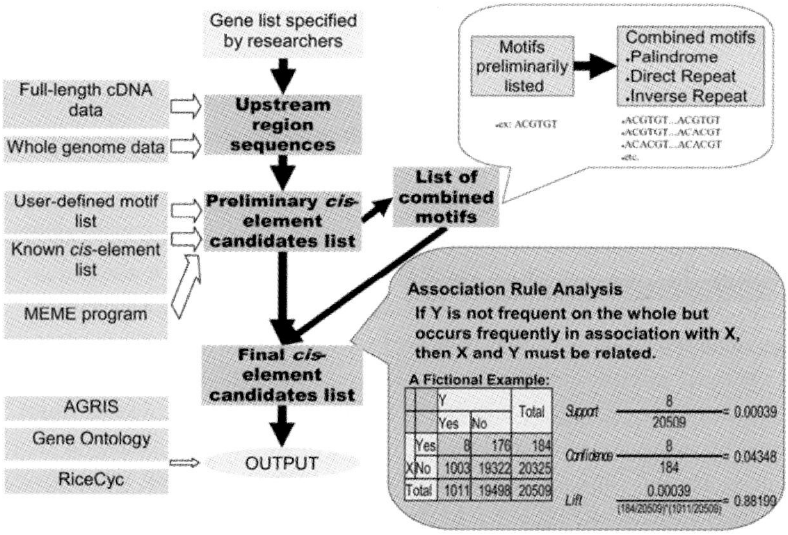

Fig. 3 Features of RiCES. (Source: Doi et al. BMC Plant Biology 2008, 8:20(59))

Validation

To test whether or not the output of RiCES was meaningful, we validated it with a list of auxin-inducible genes with known characteristics, compiled from RiceTFDB 2.0 (http://ricetfdb.bio.uni-potsdam.de/). First, Aux/IAA genes stored in RiceTFDB were applied as queries in a BLASTN search (47) of

GenBank, returning a list containing 28 rice TUs. These genes were fed into the pipeline. When the MEME program was called, the length of target motifs was set to 6, 7, or 8 bases, the number of occurrences of each motif was set to 7, 14, or 21, and the search algorithm was set to 'zoops' to check zero or one occurrence per sequence. The outputs of each option setting were merged but not otherwise filtered.

Many Aux/IAA genes are auxin-inducible (48) and contain the TGTCTC element (49). This element is commonly found in the upstream regions of auxin-responsive genes. Thus, detection of all instances of the motif by the pipeline could serve as a validation of the pipeline algorithm. The auxin-responsive element (AuxRE) containing the TGTCTC motif in some cases requires another proximal AuxRE for biological activity (44, 50). In other contexts, AuxRE functions only when it occurs with its palindromic components separated by 7 or 8 nucleotides (51). A search of AtcisDB for these motifs returned 4 showing a partial match to the record of 'PRHA binding sites', which is derived from the report of Plesch et al. (52), describing auxin-induced expression of the Arabidopsis *prha* homeobox gene. Another 4 motifs contained the TGTCTC element. The result was consistent with results of previous work, as TGTCTC was listed as a candidate in the single-motif search of Aux/IAA genes. The analysis returned 22 *cis*-element candidates with lift > 1.0. Some of these candidates were suggested by previous studies to have some kind of relationship to auxin response. For example, RAV1 was found in the promoter region of ABP, which encodes an auxin-binding protein (53). Expression of LEAFY (LFY) is affected by the auxin gradient in Arabidopsis (54). ETT is another auxin response factor (55), and LFY and ETT expression are closely correlated (45, 56). The position of a *cis*-element is important information to consider in relation to the function of the *cis*-element. For biological activity to occur, the distance of some *cis*-elements from the coding region or other collaborating elements is constrained. To this end, RiCES highlights the distribution of *cis*-element candidates. It provides tables of identified *cis*-element motifs and graphical motif maps to help researchers grasp positional relationships among the candidate elements.

The positions of the listed elements, some of which include TGTCTC, varied among upstream regions of genes, and it was hard to detect any skewed distribution of motifs. Goda et al. (57) studied the distribution of TGTCTC motifs in the genome of *A. thaliana*, and they pointed out that 25% of genes investigated had TGTCTC motifs in the upstream region within 1000 bp of the start codon, and 14% within 500 bps. Our results do not seem to conflict with theirs. TGTCTC motifs are scattered over wide regions of many plant species. It is possible that the variety of the roles of genes reflects the variety of mechanisms regulating gene expression and positions of *cis*-elements, even if the genes in question can be classified as 'auxin-responsive genes' in a larger sense. A major research concern is how to pick up *cis*-element candidates worthy of further experimentation. Computational and manual selection of *cis*-element candidates should play complementary roles to resolve this issue. It should be emphasized that *cis*-element candidates listed by RiCES are rated according to the likelihood provided by association rule analysis. On the other hand, researchers can check the significance of candidates in detail by using related information derived from several databases. The supported databases include AGRIS, Gene Ontology, and RiceCyc, as well as the map information described above. The outputs are not only easily accessible in a Web browser, but are also usable in further statistical or bioinformatics analyses, as they are also provided in XML format, which is a tagged plain-text format compatible with various computer programs. In some cases, the results of the analysis from the pre-compiled list of elements will be easily comparable with prior knowledge. In other cases involving solely *ab initio* evidence from MEME, the results of motif searches should be interpreted carefully, because the result will change considerably in accordance with the options selected. An appropriate set of motif search options should be determined each time, by trial and error. However, as described above, a motif search can find *cis*-element candidates whose sequences do not exactly match those of known *cis*-elements. Although RiCES is focused on the role of *cis*-elements in *Oryza sativa* ssp. *japonica*, the methodology can be applied easily to studies of other plant species, or of other genome sequence motifs involving gene expression regulation, such as motifs in coding regions of genes or downstream of the gene sequence. Such work can be made possible by replacing the reference dataset containing whole genes of rice with other datasets.

We have presented here a newly developed tool for searching for *cis*-element candidates in lists of genes. A case study showed the applicability of the tool. The tool is easy to use and publicly available. We expect that its use will deepen our understanding of the mechanisms that regulate gene expression in plants.

Contributions

This presentation is based on three recently published reports from our laboratory (6, 58, 59). I sincerely thank all the authors of the reports.

References

1. S.A. Goff, D. Ricke, T.H. Lan, G. Presting, R. Wang, M. Dunn, et al., *Science* **296**, 92 (2002).
2. J.Yu, S. Hu, J. Wang, G.K. Wong, S. Li, et al., *Science* **296**, 79 (2002).
3. J.Yu, J. Wang, W. Lin, S. Li, H. Li, et al. *PLoS Biol.* Feb, 3e38 (2005).
4. International Rice Genome Sequencing Project, *Nature* **436**, 793 (2005).
5. Rice Full-Length cDNA Consortium, *Science* **301**, 376 (2003).
6. K. Satoh, K. Doi, T. Nagata, N. Kishimoto, K. Suzuki, et al., *PLoS ONE* **2**, e1235 (2007).
7. J. Wu, T. Maehara, T. Shimokawa, S. Yamamoto, C. Harada, et al., *Plant Cell* **14**, 525 (2002).
8. Y. Zhou, J. Tang, M.G. Walker, X. Zhang, J. Wang, et al., Genomics *Proteomics Bioinformatics* **1**, 26 (2003).
9. J. Zhang, Q. Feng, C. Jin, D. Qiu, L. Zhang, et al., *Plant J* **42**, 772 (2005)
10. M. Seki, M. Narusaka, A. Kamiya, J. Ishida, M. Satou, et al., *Science* **296**, 141 (2002).
11. T. Imanishi, T Itoh, Y Suzuki, C O'Donovan, S Fukuchi, et al., *PLoS Biol* **2** (2004).
12. The RIKEN Genome Exploration Research Group Phase II Team and the FANTOM Consortium, *Nature* **409**, 685 (2001).
13. T. Itoh, T. Tanaka, R. A. Barrero, C. Yamasaki et al., *Genome Research* **17**, 175 (2007).
14. Arabidopsis Genome Initiative, *Nature* **408**,796 (2000).
15. H. Schoof H, R. Ernst, V. Nazarov, L. Pfeifer, H.W. Mewes, K. and F. Mayer, *Nucleic Acids Res.* **32** (database issue), D373 (2004).
16. W. M. Karlowski, H. Schoof, V. Janakiraman, V. Stuempflen and K. F. Mayer, *Nucleic Acids Res.* **31**, 190 (2003).
17. T. Tanaka, B. A. Antonio, S. Kikuchi, T. Matsumoto et al., *Nucleic Acids Res.* **36** (database issue), D1028 (2008).
18. S. Ouyang, W. Zhu, J. Hamilton, H. Lin, M. Campbell, K. Childs, F. Thibaud-Nissen, R. L. Malek, Y. Lee et al., *Nucleic Acids Res.* **35**, D883 (2007)
19. B. H. Eng, M. L. Guerinot, D. Eide and M. H. Jr. Saier, *J. Membr. Biol.* **166**, 1 (1998)
20. S. S. Pao, I. T. Paulsen and M. H. Jr. Saier, *Microbiol. Mol. Biol. Rev.* **62**, 1 (1999).
21. P. Mäser, S. Thomine, J. I. Schroeder, J. M. Ward, K. Hirschi, H. Sze, I. N. Talke, A. Amtmann, F. J. Maathuis, D. Sanders, J. F. Harper, J. Tchieu, M. Gribskov, M. W. Persans, D. E. Salt, S. A. Kim and M. L. Guerinot, *Plant Physiol.* **126**, 1646 (2001).
22. R. Sánchez-Fernández, T. G. Davies and J. O. Coleman, *J. Biol. Chem.* **276**, 30231 (2001).
23. Q. Ren and I. T. Paulsen, *PLoS Comput. Biol.* **3**, e27 (2005).

24. H. M. Wain, M. J. Lush, F. Ducluzeau, V. K. Khodiyar and S. Povey, *Nucleic Acids Res.* **32** (database issue), D255 (2004).
25. Q. Ren, K. H. Kang and I. T. Paulsen, *Nucleic Acids Res* **32** (database issue), D284 (2004).
26. J. H. Tchieu, F. Fana, J. L. Fink, J. Harper, T. M. Nair, R. H. Niedner, D. W. Smith, K. Steube, T. M. Tam, S. Veretnik, D. Wang, M. Gribskov, *Nucleic Acids Res.* **31**, 342 (2003).
27. R. Schwacke, A. Schneider, E. van der Graaff, K. Fischer, E. Catoni, M. Desimone, W. B. Frommer, U. I. Flügge and R. Kunze, *Plant Physiol.* **131**, 16 (2003).
28. M. G. Carter, T. Hamatani, A. A. Sharov, C. E. Carmack, Y. Qian, K. Aiba, N. T. Ko, D. B. Dudekula, P. M. Brzoska, S. S. Hwang and M. S. H. Ko, *Genome Res.* **13**, 1011 (2003).
29. T. R. Hughes, M. Mao, A. R. Jones, J. Burchard, et al. *Nat. Biotechnol.* **19**, 342 (2001).
30. D. D. Shoemaker, E. E. Schadt, C. D. Armour, Y. D. He, P. Garrett-Engele, P. D. McDonagh, P. M. Loerch, and A. Leonardson. *Nature* **409**, 922 (2001).
31. J. Yazaki, Z. Shimatani, A. Hashimoto, Y. Nagata, F. Fujii et al. *Physiol. Genomics*. **17**, 87 (2004).
32. J. van Helden, *Nucleic Acids Res.* **31**, 3593 (2003).
33. K. E. Holt, A. H. Millar and J. Whelan, *Plant Methods* 2, 8 (2006).
34 F. Zhao, Z. Xuan, L. Liu, M. Q. Zhang. *Nucleic Acids Res.* **33**, D103 (2005).
35. S.Y. Rhee, W. Beavis, T. Z. Berardini, G. Chen, D. Dixon, et al., *Nucleic Acids Res.* **31**, 224 (2003).
36. T. Ravasi, H. Suzuki, K. C. Pang, S. Katayama, M. Furuno, et al., *Genome Res.* **16**, 11 (2006).
37. X. Y. Ren, O. Vorst, M. W. Fiers, W. J. Stiekema, J. P. Nap, *Trends Genet* **22**, 528. (2006).
38. G. A. Wray , M. W. Hahn, E. Abouheif, J. P. Balhoff, M. Pizer, M. V. Rockman, L. A. Romano. *Mol. Biol. Evol.* **20**, 1377 (2003).
39. T. L. Bailey and C. Elkan *Proc. Int. Conf. Intell. Syst. Mol. Biol.* **2**, 28 (1994).
40. K. Higo, Y. Ugawa, M. Iwamoto and T. Korenaga *Nucleic Acids Res.* **27**, 297 (1999).
41. R. V. Davuluri, H. Sun, S. K. Palaniswamy, N. Matthews, C. Molina, M. Kurtz and E. Grotewold, *BMC Bioinformatics* **4**, 4:25 (2003).
42. P. Carmona-Saez, M. Chagoyen, A. Rodriguez, O. Trelles, J. M. Carazo and A. Pascual-Montano, *BMC Bioinformatics* **7**, 54 (2006).
43. D. Conklin, I. Jonassen, R. Aasland and W. R. Taylor, *Bioinformatics* **18**, 182 (2002).
44. T. Ulmasov, Z. B. Liu, G. Hagen and T. J. Guilfoyle, *Plant Cell* **7**, 1611 (1995).
45. T. Ulmasov, G. Hagen and T. J. Guilfoyle *Plant J.* **19**, 309 (1999).

46. M. Ashburner, C. A. Ball, J. A. Blake, D. Botstein, H. Butler, et al. *Nat. Genet.* **25**, 25 (2000).
47. S. F. Altschul, W. Gish, W. Miller, E. W. Myers and D. J. Lipman, *J. Mol. Biol.* **215**, 403 (1990).
48. J. W. Reed *Trends Plant Sci.* **6**, 420 (2001).
49. S. B. Tiwari, X. J. Wang, G. Hagen and T. J. Guilfoyle, *Plant Cell* **13**, 2809 (2001).
50. Z. B. Liu, G. Hagen and T. J. Guilfoyle, *Plant Physiol.* **115**, 397 (1997).
51. T. Ulmasov, G. Hagen, T. J. Guilfoyle, *Science* 276, 1865 (1997).
52. G. Plesch, K. Stoermann J. T. Torres, R. Walden, I E. Somssich, *Plant J.* **12**, 635 (1997).
53. Y. Kagaya, K. Ohmiya and T. Hattori, *Nucleic Acids Res.* **27**, 470 (1999).
54. T. A. Ezhova, O. P. Soldatova, A. Iu. Kalinina, S. S. Medvedev, *Genetika* **36**, 1682 (2000).
55. A. Sessions, J. L. Nemhauser, A. McColl, J. L. Roe, K. A. Feldmann and P. C. Zambryski, *Development* **124**, 4481 (1997).
56. D. L. Remington, T. J. Vision, T. J. Guilfoyle and J. W. Reed, *Plant Physiol.* 135, 1738. (2004).
57. H. Goda, S. Sawa, T. Asami, S. Fujioka, Y. Shimada and S. Yoshida, *Plant Physiol.* **134**, 1555 (2004).
58. T. Nagata, S. Iizumi, K. Satoh and S. Kikuchi, *Plant Mol. Biol.* **66**, 565 (2008).
59. K. Doi, A. Hosaka, T. Nagata, K. Satoh, K. Suzuki, R. Mauleon, M. J. Mendoza, R. Bruskiewich, S. Kikuchi. *BMC Plant Biol.* Feb 27 **8**, 20 (2008).

CHANGES OF INFLUENZA A (H5) VIRUSES BY MEANS OF ENTROPIC CHAOS DEGREE

KEIKO SATO AND MASANORI OHYA
Department of Information Sciences, Tokyo University of Science, Noda City, Chiba 278-8510, Japan

To understand how influenza A H5 viruses change and how we can classify the viruses, we applied the entropic chaos degree introduced in information dynamics to the course of sequence changes in hemagglutinin (HA1) protein of all H5 viruses. Phylogenetic analysis of HA1 amino acid sequences of H5 viruses revealed that the HPAI H5N1 viruses appeared after A/Goose/Guangdong/1/96 were different from the cluster made of the LPAI H5 viruses, the HPAI H5N2 and H5N9 viruses and the HPAI H5N1viruses before 1996. Moreover, the characteristics of the HA1 sequences of H5 viruses are discussed in this paper.

1. Introduction

Influenza A viruses have eight pieces of segmented RNA that encode 11 proteins [1]. The responses of antibodies to the two viral surface proteins hemagglutinin (HA) and neuraminidase (NA) are used to classify the different subtypes of influenza A viruses. Influenza A viruses representing 16 HA (H1 to H16) and 9 NA (N1 to N9) antigens have been detected in wild birds and poultry all over the world and each viral subtype is identified by the antigen combination (e.g., H1N1, H3N2) [1] [3]. Wild aquatic birds are considered to be natural reservoirs for all subtypes of Influenza A viruses [2][4]. The antigenic change of influenza A virus results from at least two mechanisms: the accumulation of point mutation in the gene that encodes the two proteins HA and NA, and the formation of a subtype with a novel combination of HA and NA by reassortment that two more influenza virus strains of the same or different subtypes co-infect a single host cell and they exchange the RNA segments [4]. The HA is initially synthesized as a single polypeptide precursor (HA0) that is post-translationally cleaved into two subunits HA1 and HA2. The HA1 mediates binding of the virus to target cells, and the HA2 fuses the viral membrane. On the other hand, the NA facilitates the release of progeny virus from infected cells [2] [7].

Avian influenza A virus strains from wild birds are designated as low pathogenic when they cause mild or inapparent disease, and highly pathogenic when they cause serious disease and high mortality. It is the characteristics of highly pathogenic avian influenza viruses to acquire an insertional mutation resulting in a polybasic amino acid cleavage site within the HA. To date, only subtypes containing H5 or H7 have been found in the highly pathogenic form. Although low pathogenic avian influenza (LPAI) H5 and H7 viruses also exist, these viruses transmitted to poultry change highly pathogenic avian influenza (HPAI) viruses [5] [6]. In addition, there is a report indicated that some wild aquatic birds appear to carry and disseminate the HPAI H5N1 virus strains. Since 2003, the HPAI H5N1 viruses have spread throughout Asia, Europe, Middle East, North and West Africa with outbreaks in poultry and continuing cases of human infection [6]. It is believed that these H5N1 strains have been diversified from the A/Goose/Guangdong/1/1996-like H5N1 HPAI viruses that were initially isolated from geese in Guangdong Province, China, in 1996 [1][8]. The currently H5N1 HPAI strains spread geographically, exhibit the genetic diversity and infect a variety of host species.

Here to understand how the H5 viruses change and how we can classify the viruses, we applied the entropic chaos degree introduced in information dynamics to the course of sequence changes in hemagglutinin (HA1) protein of all H5 viruses.

2. Methods

2.1. *Entropic Chaos Degree*

Entropic Chaos Degree (ECD for short) has been used to characterize the chaotic aspects of the dynamics leading sequence changes [9][10]. We briefly explain the ECD. The ECD for the amino acid sequences is given as follows [12]. After two amino acid sequences are aligned, they are symbolized as X and Y. The complete event system of X is determined by the occurrence probability p_i of each amino acid a_i and p_0 of the gap $*$;

$$(X,p)=\begin{pmatrix} * & a_1 & \cdots & a_{20} \\ p_0 & p_1 & \cdots & p_{20} \end{pmatrix}.$$

In the same way, the complete event system Y is denoted by

$$(Y,\overline{p})=\begin{pmatrix} * & a_1 & \cdots & a_{20} \\ \overline{p_0} & \overline{p_1} & \cdots & \overline{p_{20}} \end{pmatrix}.$$

The compound event system of X and Y is denoted by

$$(X\times Y,r)=\begin{pmatrix} ** & *a_1 & \cdots & a_{20}a_{20} \\ r_{00} & r_{01} & \cdots & r_{2020} \end{pmatrix}\left(\sum_{j=0}^{20} r_{ij}=p_i,\ \sum_{i=0}^{20} r_{ij}=\overline{p}_j\right),$$

where r_{ij} represents the joint probability for the event a_i of X and the event a_j of Y. We suppose the dynamics describing the change of sequence X to Y is given by a certain mapping Λ called a channel sending the probability distribution p to $\overline{p}$; $\overline{p} = \Lambda p$. It is very difficult to know the exact form of this dynamics Λ in the course of sequence changes. However, the ECD can be used to measure the complexity without knowing the exact form [11].

The ECD for the amino acid sequences is given the following formula;

$$ECD(X,Y) \equiv \sum_i p_i S(\Lambda \delta_i),$$

where $S(\cdot)$ is the Shannon entropy and $p = \sum_i p_i \delta_i$, $\delta_i(j) = \begin{cases} 1 & (i = j) \\ 0 & (i \neq j) \end{cases}$. Note that the $ECD(X,Y)$ is written as $ECD(p,\Lambda)$ when p and Λ are desired to indicate. By a simple computation, when p, $\overline{p}(= \Lambda p)$ and r are obtained in some proper means, the ECD above is represented as

$$ECD = \sum_{i,j} r_{ij} \log \frac{p_i}{r_{ij}}.$$

This chaos degree is originally considered how much chaos is produced by the dynamics Λ. Therefore it is defined

(1) Λ produces a chaos iff ECD > 0

(2) Λ does not produce a chaos iff ECD = 0.

Moreover, the chaos degree $ECD(X,Y)$ provides a certain difference between X and Y through a change from X to Y, so that the chaos degree characterizes the dynamics changing X to Y.

We apply the Rate of Entropic Chaos Degree $\left(; RECD(X,Y) \equiv \dfrac{ECD(X,Y)}{S(Y)}\right)$ to characterize and classify Influenza A H5 viruses in the sequel sections.

2.2. *Time Series Analysis*

These analyses were carried out with 1,584 HA1 amino acid sequences of all available H5 viruses (i.e., H5N1-H5N9, except H5N4) from Influenza Virus Resource of the NCBI. They are composed of nearly 323 amino acids. The HA cleavage site are not included in our analyses.

We took a HA1 sequence isolated in a certain year from HA1 sequences of the H5 viruses as a fixed sequence and evaluated the RECD for the fixed sequence with other sequences to understand the sequence changes at later successive times..

2.3. *Phylogenetic Analysis*

1,000 HA1 amino acid sequences from the H5 viruses were used in phylogenetic analyses. We reduced the number of sequences to construct phylogenetic trees easily. The difference between the sequences calculated by using the RECD. Phylogenetic trees were constructed by Neighbor Joining method based on the RECD. In addition, we constructed phylogenetic tree for 1000 HA1 amino acid sequences of HPAI H5N1 viruses.

3. Results

3.1. *Time Series Analysis*

When we took the HA1 sequence of H5N2 virus isolated in Wisconsin in 1975 as a fixed sequence, most of the HA1 sequences of all available North American H5 viruses (i.e., H5N1, H5N2, H5N3, H5N5, H5N7 and H5N9) isolated from wild birds and poultry continued to be present with a little change (RECD < 0.06) from the fixed sequence, as shown in Fig. 1(a). A few North American HA1 sequences showed a big change, but these values were not continuous. Although the HA1 sequences with small values were those of all LPAI H5 viruses isolated from wild birds and poultry, the HA1 sequences with large values were those of either HPAI H5 viruses or LPAI H5 viruses isolated from poultry. We suppose that the predominant LPAI H5 viruses in North America are transmitted to poultry, and then these HA1 sequences accumulate mutations and have a insertion of consecutive basic amino acid at the HA cleavage site. In the process like this, the HA1 sequences of the HPAI H5 viruses isolated from poultry in North America seem to have appeared; however, the HPAI H5 viruses were not perpetuated in poultry.

When we took the HA1 sequence of H5N3 virus isolated in Hong Kong in 1976 as a fixed sequence, all HA1 sequences of H5 viruses isolated in North America showed a big change (RECD > 0.07). That is, the HA1 sequences of H5 viruses isolated in North America are independent of those in Eurasia. Here we eliminated the HA1 sequences from North America, as shown in Fig. 1(b). All values of the RECD for the fixed HA1 sequence and the others were classified into two groups. Group 1 consisted of relatively small and unchanging values, but, in contrast, Group 2 consisted of relatively large and sharply increasing values. Interestingly, Group 1 resulted from the strains of many subtypes containing H5 from wild birds and poultry, whereas Group 2 resulted from the strains of HPAI H5N1 viruses circulating between wild birds, poultry, swine, humans and other animals. Although most of the strains resulting in Group 1' were LPAI H5 viruses, two HPAI H5N1 strains (A/chicken/Scotland/59(H5N1), A/turkey/England/50-92/91(H5N1)) and some HPAI H5N2 strains isolated from poultry in Italy in 1997, etc. belonged to Group 1 by the value for these HA1 sequence. We suppose that the LPAI H5 viruses existing in Eurasia are transmitted to poultry, and then these HA1 sequences have an insertion of consecutive basic amino acids at the HA cleavage site. In the simple process like this, the HA1 sequences of a few HPAI

H5 viruses isolated from poultry in Eurasia seem to have appeared; however, the HPAI H5 viruses were not perpetuated in poultry as well as the HPAI H5 viruses of North America.

(a)

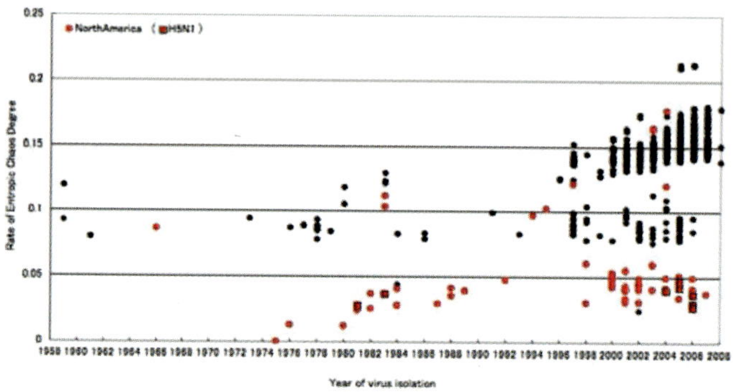

(b)

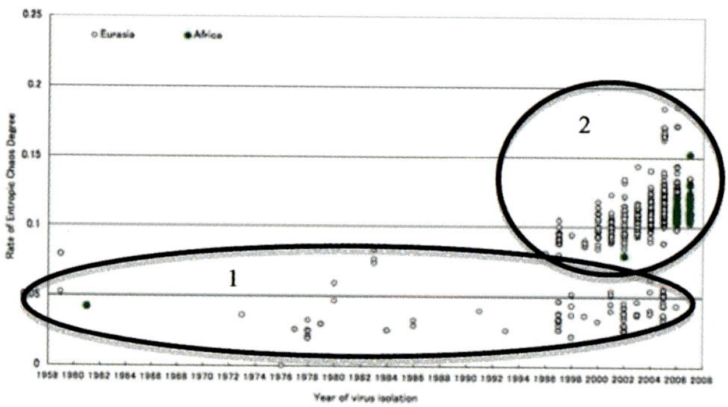

Figure 1. Each value of the RECD for the fixed HA1 sequence and the other H5 HA1 sequences is plotted on the year that each H5 virus strain was isolated. **(a) RECD for HA1 of A/blue goose/Wisconsin/711/1975/ (H5N2) and others.** When the geographical location of virus isolation is North America, its value of the RECD is represented by a red circle and a red square. Most values of the RECD represented by the red circles and squares were less than 0.06. The HA1 sequences with small values included also those of LPAI H5N1 strains (red squares). **(b) RECD for HA1 of A/duck/Hong Kong/23/1976 (H5N3) and others.** When the geographical location of virus isolation is Eurasia, its value of the RECD is represented by a white circle. The HA1 sequences of H5 viruses isolated in Eurasia were classified into two groups. The HA1 sequences of the HPAI H5N1 viruses belonged to Group 2.

Figure 2 shows the RECD for the HA1 sequence of HPAI H5N1 virus isolated from a goose in China in 1996 and the other HPAI H5N1 HA1 sequences. The HA1 sequence from the goose was close to the HA1 sequences of H5N1 viruses isolated from humans in Hong Kong in 1997. However, HA1 sequences of H5N1 viruses continued to change rapidly for about 10 years.

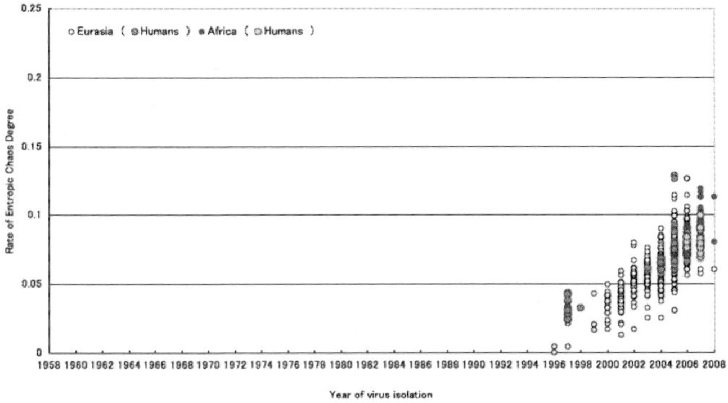

Figure 2. RECD for HA1 of A/Goose/Guangdong/1/96 (H5N1) and others. Each value of the RECD for the HA1 sequence of A/Goose/Guangdong/1/96 (H5N1) and the other H5N1 HA1 sequences is plotted on the year that each H5N1 strain was isolated. For indicating the value of the RECD, we use the colors; white for Eurasian species except humans, gray for Eurasian humans, green for African species except humans and yellow for African humans. The HA1 sequences of the HPAI H5N1 viruses continue to change rapidly and diversify.

3.2. *Phylogenetic Analysis*

Phylogenetic analysis of HA1 amino acid sequences from the H5 viruses revealed that almost all HA1 sequences of HPAI H5N1 virus strains from wild birds, poultry, swine, humans and other animals made a group as a large cluster (denoted by salmon pink in Figure 6(a)), which were phylogenetically different from HA1 sequences of LPAI H5N1 strains and the other subtypes (mauve and green in Figure 6(a)). Figure 6(b) shows that the HA1 sequences of HPAI H5N1 virus appeared in 1996 can be further divided into eight sub-clusters according to differences in species, the geographical locations and the time of viral isolation. The phylogenetic analysis indicated, as the time series analysis above, that the HA1 sequences of HPAI H5N1 virus strains isolated from humans in Hong Kong in 1997 were close to those isolated from geese in Guangdong in 1996. The HA1 sequences of H5N1 virus strains causing the infection in humans since 2003 were separated into 4 clusters. As to the HA1 sequences of LPAI H5N1 virus strains from wild birds, they were grouped into two clusters (mauve and green in Figure 6(a)) according to the continent each strain was isolated.

(a)

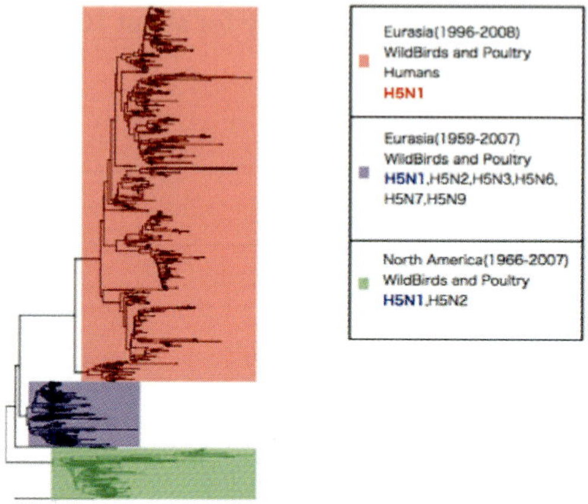

(b)

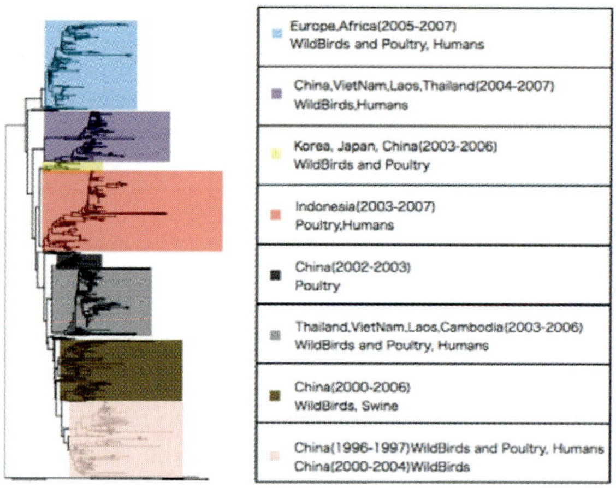

Figure 6. We listed the predominant species and subtypes among the clusters of HA1 phylogenetic tree. The red letter denotes the HPAI viruses and blue one denotes the LPAI viruses.

(a) Phylogenetic tree for HA1 amino acid sequences of H5 viruses.

The HA1 sequences of H5 viruses were clearly separated into 3 clusters. The cluster represented by salmon pink was composed the HA sequences of HPAI H5N1 virus strains. The cluster represented by mauve was composed the HA sequences of LPAI H5 virus strains in Eurasia except two HPAI H5N1 strains and some HPAI H5N2 and H5N9 strains isolated from poultry in Italy in 1997 and 1998. The cluster represented by green was composed the HA sequences of LPAI H5 virus strains in North America except some HPAI H5N2 and H5N9 strains isolated from poultry in North America and LPAI H5N2 strains isolated from chicken in Japan in 2005 and 2006.

(b) Phylogenetic tree for HA1 amino acid sequences of HPAI H5N1 viruses.

The HA1 sequences of HPAI H5N1 strains isolated from geese in Guangdong in 1996 and those isolated from humans in Hong Kong in 1997 belonged together in the same cluster (pink). The HA1 sequences of H5N1 virus strains, which cause infections in humans since 2003, were separated into 4 clusters (sky blue, mauve, salmon pink and gray).

4. Discussion

Our results indicate that the hemagglutinin (HA1) protein of avian influenza H5 viruses has three lineages: Eurasian lineage of HPAI H5N1 viruses, Eurasian lineage of LPAI H5 viruses and North American lineage of LPAI H5 viruses. A few HA1 sequences of HPAI H5 strains isolated from poultry certainly are included in two lineages for LPAI H5 viruses. We consider that these special HA1 sequences and the HA1 sequences of the LPAI H5 viruses have the same characteristic and function. The reason that the HPAI H5 viruses belonging to the lineage of LPAI H5 viruses were not epidemic is considered the HPAI H5 viruses are perpetuated in certain species of wild birds, not in poultry.

The HA1 protein of A/Goose/Guangdong/1/96, the first Asian HPAI virus of subtype H5N1, is considered as a result of mutations of the HA1 gene that encodes HA1 protein of LPAI H5 virus resident in Eurasian wild birds. That is, the spread of HPAI H5 viruses occur only if HPAI H5 viruses are transmitted to some wild birds that are considered to be natural reservoirs for LPAI viruses, and then those wild birds scatter the HPAI H5 viruses. The HA1 protein of the HPAI H5N1 indicates a different lineage from HA1 protein of H5 viruses that have existed for long time. It is interesting that although we omitted the cleavage site of HA that determine the pathogenicity, we obtained the results above. These results suggest that not only the cleavage site but also the HA1 protein have influence on the decision of pathogenicity.

We are sure that the HA1 protein of the currently circulating HPAI H5N1 strains has been derived from the gene that encodes the HA1 protein of A/Goose/Guangdong/1/96. In addition, according to our time series analysis for the Rate of Entropic Chaos Degree, as shown in Figure 2, these HA1 sequences exist as a result of change with a certain constant width and a rather sharp tangent. We conclude that the Entropic Chaos Degree can be a measure to characterize and classify the H5 viruses.

References

[1] Björn Olsen, Vincent J. Munster, Anders Wallensten, Jonas Waldenström, Albert D. M. E. Osterhaus and Ron A. M. Fouchier, Global patterns of influenza a virus in wild birds., Science, 312(5772): 384-388, Review (2006)

[2] Ron A. M. Fouchier, Vincent Munster, Anders Wallensten, Theo M. Bestebroer, Sander Herfst, Derek Smith, Guus F. Rimmelzwaan, Björn Olsen, and Albert D. M. E. Osterhaus, Characterization of a Novel Influenza A Virus Hemagglutinin Subtype (H16) Obtained from Black-Headed Gulls, J. Virol. 79(5):2814-22 (2005)

[3] Scott Krauss, Caroline A Obert, John Franks, David Walker, Kelly Jones, Patrick Seiler, Larry Niles, S. Paul Pryor, John C Obenauer, Clayton W Naeve, Linda Widjaja, Richard J Webby, and Robert G Webster, Influenza in Migratory Birds and Evidence of Limited Intercontinental Virus Exchange, PLoS Pathog., 3(11): e167: 1684-1693 (2007)
[4] Martha I. Nelson and Edward C. Holmes, The evolution of epidemic influenza, Nature Reviews Genetics, 196-205 (2007)
[5] Vivien G. Dugan, Rubing Chen, David J. Spiro, Naomi Sengamalay, Jennifer Zaborsky, Elodie Ghedin, Jacqueline Nolting, David E. Swayne, Jonathan A. Runstadler, George M. Happ, Dennis A. Senne, Ruixue Wang, Richard D. Slemons, Edward C. Holmes and Jeffery K. Taubenberger, The Evolutionary Genetics and Emergence of Avian Influenza Viruses in Wild Birds, PLoS Pathog., 4(5): e1000076: 1-9 (2008)
[6] B. Pattnaik, A. K. Pateriya, R. Khandia, C. Tosh, S. Nagarajan, S. Gounalan, H. V. Murugkar,
B. P. Shankar, N. Shrivastava, P. Behera, S. Bhagat, J. S. M. Peiris and H. K. Pradhan, Phylogenetic analysis revealed genetic similarity of the H5N1 avian influenza viruses isolated from HPAI outbreaks in chickens in Maharashtra, India with those isolated from swan in Italy and Iran in 2006 , CURRENT SCIENCE, 96(1): 77-81 (2006)
[7] Sunil K. Lal and Vincent T.K. Chow, Avian Influenza H5N1 Virus: An Emerging Global Pandemic, Emerging Viral Disease of Southeast Asia, 4: 59-77 (2007)
[8] Xiyan Xu, Kanta Subbarao, Nancy J. Cox and Yuanji Guo, Genetic Characterization of the Pathogenic Influenza A/Goose/Guangdong/1/96 (H5N1) Virus: Similarity of Its Hemagglutinin Gene to Those of H5N1 Viruses from the 1997 Outbreaks in Hong Kong, 261(1): 15-19 (1999)
[9] Masanori Ohya, Information Dynamics and Its Applications to Optical Communication Processes, Lecture Note in Physics, Vol.378, 81-92 (1991)
[10] Ohya M, Complexities and their application to characterization of chaos, International Journal of Theoretical Physics, 37(1): 495-505 (1998)
[11] Kei Inoue, Masanori Ohya and Keiko Sato, Application of chaos degree to some dynamical systems, Chaos, Solitons and Fractals, 11(9): 1377-1385 (2000).
[12] K. Sato and M. Ohya, Analysis of the disease course of HIV-1 by entropic chaos degree, Amino Acids, 20: 155-162 (2001)

BASICS OF GENOME SEQUENCE ANALYSIS IN BIOINFORMATICS
~ ITS FUNDAMENTAL IDEAS AND PROBLEMS ~

TOMONORI SUZUKI AND SATORU MIYAZAKI

Department of Medicinal and Life Science, Faculty of Pharmaceutical Sciences, Tokyo University of Science, 2641 Yamazaki, Noda 278-8510, Japan

The genome sequences are one of the most fundamental data among various omics analyses. So far, basic bioinformatics tools have developing to treat genome sequences. First step of genome sequence analysis is to predict or assign "genes" on genome sequences. In the case of Eukaryotes, we can identify genes by use of full length cDNA sequences with local alignment tools such as ssearch, blast and fasta, etc. However, it is difficult to catch mRNAs (transcripts) in Prokaryotes. Therefore, computational prediction for gene identification is first choice to start genome sequence analysis. In this review, we pick up methods for computational gene prediction first. Once genes are predicted, next step is to functions for proteins or RNAs encoded on a gene. Then, how we can define the distance between gene sequences is very important for the further analysis. So, we describe the basics of mathematical concept for gene comparison. And we also introduce our novel concept for biological sequence comparisons for the view point of informational theory.

In the post genome era, many researchers are very interested in not only gene functions but also the gene regulations whose information is also on genome sequences. *Cis*-regulatory elements, however, is too short to find some mathematical rules. Therefore, computationally predicted *cis*-elements tend to include many false-positives. To reduce the ratio false-positives, we need reliable database of set of *cis*-regulatory elements called *cis*-regulatory modules for a gene. So, we are trying to develop the *Cis*-Regulatory Elements Module Reference Database. In the third section, we introduce you the procedure to construct the *Cis*-Regulatory Elements Module Reference Database and its user interfaces.

1. Methods for Computational Gene Prediction

The most popular tools to predict genes in Prokaryote is the grimmer implemented by Mark Borodovsky *et al.* [1, 2]. In grimmer, Markov model is used. In prokaryotic genes, there are no introns. Therefore, the structure of gene is simpler than Eukaryote ones so that genes are occupied by protein coding region. And a protein coding region is consisted of codons. It suggests that prokaryotic gene might be correlated with the usage of the genetic code. This may remind us Markov model double in numeric order of alphabet for DNA or Protein. As you know, order of amino-acids determines the higher structure of protein and function. We can expect that there exist particular rule for the ordering of codons or amino-acids. This idea introduce us that occurrence of a base might be influenced by the previous base. And inter genetic regions (non-genes) are independent of genes in general. Therefore it is natural to examine the Markov property in the genome sequence. General procedure of gene finding by Markov Model is as follows;

- Let us suppose that occurrence of bases in gene follow k-th order Markov process.
- For a given genome, correct the set of biologically determined gene sequences (learning set).
- In learning set, divide the sequences into the "word" of length k+1 by k+1 size window scanning. And calculate frequency of each word.
- By the frequencies determined above, we reconstruct the transition probability matrix of k-th order Markov process for a given genome.
- For each sequence, we calculate occurrence probability of the sequence by use of the transition probability constructed the previous term.

Fig.1 shows the summary of above idea.

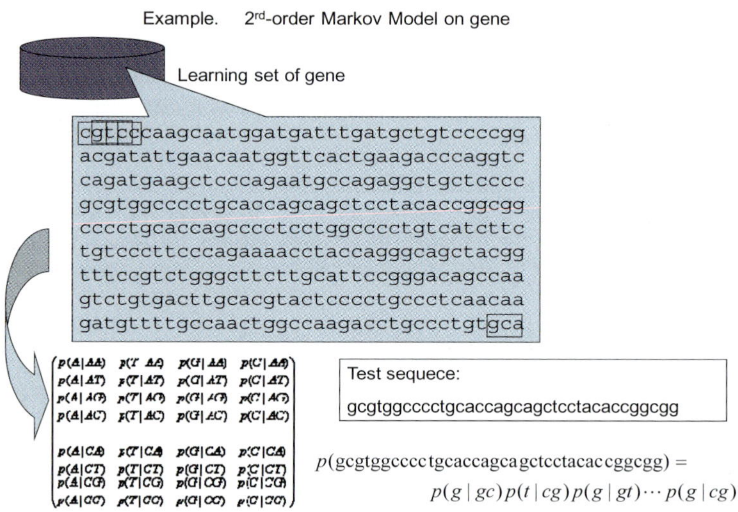

Figure 1. Summary of a procedure for gene prediction with Markov Model double.

We can calculate the possibilities of a gene as a probability $p()$ multiplying the series of transition probabilities along a give biological sequence.

2. Mathematical Measures between Two Biological Sequences

Once genes are identified, next step is to assign the function of each gene. In order to predict functions for given gene, we have to know how to compare biological sequences from the view point of biology. So far, evolutional distances or biochemical distances have been proposing. Because, active sites are conserved evolutionally in general and we can see the common "motif" sequence among family proteins whose proteins share common bio-chemical activities. However, 30-40% of genes have some unique property that could not have homologues in other organisms. In order to compare two biological sequences with no-identities, we need another type of a mathematical distance. We think that the definition of "information" on a biological sequence might bring a new approach for the sequence comparison. In this section, we review traditional evolutional distance and introduce our novel distance on biological sequences by use of Shannon and mutual entropy.

2.1. *Evolutional Distance* [3]

Let us consider flowing two biological sequences A and B of ten letters.

A:GTTCTAGTCC

B:ATTCTTGTCT

At a glance, you may think the distance between above two is 3/10. 3/10 is a simple substitution rate between two sequences. However, you may also image the situation described at Fig. 2. Almost all available biological sequences are current ones. Point mutations are accumulated by the progress from ancestral sequence to current one. And we should take some hidden substitutions. In Fig. 2, the accumulation of 6 point mutations is considered from a common ancestor sequence to current sequences. And we should also notice that divergence time between two current sequences will be 2T time if we take T time from common ancestor to current species.

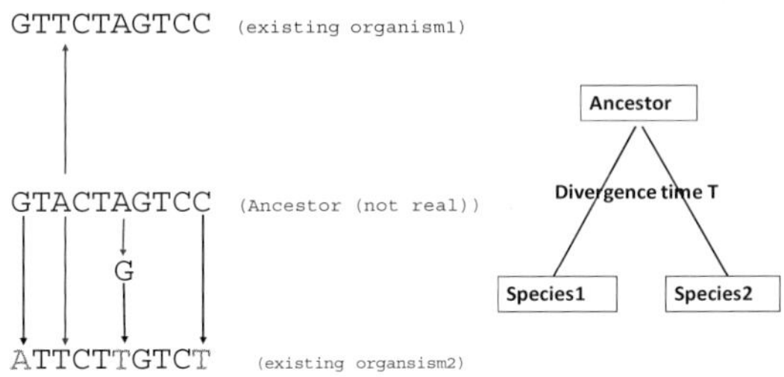

Figure 2. Divergence time and accumulated point mutations between two current sequences
We have a possibility that 6 substitutions in 2T periods might be occurred.

To handle the hidden substitution, we consider that the substitution ratio of each site follow a Poisson distribution with mean value of λt. Based on this idea, Jukes-Cantor model [4] was established.

$$p(X = k) = p(k) = (\lambda t)^k \frac{e^{-\lambda t}}{k!}$$

Here λ represent the substitution ratio in a unit time at a site. Then λt represent the mean value of total number of substitutions in time t at a site. Therefore, the probability for no substitution at the site is calculated by

$$\begin{aligned} p(X = 0) &= (\lambda t)^0 \frac{e^{-\lambda t}}{0!} \\ &= e^{-\lambda t} \approx 1 - \lambda t \end{aligned}$$

Next, let us define $Pij(t)$ as the probability base i change base j after time t. Then we can derive the following equation (1).

$$p_{ij}(t+dt) = p_{ij}(t)(1-\lambda dt) + (1-p_{ij}(t))\frac{\lambda dt}{3} \quad (1)$$

First term of right side in equation (1) represents the probability i changed j at time t and no substitution occurred in next dt time. Second term represents the probability i changing any acids excepting j and will change into j in next dt. We give a sample of transition for $P_{AC}(t+dt)$ in Fig.3 .

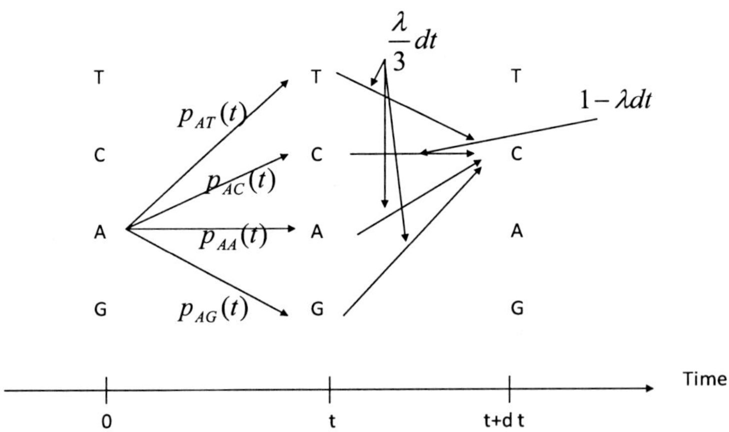

Figure 3. Transition probability and basics of a differential equation for Jukes-Cantor model.

Equation for $P_{AC}(t+dt)$ is as follows;

$$p_{AC}(t+dt) = p_{AC}(t)(1-\lambda dt) + (1-p_{AC}(t))\frac{\lambda dt}{3}$$

Then we can get the following differential equation from (1).

$$\frac{d}{dt}p_{ij}(t) = \frac{\lambda}{3} - \frac{4}{3}\lambda p_{ij}(t)$$

$$\textit{With inital values,}\ \ p_{ii}(0) = 1, \quad p_{ij}(0) = 0 \quad (i \neq j)$$

From above equation, we get the following transition probabilities *Pii*(t) and *Pij*(t).

$$p_{ii}(t) = \frac{1}{4} + \frac{3}{4}\exp(-\frac{4}{3}\lambda t)$$

$$p_{ij}(t) = \frac{1}{4} + (1-\exp(-\frac{4}{3}\lambda t))(i \neq j) \qquad (2)$$

And we calculate the expected difference $d(t)$ between after time t by use of equation (2) as follows,

$$d(t) = \sum_i \frac{1}{4}\sum_{i \neq j} p_{ij}(t) = \frac{3}{4}(1-\exp(-\frac{4}{3}\lambda t))$$

$$\lambda t = -\frac{3}{4}\log(1-\frac{4}{3}d) \qquad (3)$$

(3) gives us the expected number of substitution by Juke-Cantors model. According to Juke-Cantors model, expected substitution for sequence A and B will be

$$d = 2\lambda t = -\frac{3}{4}\log_e(1-\frac{4}{3}\times\frac{3}{10}) = 0.38311$$

As almost same manner mentioned above, some researchers have given equations to calculate the expected substitutions. Among them, let us introduce you Kimura's 2 parameter models, which is one of the most famous models in molecule biologists [6, 7].

$$d = 2\lambda t = -\frac{1}{2}\log_e[(1-2P-Q)\sqrt{1-2Q}]$$

Here, P and Q are the number of "transitions" and "transversions", respectively. By use of Kimmura's 2 parameter model, we can calculate the expected number of substitutions for sample A and B sequences is

$$d = 2\lambda t = -\frac{1}{2}\log_e[(1-2\times\frac{2}{10}-\frac{1}{10})\sqrt{1-2\times\frac{1}{10}}]$$

$$= -\frac{1}{2}\log_e[(0.5)\sqrt{0.8}] = 0.4023$$

2.2. *Informational Approach to Compare Two Biological Sequences*

As a novel measure to compare between two biological sequences, M. Ohya defined so-called Entropy Evolutional Ratio (EER) [7, 8].

The EER between any two sequences of X, Y is given by following equation using mutual information (*I*) and entropy (*S*):

$$EER(X;Y) = \frac{1}{2}\left(\frac{I(X;Y)}{S(X)} + \frac{I(X;Y)}{S(Y)}\right)$$

here, entropy (*S*) and mutual information (*I*) is calculated as follows.

$$S = -\sum_{i=A,T,G,C} P_i \log_2 P_i$$

$$I(X;Y) = \sum_{\substack{i=A,T,G,C \\ j=A,T,G,C}} P_{ij} \log_2\left(\frac{P_{ij}}{P_i P_j}\right)$$

and in this case, EER takes $0 \le \text{EER} \le 1$.

EER is a sort of the normalized mutual information quantity, and mutual information represents degree of relevance between two sources of information. Thus, if the EER value between two sequences of X and Y to compare is near to 0, it means there is not relation for appearance of their bases. If the EER value is near to 1, it means there is dependency for the appearance of bases which compose sequence X and Y. Fig. 4 shows an example of calculation for a sequence X and Y.

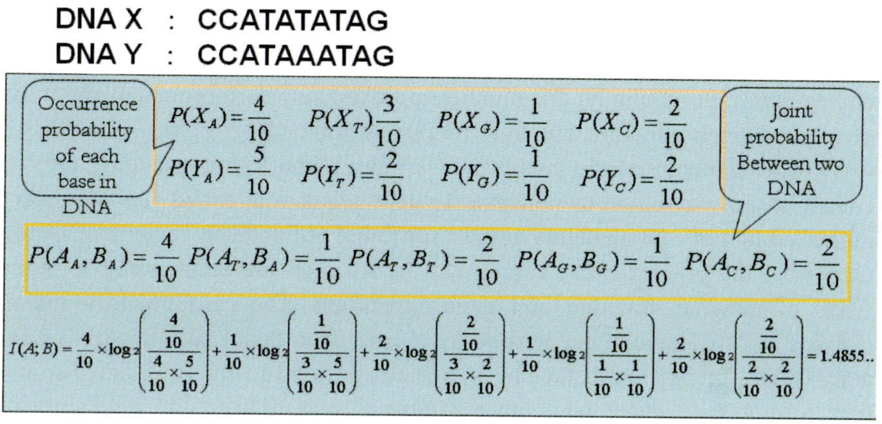

Figure 4. Mutual Entropy between DNA sequences.

If given two sequences are resemble, EER between them will be high. This property of EER seems to be as same as Evolutional distance or similarity explained section 2-1. However, EER is an idea based on "Information transition", therefore, EER will take high score if we can see some rules between sequences. In Fig.5, we described a typical case in which EER and evolutional measure take an opposite value.

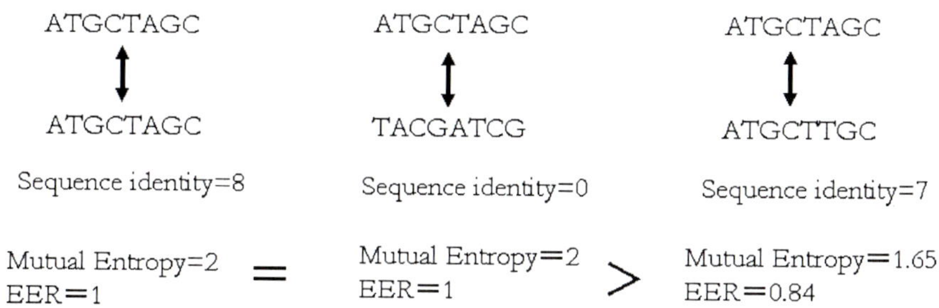

Figure 5. Difference between identities and EER (Mutual entropy).

3. Construction of the *Cis*-Regulatory Elements Module Reference Database and Its Usage

The Human Genome Project started in 1991, and the draft genome sequence was reported [9]. Three years later, this project was finished [10], and they reported that the human genome seems to encode only 20,000 to 25,000 protein-coding genes. Surprisingly, this number approximately corresponds to about 3% of whole human genome. The other region is non-coding region, which has a wide distribution in whole genome. It is thought that the non-coding regions contain a specific region responsible for gene regulation, called "*cis*-regulatory region". Most of *cis*-regulatory region are found at upstream regions of genes. Currently, the *cis*-regulatory region and its function are emphasized in study fields of genome structure and gene expression. The *cis*-regulatory region consists of "*cis*-element" recognized by transcription factor such as AP-1, Sp-1 and NF-kB. Transcription factors recognize and bind to the *cis*-elements, and then form "*cis*-module" with other components, leading to the regulation of gene expression. The *cis*-elements have been discovered by many biochemical experiments, which have been submitted to databases such as TRANSFAC [11] and JASPAR [12]. However, it is difficult to discover the all *cis*-elements distributed in whole genome with biochemical experiments. Therefore, a bioinformatical approach is very undoubtedly important and effective. Thus far, *cis*-element prediction algorithms have been developed [13]. However, because the *cis*-elements consist of 4-20 bp of short sequence, predicted results often contain false-positive. We believe reference database of *cis*-elements required for assessment of false-positive results. Here, we constructed the "*Cis*-Regulatory Elements Module Reference Database", which provides for information on *cis*-element distribution and location on whole genome.

3.1. *Important Consideration for Cis-element Information Submitted on Public Databases*

Currently, a large number of experimentally obtained sequence data are submitted on International Nucleotide Sequence Database Collaboration (INSDC) (http://www.insdc.org/page.php?page=home). The INSDC have been developed and maintained collaboratively between DNA Data Bank of Japan (DDBJ), European Molecular Biology Laboratory (EMBL) and GenBank. These databases are a collection of entry which is the unit of the data. The flat file of submitted entry includes sequence, submitters information, references, source organism, feature information and so on. The feature information includes biological features of a submitted sequence data; coding sequence for protein (CDS), *cis*-element information and so on. However, as shown Fig. 6, even if same gene information, the description depends on submitters. In addition, most of data are not mainly focused on *cis*-element, thus the descriptions of *cis*-element are defective. Therefore, the first step to construct the *Cis*-Regulatory Elements Module Reference Database is merger of submitted *cis*-element informations. Hereinafter, we present the process of construction of the *Cis*-Regulatory Elements Module Reference Database.

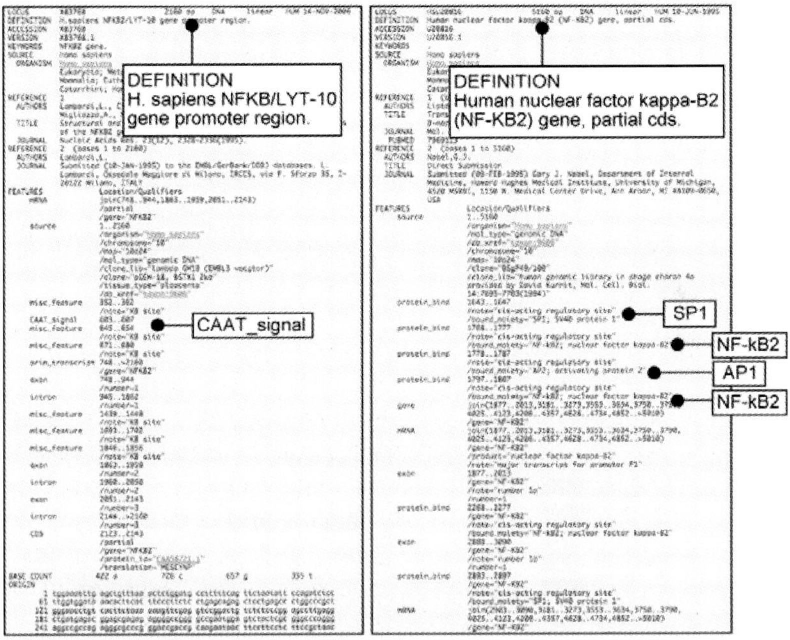

Figure 6. Example of DDBJ flat files described same gene.

3.2. *Data Sources and Data Collection*

In this work, the database responsible for five organic species (Homo sapiens, Mus musculus, Rattus norvegicus, Drosophila melanogaster and Saccharomyces serevisiae) was constructed. Data files contained *cis*-element informations were obtained from the DDBJ (http://www.ddbj.nig.ac.jp/) which is one of the three members of INSDC. The number of collected data files is shown as Table 1. Genome sequences of five organic species were obtained from Ensemble database (http://www.ensembl.org/index.html), which contained location of gene loci (Fig. 7).

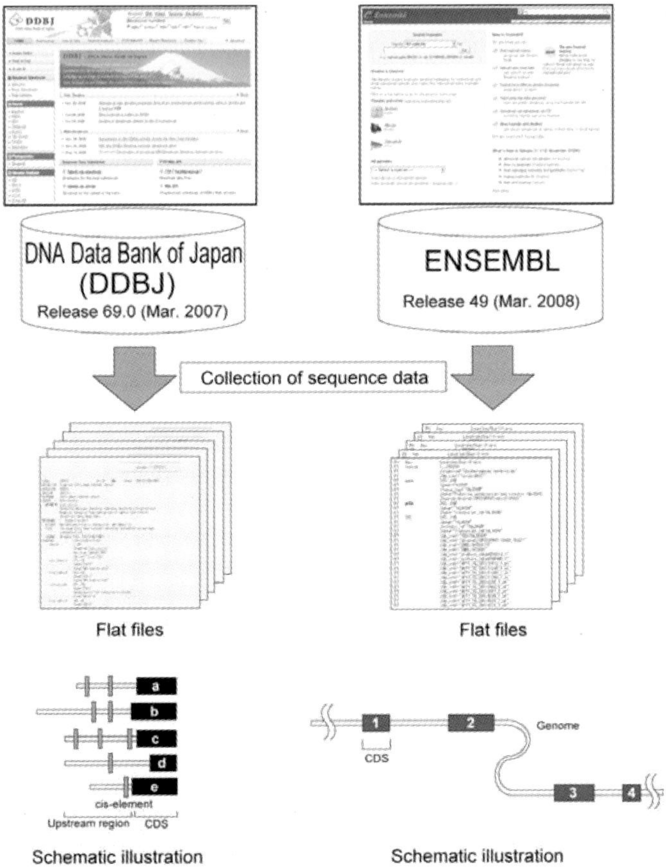

Figure 7. Data collection. Schematic illustrations represent informations on CDS, upstream region and *cis*-element described in flat files.

Table 1. The number of collected data files containing *cis*-element information.

Organic species	The number of collected data files
Homo sapiens	5,251
Mus musculus	2,879
Rattus norvegicus	848
Drosophila melanogaster	878
Saccharomyces cerevisiae	889

3.3. *Identification of CDS Location on Genome*

Amino acid sequences of CDSs on genomes (genome/CDSs) were extracted from genome sequence data, and stored in "Temporally Sequence Database" (Fig. 8). Amino acid sequences of CDSs of DDBJ entries (entry/CDSs) were extracted from DDBJ data, and compared with Blast database using ssearch program, leading to the identification of entry/CDS locations on genomes. Due to this process, it is able to specify the rough locations of entry/CDS upstream region (entry/USR) on genomes.

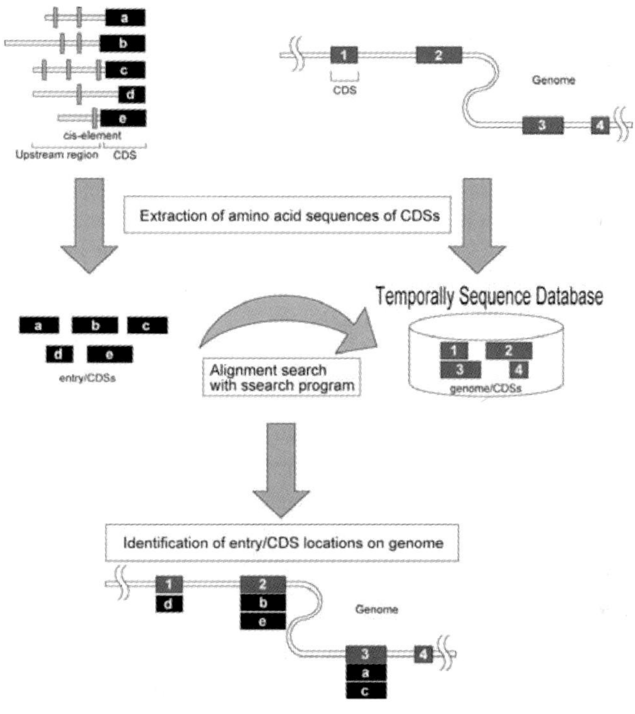

Figure 8. Overview of identification of entry/CDS location on genome.

3.4. *Identification of Upstream Region Location on Genome*

DNA sequences of CDS upstream regions on genomes (genome/USRs) were extracted from genome sequence data, and stored in "Mini-ensembl Database" (Fig. 9). The genome/USR length was according to maximum length of corresponding entry/USR. DNA sequences of entry/USRs were extracted from DDBJ data, and compared with corresponding genome/USRs stored in Mini-ensembl Database using ssearch program, leading to the identification of entry/USR locations on genomes. Then, each *cis*-element location on genomes was identified. Due to this process, *cis*-element informations of different entries were merged and mapped on genomes.

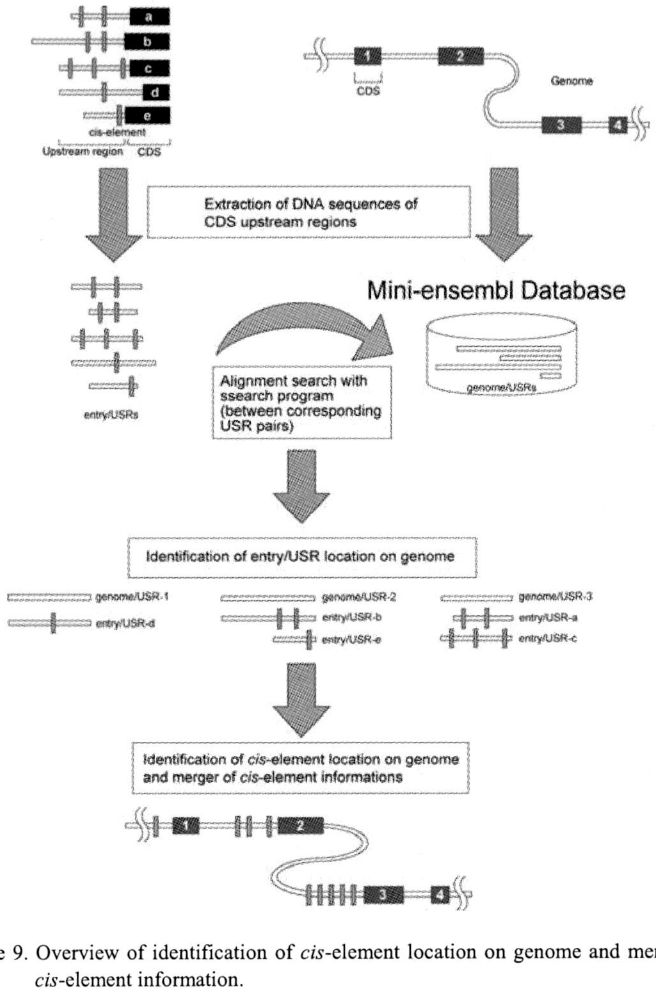

Figure 9. Overview of identification of *cis*-element location on genome and merger of *cis*-element information.

3.5. *Explanation of User Interface and the Cis-Regulatory Elements Module Reference Database Flat File Format*

The database provides *cis*-element information of each gene upstream region so that researchers are able to utilize the information. The database is freely available on the web (http://www.pharmacoinformatics.jp/cis/). Fig. 10 illustrates how the *cis*-element information is displayed. In addition, users are able to download flat files. In the case of yeast database, there are about 300 files by gene. The items of flat file are following;

(1) CIS_MODULE_ID: A unique ID of the entry in our database.

(2) DEFINITION: Information of related gene.

(3) SOURCE: DDBJ/EMBL/GenBank accession number of referred entry data.

(4) LAST_UPDATE: The date of last release.

(5) FEATURES: *Cis*-element information. In this item, the location of *cis*-element region and supplementary information of *cis*-element are described.

(6) ORIGIN: Sequence of gene upstream region.

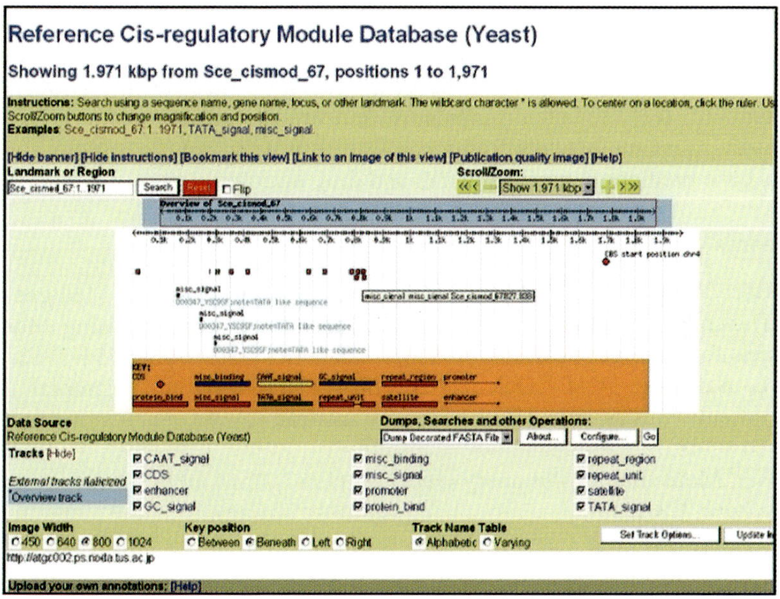

Figure 10. User interface of the *Cis*-Regulatory Elements Module Reference Database.

3.6. *Conclusions and Future Works*

In this review, we introduced you the *Cis*-Regulatory Elements Module Reference Database, which was constructed through several processes. It might be thought that upstream region sequence contained *cis*-element information should have directly aligned to genome sequence at first step. However, this simple method has problems in calculation cost and data reliability. Because genome sequences are far larger than gene sequences studied individually, the simple method requires quite high calculation cost, and that is problem for data update. In addition, most of upstream region contain repeating sequence and similar sequence, therefore the simple method might cause miss-alignment errors. In our method, upstream region of genome, which should be aligned with upstream region containing *cis*-element information, are specified, leading to reduction of calculation cost and improvement of data reliability.

Currently, there are several available databases on transcription factors and their DNA binding sites [11]. However, *cis*-elements are too short to find some mathematical rules, therefore computationally predicted *cis*-elements tend to include many false-positives. Our database will contribute to reduce the false positive ratio. Many researchers are gradually interested in the gene regulations, and analyze them with microarray analysis which provides information about global gene regulation. As an applied study, we combine our database and microarray analysis, and analyze the gene regulation mechanisms through the viewpoint of features of *cis*-elements and transcription factors. The *cis*-Regulatory Elements Module Reference Database will apply to various studies such as gene prediction, gene networks and drug discovery.

References

1. M. Borodovsky, J. D. Mclninch, E. V. Koonin, K. E. Rudd, C. Médigue and A. Danchin, Detection of new genes in a bacterial genome using Markov models for three gene classes, (1995), *Nucleic Acids Res.*, **23**, 3554-3562.
2. S. L. Salzberg, A. L. Delcher, S. Kasif and O. White, Microbial gene identification using interpolated Markov models, (1998), *Nucleic Acids Res.*, **26**, 544-548.
3. M. Nei, Molecular Evolutionary Genetics, (1987), *Columbia University Press, New York*, USA.
4. T. H. Jukes and C. R. Cantor, Evolution of protein molecules. In H. N. Munro, edt., Mammalian Protein Metabolism, Vol.III, (1969), *Academic Press, New York*, pp 21-132.
5. M. Kimura, A simple method for estimating evolutionary rates of base substitutions through comparative studies of nucleotide sequences, (1980), *J. Mol. Evol.*, **16**, 111-120.

6. M. Kimura, Estimation of evolutionary distances between homologous nucleotide sequences, (1981), *Proc. Natl. Acad. Sci.*, **78**, 454-458
7. C. E. Shannon, A mathematical theory of communication, (1948), *Bell System Technical Journal*, **27**, 379-423 and 623-656
8. M. Ohya, Information theoretical treatment of genes, (1989), *Trans. IEICE*, **72**, 556-560.
9. J. C. Venter, M. D. Adams, E. W. Myers et al., The human genome, (2001), *Science*, **291**, 1304-1351.
10. International human genome sequencing consortium, Fishing the eucharomatic sequence of the human genome, (2004), *Nature*, **431**, 931-945.
11. E. Wingender, A. E. Kel, O. V. Kel, H. Karas, T. Heinemeyer, P. Dietze, R. Knüppel, A. G. Romaschenko and N. A. Kolchanov, TRANSFAC, TRRD and COMPEL: towards a federated database system on transcriptional regulation, (1997), *Nucleic Acids Res.*, **25**, 265-268.
12. J. C. Bryne, E. Valen, M. H. Tang, T. Marstrand, O. Winther, I. da Piedade, A. Krogh, B. Lenhard and A. Sandelin, JASPAR, the open access database of transcription factor-binding profiles: new content and tools in the 2008 update, (2008), *Nucleic Acids Res.*, **36**, D102-106.
13. G. G. Loots and I. Ovcharenko, rVISTA 2.0: evolutionary analysis of transcription factor binding sites, (2004), *Nucleic Acids Res.*, **32**, W217-221.

A BASIC INTRODUCTION TO GENE EXPRESSION STUDIES USING MICROARRAY EXPRESSION DATA ANALYSIS

DIERK WANKE*, JOACHIM KILIAN
ZMBP Pflanzenphysiologie, Universität Tübingen
Auf der Morgenstelle 1, D-72076 Tübingen, Germany
**E-mail: dierk.wanke@zmbp.uni-tuebingen.de*

Keywords :*Multidimensional gene expression datasets, transcriptional regulation, AtGenExpress.*

1. Introduction

Assigning functional attributes to any kind of DNA region is still a challenging task for biologists and bioinformation scientists. With the help of new technologies, the speed of DNA sequencing is continually increasing and more and more sequenced eukaryote genomes have become accessible via public databases. Only a little amount of this DNA represents genes that are transcribed into RNA, and not all of this RNA encodes for functional proteins. Today, it is evident that there are many RNAs of regulatory function as well. Gene expression studies are performed to gain insight into the process of transcription and its regulation. Microarray hybridization experiments, where several thousands of gene transcripts can be monitored simultaneously, have especially been helpful. This data can be used to generate models of gene expression changes that explain e.g. the flow of information from incoming signals to downstream responses. Their ease of use, their reliability and the decreasing cost of the arrays themselves have resulted in a wide acceptance of microarray based studies amongst various fields of research. As a consequence, the amount of data produced is exponentially increasing. And hence, more and more very large, complex or multidimensional microarray expression experiments are being conducted. At present, there is no agreement on a standard for data analysis. Moreover, suitable solutions are lacking for the successful evaluation of the more complicated experimental designs.

While the amount of raw bio-information is exponentially increasing, the computational efforts taken to assign meaningful attributes or extracting robust models of how and when a gene is regulated lags far behind.

2. Transcription: From genes to RNA

Now that complete genome sequences from many prokaryotes and eukaryotes are available [1], the analysis of this data and its combinatorial use for organism based research will be one of the major challenges during the upcoming decades. Our understanding of how genetics, development and evolution functions is greatly enhanced by the investigation of sequence polymorphisms and the deciphering of traits [2].

An unambiguous terminology has evolved to link regions of DNA with the physical units of the functional molecules it encodes (Fig. 1). DNA regions that serve as a template for transcription into RNA are termed genes. A portion of all the RNAs produced within the cells are also used as templates: They are translated and encode for the unfolded primary amino acid chain, which folds and maturates into proteins. However, the majority of the RNAs transcribed is non-coding and have other functions [3,4] A prominent function is to fold and aggregate into higher order Ribosomes, huge entities composed of RNAs and proteins, which mediate the translation of RNA. This means, RNA and proteins are cooperatively involved in reading and translating the genetic code of RNA-bases into the nascent proteins. While this process based on "*DNA makes RNA, makes protein*" is the basis of the "*central dogma of molecular biology*" [5], we know since the year 2000 that there is more to non-coding RNA [6]: The importance of regulatory RNAs in development by a process termed "*RNA mediated RNA degradation*" has been shown [7]. To date many, especially small RNAs have been proven for their functionality in regulating both gene expression (transcription) and RNA stability [3,6,8]. The "*central dogma of molecular biology*" places DNA at center stage - We know today that this view is no longer true. RNA appears to be the key source of genetic information and DNA constitutes its unit. While the "*central dogma*" implies unidirectionality and everything is based on DNA, we have learned throughout the last few years that the flow of information is reversible and RNA can modify DNA in a stable, genetically inheritable manner[3,4].

A modified, more suitable dogma would be:

"All genes are transcribed into RNA - some of the RNAs encode for Ribosomes, some for regulatory RNA, some for proteins and some might

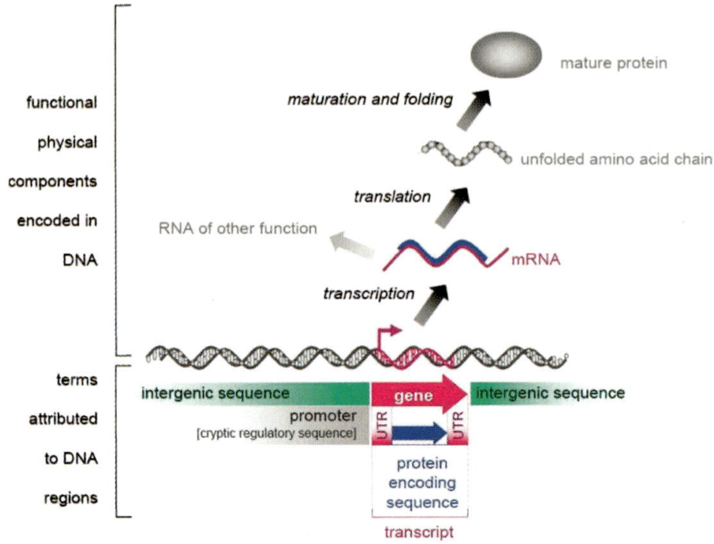

Figure 1. Schematic overview on terms and information derived and encoded in the eukaryote DNA sequence. Any region of DNA transcribed into RNA is termed gene. Many, but not all, RNA transcripts contain the signatures for translation and encode for proteins. All protein coding sequences are translated into an unfolded peptide sequence of covalently bonded amino acids. During maturation, this nascent peptide chain correctly folds into the functional protein, which usually is the conformation of lowest energy. Gene sequences that are transcribed, but which are not translated into proteins are so called untranslated regions (UTRs). The regulation of transcription and the amount of RNA is controlled by the promoter region. This sequence of cryptic DNA motifs is essential for initiation of transcription and directs the orientation in which genes are transcribed. The majority of genomic DNA is intergenic sequence of mostly unknown function. It is filling the spaces between the genes and must contain a variety of information, e.g. on how DNA is organized into higher order chromosomes.

have so far unknown functions."

Although, in colloquial language, genes are known as the units of the genetic make-up, which is inherited from parents to their children, they only represent a small fraction of the informative DNA sequence. The vast majority of genomic DNA is intergenic, non-coding sequence of mostly unknown function (Fig. 1). It fills the spaces between the transcribed genes and contains a variety of cryptic information, e.g. on how DNA is organized into higher order chromosomes. Parts of this intergenic sequence encode the cryptic information on when and how strongly a gene should be expressed. These regulatory sequences are located in close proximity to genes and contain a promoter region that directs the orientation in which genes are transcribed into RNA (Fig. 1) [9].

While proteins are encoded by the genetic code in an unequivocal manner, transcription is controlled by regulatory signals encrypted within a more extended "code". Especially the interactions of transcription factors, proteins that bind to certain DNA sequences usually not longer than six base pairs, play a crucial role in switching gene expression on or off. These DNA sequences of regulatory functionality (cis-regulatory elements) are usually too short and too degenerate to be investigated by statistical means [9,10]. Therefore, it is not yet known how transcription factors efficiently recognize their correct binding consensus within a regulatory sequence amongst the vast amount of non-specific DNA-sequences arbitrarily harboring similar motifs.

3. Microarray expression analysis: Quantification of RNA amounts

Microarrays constitute an effective means to gain insight into gene expression on a broad scale and enable the researcher to analyze the transcripts of thousands of genes simultaneously. The entire population of transcribed mRNAs present inside a cell or a population of cells is called *transcriptome* - analogous to the term *genome*, for the entire amount of genetic material per cell.

The method employs RNA extraction, processing and amplification that results in many small fragments of copy-RNAs (Fig. 2). The resulting samples can be matched to immobilized complementary single stranded probes, spotted in arrays onto a glass surface - the inner core of any microarray. This hybridization event can be read out and quantified using fluorescent dyes. The strength of the fluorescent signal is proportional to the amount of specifically bound RNAs [11,12].

In standard experiments, microarrays are used to quantify gene expression changes over time or under different conditions (Fig. 2). After the normalization of each microarray, discrete values can be assigned to any of the probes present and be used to infer gene expression changes.

To date, microarrays exist that represent entire genomes in one-base pair resolution [13,14]. These can be used to search for single nucleotide polymorphisms or yet unknown genes/RNAs.

There are initial steps taken to facilitate microarray technology for whole genome sequencing in future [13].

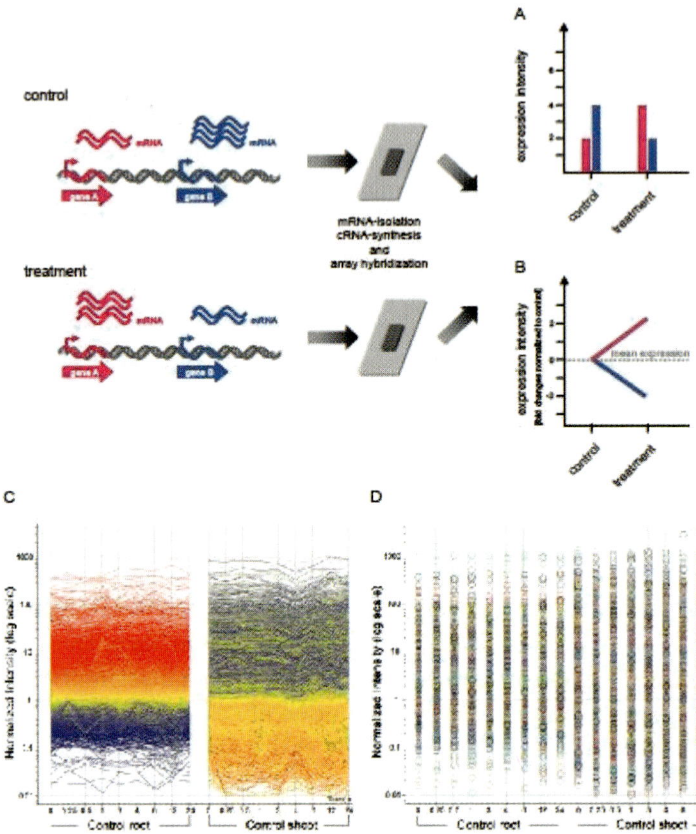

Figure 2. Scheme for mRNA expression profile measurements using microarray hybridization technology. Changes in mRNA-abundance can be quantified by using standardized mircoarrays. After mRNA-extraction and processing, the resulting samples can be matched to immobilized probes, spotted in arrays onto a glass surface. This hybridization event can be read out and quantified using fluorescent dyes. For reasons of comparison, the changes in expression states are displayed for the different conditions, here "control" and "treatment". The example given, depicts the genes A (pink) and B (blue), which transcript abundance changes after treatment respective to the untreated control (ground state). It can be seen, that gene A shows a two-fold induction in the treated condition, whereas gene B is repressed and its transcript amounts decrease by two-fold. **(A)** Absolute values for transcript abundance for two genes displayed as bar diagram. **(B)** Expression values of genes A and B normalized to the transcript abundance respective to the control condition. The values of independent measurements are color coded for different genes and connected by a line. This illustration is frequently chosen to support visualization of many genes over many conditions. An example is given in **(C)** and **(D)**, which display identical expression information for 23747 Arabidopsis genes: In **(C)** expression values have been connected by lines and the genes are color coded respective to their transcript abundance. Similarities and differences in expression trajectories are clearly visible for the two tissues tested. The identical information in **(D)** does not provide any obvious pattern in expression to the reader, although each data is color-coded for the respective genes.

Although microarrays constitute a certain standard analysis method in some areas of research, there are many obstacles to overcome, many of which have been ignored by the community of microarray users. All of these obstacles create background noise and thereby reduce the amount of informative signals [15]. Three different sources of background noise exist: technical variability, biological variation and the effects caused by the operating scientist. The resulting noise within a microarray experiment is due to reasons that can be associated with these three main causes for variability [16,15]

- variability of the chip (technical)
- variability of the replicates (technical & biological & scientist)
- variability in preparation of the labeled RNA (scientist)
- variability of the organism / individual / tissue (biological)
- variability of the organism / individual / tissue due to variability in the experimental setup (scientist)

While these five sources of noise in the dataset can be avoided or at least reduced, the effects of cross-hybridization, which is detecting transcripts that are encoded by another gene, or of optical noise, which is the effect of neighboring fluorescing probes during the read-out process, are both unavoidable [11,17,18]

4. AtGenExpress: *Arabidopsis thaliana* as a model organism for expression profile analysis

One of the most studied and best understood eukaryotic model organisms to date is the flowering plant Arabidopsis thaliana [19]. The DFG-funded Arabidopsis Functional Genomics Network (AFGN) AtGenExpress project is a multinational attempt which is designed to uncover the transcriptome of Arabidopsis with the help of microarrays. The entire AtGenExpress setup contains 41 different experimental core-conditions which constitute 1295 independent microarray hybridization events [20]. At the end of 2005 these core-conditionshave been completed and made publicly available [21,20]. These core-conditions comprise a developmental baseline, biotic and abiotic stress treatments, hormone application, growth under different light qualities and nutrient availability [21,22,23,20]

All the microarray expression data used here to provide a basic introduction for the reader is derived from the AtGenExpress.

Although all the work focuses solely on a plant as a model organism, the experimental design is universal and plainly demonstrates the complexity of all the microarray experiments conducted.

Moreover, the experiments done with Arabidopsis contain the lowest background noise of all present day microarrays and hence serve as a good basis for the development of novel evaluation methods.

5. Displaying and analyzing microarray expression data

5.1. *Displaying expression data*

To provide a good visual image of gene expression changes over the many conditions of a microarray experiment, the best way is to color code each of the genes based on expression and to connect its independent conditions by a line. This illustration is frequently chosen to support visualization of many genes over many conditions (Fig. 2).

5.2. *Biologically simple but mathematically complex experimental design*

Although the expression values are treated as independent numerical mathematical integers, they can not be evaluated without considering the experimental design that follows a more biological logic.

Biologically simple experiments can be difficult to analyze, because of their mathematical complexity.

As microarray data can only be compared within one experiment, researchers tend to perform large and complex experiments. Many of them ignore the problem that suitable methods for analyzing them are still lacking [16,? 15]. To date, there are no standards employed for data evaluation and the available programs are only capable of analyzing on up to three-dimensional datasets.

The simplest experiments contain expression data from conditions of equal weight (Fig. 2 A). This kind of *gene* x *condition* data usually lacks a control treatment and is purely descriptive rather than analytic. A good example is the examination of gene expression at different development stages, e.g. sampling individuals at different age [22].

Expression profiles of a higher dimensional order are more complicated:

A simple three-dimensional dataset is used to compare the gene expression of at least two sets of logically interconnected conditions. The example given in Fig. 3B depicts a *gene* x *condition* x *time* matrix. Here, gene expression is followed over time to compare the effects of five hormone treatments with their respective control [22].

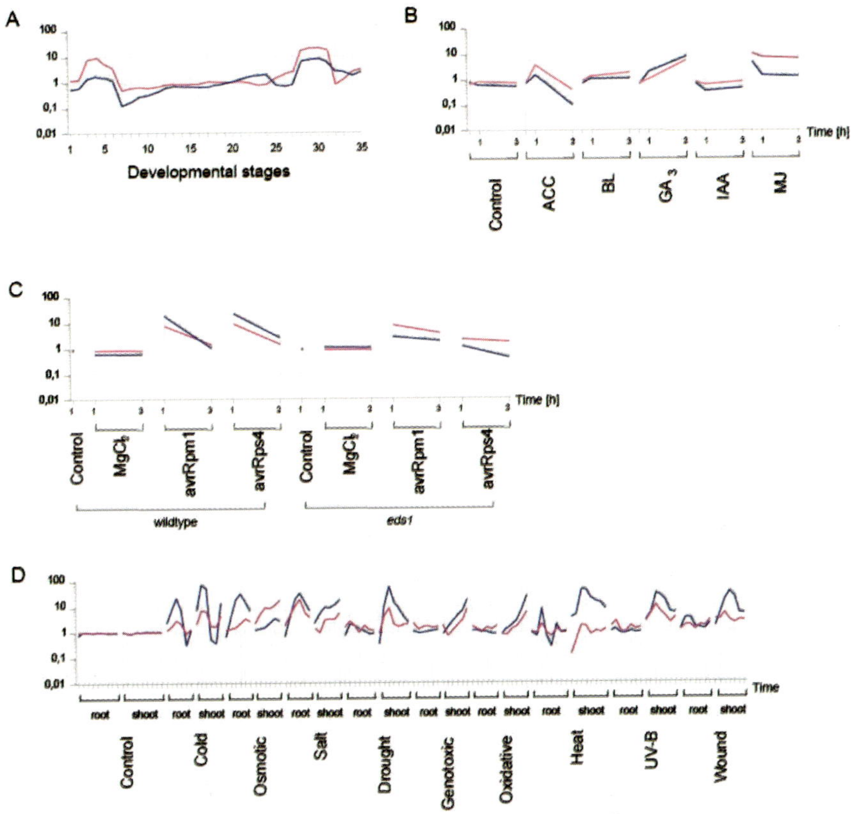

Figure 3. Examples of experimental microarray expression profile design. All examples display the highly correlated expression profiles of two genes to provide preliminary insight into the problems microarray expression profile analysis. **(A)** two-dimensional dataset of genes examined under various conditions (developmental stages) of equal weight [22]. **(B)** simple three-dimensional dataset of a *gene* x *condition* x *time* matrix [23]. When multidimensional datasets are evaluated with two-dimensional analysis algorithms, the trajectories of at least one of the dimensions get lost. It will not be revealed that the genes follow the same gene expression trajectory, but in a time and condition independent way. **(C)** complex three-dimensional expression profile of a *gene* x *condition* [*condition*] x *time* experiment [24]. A mutant and a wildtype are compared respective to their responsiveness towards two treatments [*avrRpm1* and *avrRps4*]. Two nested controls (control and mock treatment [$MgCl_2$]) are needed for proper comparison and normalization. Furthermore, the timeline is broken in trajectories of 1 and 3 hours for treatments and a 1 hour only sample as untreated control condition. **(D)** four-dimensional dataset of *gene* x *condition* x *organs* x *time* matrix [21], which can not sufficiently be analyzed by present day methods.

Another example of this experimental design comes from clinical research, where sampling was performed on patients that are suffering from a disease, before and after a medical treatment in comparison with a control group.

Evaluation of such multidimensional datasets with two-dimensional analysis algorithms results in losing at least one of the dimensions.

Specialized algorithms are needed, otherwise information on genes that follow the same expression trajectories, but in a time and/or condition independent way, gets lost [25,16]. There are two types of complex three-dimensional expression profiles that can not be analyzed by present day methods:

The first type has a so called nested composition, where the conditions are logically connected with a higher order of information (*condition* [*condition*]). Usually, the experimental design contains more than only one control for treated or non-treated populations. Examples for this kind of experiments are studies in medical research, where two groups of persons (healthy and diseased) are non-treated, treated with a placebo (mock) and with medication. Here, healthy and diseased populations create a higher order condition that includes three sub-conditions each.

The second type of complex three-dimensional expression profiles has a broken time line. An example is an experiment that monitors gene expression in treated and non-treated populations over time, with one single sample that serves as a control for both populations ahead of the treatment ([*gene* x *condition*] [*gene* x *condition* x *time*]). Still, these complex compositions are three-dimensional experiments as the information within one of the dimensions is of equal weight.

The example given in Fig. 3 C combines both types of complex three-dimensional expression datasets [24]. It is composed of a *gene* x *condition* [*condition*] x *time* matrix, with a broken time line for the controls: Mutant and wild type plants are compared with respect to their responsiveness towards two treatments [*avrRpm1* and *avrRps4*]. Two nested controls (control and mock treatment [$MgCl_2$]) are needed for proper comparison and normalization. Furthermore, the basic timeline of 1 and 3 hour samples for the treatments is broken by a 1 hour only sample in the untreated control condition.

Microarray expression datasets of four or more dimensions can not be sufficiently analyzed, because suitable methods are lacking that allow to follow gene expression trajectories over all dimensions.

To date, there are only a few existing four-dimensional microarray expression datasets, but one can foresee that this number will drastically increase in the future. The best studied four-dimensional experiment to date is given as an example in Figure 3 D: Arabidopsis plants have been treated with nine different abiotic stresses. Samples are taken at 6 points in time from two tissues (roots and shoots) [21]. This experimental design constitutes a *gene* x *condition* x *time* x *tissue* matrix.

So far, the analysis of n-dimensional datasets is carried out by simplifying the questions addressed [25,16]. For example, which genes exhibit a time dependent expression or are regulated during one of the stresses? To date, a complex analysis that considers all four dimensions is not possible.

5.3. *Mining for meaningful data*

The most obvious information gained by microarray experiments is on the genes that exhibit changes in transcript abundance under inducing conditions over the background compared with the control conditions cite6, 8. Besides that, there are three different kinds of meaningful data that are frequently inferred and searched for in microarray expression experiments.

- genes with highly similar expression trajectories over all conditions (co-regulated genes)
- genes that are involved in specific biological processes
- genes that might be controlled by similar regulatory sequences

Co-regulated genes are most informative as they might be involved in the same biological processes and have a higher chance of being regulated by similar regulatory sequences [26,27,16]. Hence, the focus of gene expression data analysis is on the identification of genes that form clusters of co-regulated gene modules.

While simple correlation algorithms sufficiently support the analysis of co-regulated gene modules in two-dimensional microarray experiments, novel types of approach had to be developed in the past to find modules in three- or multi-dimensional datasets [25,16,15].

Recently it was shown that four-dimensional datasets can preliminarily be analyzed, by examining the fourth-dimension independently [25,16]. Thus, a large 4D experiment needs to be broken up into several experiments of three-dimensions.

However, the quality of information gained from experiments strictly depends on the variability of the signals [as discussed in Section 3], the complexity of the experimental design and the infrastructure supporting the data analysis.

6. Conclusions & Outlook

The present manuscript provides an insight into how microarray experiments are performed, their value for life-scientists and the problems arising from the different experimental designs.

Many biologists are unaware of the downstream problems caused by their experimental design and ignore the complexity of the datasets. Many of the complex experiments have been evaluated with two-dimensional analysis procedures, which only allow examination of the dramatic expression changes over the background. More subtle, but nonetheless informative, changes in gene expression can not be detected that way.

As experiments of n-Dimensions or of a complex, nested design can not be sufficiently evaluated, novel evaluation procedures need to be developed [25,16,15]. Moreover, methods have to be established that allow the comparison of independent microarray experiments from different researchers.

On the one hand these urgent calls for new methods and cautious remarks by computer scientist stay unheard. On the other hand biologists plan even more complex microarray experiments, although the developers of microarray analysis tools lag far behind. Thus, many researchers develop tools on their own, decreasing the possibility of generating comparable data.

A major drawback in the analysis of biological data is that populations do not follow distinct state changes, but respond in a bell shaped frequency distribution. There is a frequency distribution-like predisposition for each of the individuals to behave in a certain manner, resulting in a frequency distribution-like response.

This is the same for gene expression: There is a certain possibility for a gene to change its expression state after a stimulus. Hence, this change will possibly affect the gene expression of other genes or feed back on its own expression. Modules of co-regulated genes are established, which enable the cell to integrate information and respond in a concerted manner. It remains to be investigated whether the strength of gene expression changes of a few genes or the functional composition of the co-regulated gene modules translates into a specific frequency distribution-like response of the individual cells.

These undefined state changes make an evaluation of biological data difficult for mathematicians and information scientists.

References

1. A. G. Cridge, L. L. Major, A. A. Mahagaonkar, E. S. Poole, L. A. Isaksson and W. P. Tate, Nucleic Acids Res 34, 1959 (2006).
2. J. R. Gibbs and A. Singleton, PLoS Genet 2, p. e150(Oct 2006).
3. J. Brosius and H. Tiedge, RNA Biol 1, 81(Jul 2004).
4. Z. Pezer and D. Ugarkovic, Semin Cancer Biol 18, 123(Apr 2008).
5. F. Crick, Nature 227, 561(Aug 1970).
6. B. J. Reinhart, F. J. Slack, M. Basson, A. E. Pasquinelli, J. C. Bettinger, A. E. Rougvie, H. R. Horvitz and G. Ruvkun, Nature 403, 901(Feb 2000).
7. M. Metzlaff, Biol Chem 383, 1483(Oct 2002).
8. K. Huppi, N. Volfovsky,M.Mackiewicz, T. Runfola, T. L. Jones, S. E.Martin, R. Stephens and N. J. Caplen, Semin Cancer Biol 17, 65(Feb 2007).
9. K. W. Berendzen, K. Stber, K. Harter and D. Wanke, BMC Bioinformatics 7, p. 522 (2006).
10. D. Walther, R. Brunnemann and J. Selbig, PLoS Genet 3, p. e11(Feb 2007).
11. G. Liu, A. E. Loraine, R. Shigeta, M. Cline, J. Cheng, V. Valmeekam, S. Sun, D. Kulp and M. A. Siani-Rose, Nucleic Acids Res 31, 82(Jan 2003).
12. J. C. Redman, B. J. Haas, G. Tanimoto and C. D. Town, Plant J 38, 545(May 2004).
13. D. Gresham, M. J. Dunham and D. Botstein, Nat Rev Genet 9, 291(Apr 2008).
14. J. M. Johnson, S. Edwards, D. Shoemaker and E. E. Schadt, Trends Genet 21, 93(Feb 2005).
15. D.Wanke, J. Kilian, J. Supper, K.W. Berendzen, A. Zell and K. Harter, QPPQ: QUANTUM BIO-INFORMATICS - From Quantum Information to Bio-Informatics, Quantum Probability and White Noise Analysis Vol. 21 (World Scientific, 2008), ch. The Analysis of Gene Expression and Cis-Regulatory Elements in Large Microarray Expession Datasets., pp. 294 – 314.
16. J. Supper,M. Strauch, D.Wanke, K. Harter and A. Zell, BMC Bioinformatics 8, p. 334 (2007).
17. Z. Wu, R. Irizarry, R. Gentleman, F. M. Murillo and F. Spencer, A Model Based Background Adjustment for Oligonucleotide Expression Arrays, Johns Hopkins University Dept. of Biostatistics Working Paper Series 1001, Berkeley Electronic Press (July 2004), available at http://ideas.repec.org/p/bep/jhubio/1001.html.
18. J. Quackenbush, Nat Genet 32 Suppl, 496(Dec 2002).
19. Arabidopsis Genome Initiative, Nature 408, 796(Dec 2000).
20. L. N. u. D. W. Thomas Altmann, GenomXpress 3, 13 (2004).
21. J. Kilian, D. Whitehead, J. Horak, D. Wanke, S. Weinl, O. Batistic, C. D'Angelo, E. Bornberg-Bauer, J. Kudla and K. Harter, Plant J 50, 347(Apr 2007).
22. M. Schmid, T. S. Davison, S. R. Henz, U. J. Pape, M. Demar, M. Vingron, B. Schlkopf, D. Weigel and J. U. Lohmann, Nat Genet 37, 501(May 2005).
23. J. L. Nemhauser, F. Hong and J. Chory, Cell 126, 467(Aug 2006).

24. M. Bartsch, E. Gobbato, P. Bednarek, S. Debey, J. L. Schultze, J. Bautor and J. E. Parker, Plant Cell 18, 1038(Apr 2006).
25. M. Strauch, J. Supper, C. Spieth, D.Wanke, J. Kilian, K. Harter and A. Zell, Journal of Integrative Bioinformatics 4(1), p. 54 (2007), Online Journal: http://journal.imbio.de/index.php?paper id=54.
26. H. K. Lee, A. K. Hsu, J. Sajdak, J. Qin and P. Pavlidis, Genome Res 14, 1085(Jun 2004).
27. L. L. Elo, R. Lahesmaa and T. Aittokallio, Journal of Integrative Bioinformatics 3(2), p. 33 (2006), Online Journal: http://journal.imbio.de/index.php?paper id=33.

INTEGRATING BIOLOGICAL PERSPECTIVES: A QUANTUM LEAP FOR MICROARRAY EXPRESSION ANALYSIS

DIERK WANKE* [1], JOACHIM KILIAN[1], ULRICH BLOSS[1], ELKE MANGELSEN[2], JOCHEN SUPPER[3], KLAUS HARTER[1] AND KENNETH W. BERENDZEN[1]

[1]*ZMBP Pflanzenphysiologie, Universität Tübingen*
Auf der Morgenstelle 1, D-72076 Tübingen, Germany
**E-mail: dierk.wanke@zmbp.uni-tuebingen.de*
[2]*Department of Plant Biology and Forest Genetics, The Swedish University of Agricultural Sciences (SLU)*
P.O. Box 7080, SE-750 07 Uppsala, Sweden
[3]*Wilhelm-Schickard-Institut für Informatik, Universität Tübingen*
Sand 1, D-72076 Tübingen, Germany

Biologists and bioinformatic scientists cope with the analysis of transcript abundance and the extraction of meaningful information from microarray expression data. By exploiting biological information accessible in public databases, we try to extend our current knowledge over the plant model organism *Arabidopsis thaliana*. Here, we give two examples of increasing the quality of information gained from large scale expression experiments by the integration of microarray-unrelated biological information: First, we utilize Arabidopsis microarray data to demonstrate that expression profiles are usually conserved between orthologous genes of different organisms. In an initial step of the analysis, orthology has to be inferred unambiguously, which then allows comparison of expression profiles between orthologs. We make use of the publicly available microarray expression data of Arabidopsis and barley, *Hordeum vulgare*. We found a generally positive correlation in expression trajectories between true orthologs although both organisms are only distantly related in evolutionary time scale. Second, extracting clusters of co-regulated genes implies similarities in transcriptional regulation via similar *cis*-regulatory elements (CREs). *Vice versa* approaches, where co-regulated gene clusters are found by investigating on CREs were not successful in general. Nonetheless, in some cases the presence of CREs in a defined position, orientation or CRE-combinations is positively correlated with co-regulated gene clusters. Here, we make use of genes involved in the phenylpropanoid biosynthetic pathway, to give one positive example for this approach.

Keywords : *Gene expression data, cross species transcription profiling,* *comparative microarray analysis, cis-regulatory elements, AtGenExpress*

1. Introduction

One of the best studied eukaryotic model organisms to date is the flowering plant *Arabidopsis thaliana.* Its physiological responses and its lifecycle have been studied for more than 100 years. With only five chromosomes, the overall complexity of its genome is low and very early the genetic potential was recognized by investigations on naturally occurring mutants. Since the first publication of its entire genomic sequence in the end of the year 2000, Arabidopsis has risen to one of the best understood eukaryote model organisms with the best annotated genome sequence of all eukaryotes available [1]. Moreover, the genome sequence serves as a good basis to investigate also distantly related organism, as comparative analyses demonstrated for protein coding genes that 35 % of all Arabidopsis with a known function have orthologs in animal counterparts. That number even increases within the plant kingdom.

The DFG-funded Arabidopsis Functional Genomics Network (AFGN, http://www.uni-tuebingen.de/plantphys/AFGN/) AtGenExpress project is designed to uncover the transcriptome of *Arabidopsis thaliana* with the help of microarrays. Microarray technology constitutes an effective means to gain insight into gene expression on a broad scale and enables the researcher to analyze the transcripts of thousands of genes simultaneously [2]. There are 41 different AtGenExpress core-conditions that were completed and made publicly available at the end of 2005. These core-conditions comprise a developmental baseline, biotic and abiotic stress treatments, hormone application, growth under different light qualities and nutrient availability [3,4,5]. Collectively, the community has access to 1295 AtGenExpress gene expression profile datasets (TAIR: http://www.arabidopsis.org).

While *Arabidopsis thaliana* is a modern dicot plant of little agricultural importance, the evolutionary distantly related grasses are of a high economical value worldwide [6]. A major drawback for genetic approaches in grasses lies in the complex genomes of cereal crops that has multimerized and duplicated during the breeding and inbreeding processes for domestication. The small size and relatively simple organization of the genome has been in favor of rice as the second fully sequenced eukaryote plant species. Besides rice, great efforts have been made to uncover genomic information of maize or wheat [7,8], but resigned from these approaches due to the complex organization of the cereals genomes.

However, the cereal crop barley is also of high importance for food, feed and brewing [6]. After maize, rice and wheat, barley ranked as number four in cereal production in 2005 [6]. *Hordeum vulgare* is cultivated world-wide and under diverse environmental conditions, due to its wide ecological potential in sustaining and acclimating to a variety of abiotic stresses [7]. For example, barley is grown on soils and in altitudes that are unsuitable for wheat and oat. The genome of barley is simple in organization and has been estimated to be of 5000 Mbp in size. Unfortunately, despite ongoing sequencing projects, it has not been fully sequenced yet. However, the GenBank EST dataset (http://www.ncbi.nlm.nih.gov/) contains almost 440,000 barley ESTs and provides a valuable resource for bioinformatics analyses. On the basis of this information the Affymetrix 22K Barley1 GeneChip microarray was released on the market in 2004, which represents probes for roughly 22,000 genes [9]. Nowadays, a number of barley microarray experiments are accessible via public databases [10], which allows gene expression analyses of barley genes.

Most informative for biologists is the analysis of co-regulated genes [11]. These are genes with highly similar expression trajectories over different experimental conditions (Fig.1). Those are believed to constitute the best candidate genes that might likely fulfill a common function [12,11]. Either they encode for proteins that act in the same signaling cascades, are enrolled in the same enzymatic pathways or represent interacting partners that form higher order complexes. Additionally, genes with similar expression profiles in microarray experiments might be regulated by the same *cis*-regulatory elements (CREs) (Fig.2). Thus, combinatory investigations of co-regulated genes from array data and subsequent promoter analysis have in some cases resulted in the identification of novel motifs or of already known functional CREs of putative regulatory function [13,14].

Although single elements have been found to be sufficient for regulating specific responses upon simple stimuli in synthetic promoters, there is evidence that functional *cis*-elements occur in clusters rather than as isolated singletons. In this context it has been shown that spacing between two *cis*-elements and their orientation towards each other is essential for both DNA binding and transactivation [15,16].

The analysis of gene expression profiles for the identification of clusters of genes that are co-regulated under several experimental conditions with the intention of discovering putative CREs orchestrating gene regulation is still a challenging task for biologists and information scientists.

Here, we give two examples of increasing the quality of informa-

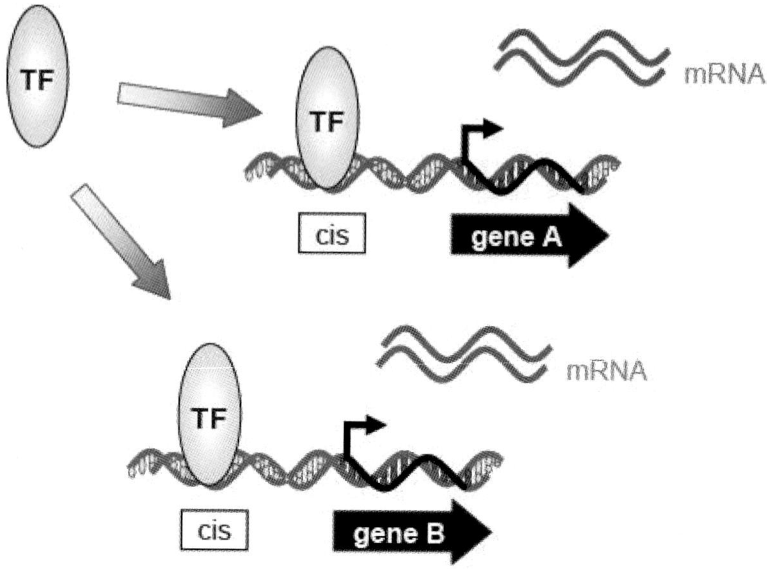

Figure 1. Scheme for co-regulation of two genes mediated by one transcription factor. A transcription factor (TF) binds to the *cis*-regulatory element present in the 5 regulatory region of the two genes A and B. The proximal and distal regions important for the regulation of gene expression are the promoters. Here, binding of TF results in the same amount of transcribed mRNA.

tion gained from large scale expression experiments by the integration of microarray-unrelated biological information. We show that the integration of phylogenetic information, which is the evolutionary distance between two sequences, provides a solid basis for the analysis of protein functions using microarray expression data. This method extends our knowledge over the fully sequenced model organism to infer protein function in distant relatives.

The second part of our work describes a cluster of CREs responsible for concerted expression and co-regulation of biosynthetic pathway genes. Different as many approaches before, the initial investigation was not based on co-expression of genes, but on information of the CREs.

2. Material and Methods

2.1. *Arabidopsis thaliana Gene Expression Data*

AtGenExpress data used in this article has been described by Kilian *et al.*[4] and Goda *et al.*[5]. The AtGenExpress abiotic stress dataset used in this analysis comprises the Affymetrix CEL file accession numbers at TAIR (www.arabidopsis.org): ME00325, ME00326, ME00327, ME00328, ME00329, ME00330, ME00338 and ME00339; the AtGenExpress plant hormone dataset used in this analysis comprises the accession numbers at TAIR (www.arabidopsis.org): ME00333, ME00334, ME00336, ME00337, ME00343, ME00344; AtGenExpress developmental baseline dataset used in this analysis comprises the Affymetrix CEL file accession numbers at TAIR (www.arabidopsis.org): ME00319. All files have been imported into GeneSpring software version 7.3.2 (Agilent, Boeblingen, Germany). The arrays have been adjusted for background optical noise using sequence dependent robust multi-array averaging (GC-RMA) software and normalized using quantile normalization. The quality of the replicates has been tested by performing a least-square regression analysis. From the resulting signal intensities, fold change values have been calculated.

2.2. *Hordeum vulgare Gene Expression Data*

Publicly available microarray data of the Affymetrix Barley1 GeneChip has been used. The annotations of the Barley1 GeneChip have been updated by the HarvEST database (http://harvest.ucr.edu/; version 1.51) [10,8]. The Affymetrix CEL files of the barley developmental baseline dataset used in this analysis comprises the accession number BB3 at BarleyBase [10] (http://www.plexdb.org/). A per-Chip normalization to the median has been applied to obtain comparability. The arrays have been adjusted for background optical noise using sequence dependent robust multi-array averaging (GC-RMA) software and normalized using quantile normalization. From the resulting signal intensities, fold change values have been calculated.

2.3. *Phylogenetic Analysis*

Multiple sequence alignments have been conducted on Arabidopsis, barley and rice protein sequences with ClustalW (http://www.ebi.ac.uk/clustalw/). Default settings were used to calculate the trees files. Phylogenetic trees were drawn using TreeView software version 1.6.6 (http://taxonomy.zoology.gla.ac.uk/rod/treeview.html).

2.4. *Cross-Species Gene Expression Analysis*

The principle approach used in this article has been described by Mangelsen *et al.* [17]. The identification of suitable barley genes for cross-species gene expression analysis has been performed with BLAST searches against the Barley1 GeneChip probe set assemblies on the Affymetrix Netaffx website (http://www.affymetrix.com/analysis/index.affx).

Expression data were extracted for homologous plant organs (1 - 10) for barley and Arabidopsis, respectively: 1 (coleoptile/cotyledon), 2 (mesocotyl/hypocotyl), 3 (radicule/seedling roots), 4 (leaf/leaf), 5 (root/root), 6 (inflorescence/inflorescence), 7 (anthers/stamen), 8 (caryopsis 5dpa/seed stage 6), 9 (caryopsis 10dpa/seed stage 8) and 10 (caryopsis 15dpa/ seed stage 10).

The correlation for gene expression levels between corresponding organs from barley and Arabidopsis were analyzed by calculating the Pearson correlation of the normalized signal intensities with regular spread sheet analysis. As a control, signal intensities of 100 randomly chosen probesets of the Barley1 GeneChip were compared to 100 randomly chosen Arabidopsis genes represented by the ATH1 GeneChip. For both comparisons, the average correlation was calculated. Students t-test was used to test the statistical significance of the difference in correlation of the datasets.

2.5. *Simple Correlation Analysis and Visualization*

For reasons of comparability, all gene expression profiles are visualized with GeneSpring software version 7.3.2 (Agilent, Boeblingen, Germany) as line graphs. Clusters of co-regulated genes are identified by simple correlation analysis of gene expression trajectories with Pearson Correlation, which is implemented in the GeneSpring software package.

2.6. *Cis-regulatory Elements*

Scripts used for the position analysis of Position Weight Matrices (PWM) by frequency distribution curves are part of the Motif Mapper open-source script
collection [15] (`http://www.zmbp.uni-tuebingen.de/PlantPhysiology/ResearchGroups/harter/berendzen/programs.html`), which was first released online in 2004. Frequency distribution curves were rooted to the translation initiation codon ATG; all curves were drawn with the root (ATG) pointing to the right. The sequence logos for the P-, A- and L-boxes have been composed with WebLogo (http://weblogo.berkeley.edu/) from 23, 19 and 21 previously identified functional elements, respectively. The output of REG_Pro_Point_Mapper v4.0.0 script has been displayed with regular spread sheet analysis programs. For each rooted dataset the

raw number of motifs was plotted in respect to their exact position.

2.7. *Variance based Comparison of Gene Expression Datasets*

The principle approach uses the variances in gene expression to compare two independent expression profile datasets. Expression estimates of the abiotic stress [4] and the hormone treatment [5] experiments of the AtGen-Express project have been calculated with GC-RMA. Only treatments for ABA, ACC, BL, GA, IAA and MJ have been considered for the hormone dataset. A per-gene Variance Analysis (VA) has been computed on the expression trajectories for both datasets independently [18,19]. The Variance has been calculated based on all genes as a sample instead of a universe, which increases the level of significance. Both, GC-RMA and VA are algorithms and procedures implemented in R (http://cran.r-project.org/). Genes have been considered differentially expressed, if their expression variances differ by more than a 3-fold from their mean expression trajectories in both datasets under at least one condition.

3. Results

One of the challenging tasks in the analysis of microarray expression data is the extraction of meaningful information [20,11]. Most informative are co-regulated genes, which exhibit highly similar expression trajectories over different experimental conditions [21,20,11]. It has been shown that co-regulated genes do either function in similar pathways (Fig.1) or are controlled by the same regulatory promoter elements (Fig.2).

Problems in the identification of co-regulated genes are manifold. However, the most difficult is to assign significance over experimental noise and technological variability. We have overcome these problems by the integration of microarray-unrelated biological information.

The first approach is based on the assumption that the overall similarities in gene expression increase the closer different species are related to each other. Figure 3 gives an example for the expression of two genes in four different species. The two phylogenetic closely related species A and B exhibit almost identical expression intensities (Fig.3 A). The genes A or B are transcribed to a similar extend and a similar amount of mRNAs are synthesized. If a gene is in direct relationship with an other gene in a different species, it is termed ortholog. In a comparison of more distantly related species, for example species A and C, evolutionary time resulted in

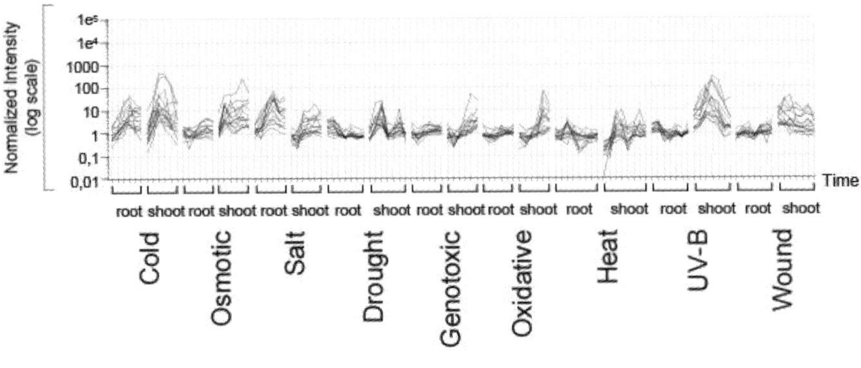

Figure 2. Line graph representing gene expression profiles of co-regulated genes under diverse stress treatments. Cluster of 32 co-regulated genes during nine different stresses of the AtGenExpress abiotic stress dataset. The lines represent expression trajectories for each gene in a *time* x *condition* x *tissue* dependent manner. The average gene expression intensities are shown. The example given depicts genes upregulated during cold and UV-B light stress.

a higher mutation rate and consequently in sequence divergence. Due to the different selective pressures, mutations within the orthologous promoter sequences are much more likely than mutations within the protein coding region [15,20]. Hence, mutations accumulating in the regulatory sequences of the promoters will account for differences in the expression of genes, while the gene function of orthologs is likely to be maintained [15,22,20].

The phylogenetic divergence between the species is calculated on the basis of DNA or protein sequence information and can be used to draw a tree of evolutionary distance [17]. As phylogenetic data is independent of and unrelated to expression signatures measured with microarray technology, it adds important extra information for the analysis. The more expression data of different species is available in public databases the more it can be used to support microarray data analysis. Besides the direct enhancement of microarray data evaluation, the evolutionary analysis of single or multiple gene expression trajectories becomes imaginable (Fig.3 B). There is expression data of close relatives, but information on distantly related species is lacking in most of the cases.

In more distantly related species, the chance is higher for other effects on the genome, which could perturb the analysis. Besides small mutations that affect the changes in DNA sequences in a more gradual manner, gene duplications and rearrangement result in erratic and sudden transitions [1,23,17].

In cases of gene duplications, it might become difficult to assign proper orthology and both duplicates have to be taken into account. Therefore, a phylogenetic analysis is inevitable to conclude on an unambiguous gene orthology [23,17].

On the other hand, the more distantly two species are related the higher is the chance to regain a defined sequence due to back-mutations. As a consequence, very distant related sequences appear closer related than they actually are. This effect is called "long branch attraction" [24,25]. Integration of data on the relative distances in gene expression could contribute in correcting the evolutionary time.

Public expression databases exist that provide data for Arabidopsis thaliana and of the distantly related grass species rice and barley. However, the paradigm of our approach relies on the unambiguous assignment of orthology.

A previous analysis on transcript abundance of Arabidopsis and rice genes revealed only few similarities between gene sequences and expression [26,13]. Nevertheless, highest similarities have been found restricted to genes of primary metabolism, while similar genes encoding for transcription factors have not been found similarly expressed [26,27].

In contrast, we have shown that this is not generally true: We assigned true orthology for WRKY transcription factors in Arabidopsis and barley [17]. Previous analysis has been shown that WRKY transcription factors are fast evolving, which means that their gene sequence is rapidly changing [28]. However, we found that true orthologs of Arabidopsis and barley WRKY genes have highly similar expression trajectories. This is in vast contrast to what has been assumed before.

Here, we extend our analysis to other transcription factor families, to support the data we have found for WRKY proteins. Additionally, we have included genes of the primary metabolism. Figure 4A gives examples for the well known Arabidopsis transcription factor genes AGAMOUS, KNAT1, HY5 and AtWRKY56. These genes are involved in flower and leaf development, light signal transduction and pathogen defense [29,30,31,32,33] and are compared to their barley HvAG, HvKn1, HvHY5 and HvWRKY13 orthologs.

Their expression trajectories are shown over ten different developmental conditions and Pearson correlation coefficients r have been calculated on the basis of normalized signal intensities for each of the orthologous gene pairs. Most of the gene pairs that have been analyzed this way display a positive correlation coefficient. This is indicative of evolutionary conserved function and conserved expression pattern. For example, the expression trajectories of the orthologous pair AGAMOUS and HvAG1 are highly similar and exhibit the highest correlation coefficient (r = 0.90) of all transcription factor genes found (Fig.4A). This can be explained by the inevitable function of AGAMOUS-like proteins in flower development. Nonetheless, this is surprising as several genes have been identified in the Arabidopsis genome with high similarities to AGAMOUS. Therefore, all these similar genes could possibly substitute AGAMOUS function and act redundantly in flower patterning. In Arabidopsis the AGAMOUS gene is located inside a duplicated locus and the SHP1 gene is its duplicate [29]. Consequently, it is impossible to deduce orthology unambiguously and it could be possible that SHP1 is the real ortholog to HvAG1. However, the integration of expression data and its high correlation coefficient revealed that only AGAMOUS and HvAG1 constitute orthologs.

We have shown that the combined analysis on sequence and expression similarities is beneficial and highly informative for both approaches. The combined examination of the orthologous pairs of the Group 2c WRKY genes and the Knotted-1 like homeobox genes has been advantageous for phylogeny and cross-species expression analysis.

In contrast, the AtHY5 gene has been positioned between two clades in a tree of HY5-like bZIPs [30]. This implies that true orthologs can be found at either side of the tree branches. The expression comparison of Arabidopsis AtHY5 with a barley gene termed HvHY5 revealed a negative correlation coefficient (r = -0.46). This suggests that HvHY5 might not be the true ortholog. On the other hand, we can not exclude the possibility that both genes are still true orthologs, which evolved different functions in the distantly related plant species.

The second group of genes is involved in primary metabolism, development and hormone signaling (Fig.4B), but the same difficulties in the assignment of orthology have been observed. The correlation coefficients found have all been positive, which indicates a higher level of expression conservation compared to the tested transcription factor genes. Moreover, all tested Arabidopsis genes do have a high similar duplicate that possibly can take over the gene function. One can speculate that only one of these genes is essential for plant life and that the high similarity is a consequence of very recent gene duplication events.

Further investigations will have to elucidate whether or not the conservation does extend over the similarities in sequence or gene expression. As similar expression trajectories have been found between the orthologs, one also has to speculate on conserved signaling cascades underlying and governing the up- or downregulation of gene transcription. Maybe, similar expression trajectories are caused by similar transcription factors conserved between the different species binding to similar DNA-elements in the promoter sequences. Keeping in mind, genes with similar expression profiles in microarray experiments might be regulated by the same *cis*-regulatory elements (CREs) (Fig.2). Berendzen *et al.*[15] have found that in eukaryotes many regulatory motifs are conserved between the species and most are located close to the gene sequence. Comparative sequence analyses of orthologous promoters from different species can provide clues for the identification of putative functional CREs, which are retained due to a higher selective pressure in favor of the motifs function [15]. This approach has successfully been used in the last years and termed phylogenetic footprinting [35,36]. Unfortunately, phylogenetic footprinting has only succeeded for sequences of closely related species [37,36].

Identification of putative CREs has been successfully performed using probabilistic analysis of CRE presence in a regulatory sequence [15,16]. That is a higher relative frequency of a DNA-motif in a regulatory sequence than expected from a reference sequence.

However, we combined the probabilistic analysis with a phylogenetic footprinting approach to infer functional combinations of CREs. Therefore, we have chosen the well known phenylpropanoid metabolism genes PAL and CHS, which have been shown regulated via the evolutionary conserved P-, A- and L-box DNA elements [37,38].

PAL and CHS are genes that function in different steps of the phenylpropanoid biosynthetic pathway [39], however, the P-, A- and L-boxes are present in the promoters of both genes and allow the concerted regulation of Chalcone biosynthesis (Fig.5B).

We first have collected known functional sequence information on all three boxes to compose position weight matrices (PWM), which subsequently allows deriving motif consensi. In figure 5A the consensi of the three boxes are depicted as weblogos. The size of each of the nucleotides is proportional to the conservation within the DNA-motif. To gain additional information on each of the three CREs, frequency distribution curves are conducted with motif mapper (Fig.5A). With the use of frequency distribution curves for position analysis of CREs, it is possible to get a more detailed insight into where an element appears to function [15].

Distribution curves on the three CREs under investigation revealed that the A-box is the most frequent. On the basis of the maximal frequency peak found in the distribution curves, the P-box is more distantly positioned than the A- or L-boxes. Integrating the positional information in our motif searches identified about 1000 promoters in the Arabidopsis genome that share the orientation of all three motifs towards each other and their position in respect of the ATG. A detailed analysis of these genes has revealed their functional involvement in several biosynthetic pathways.

We next have asked whether or not other genes involved in the phenylpropanoid pathway do have this precise arrangement of P-, A- and L-boxes in their promoters. To our great surprise, many of the genes involved in the anthocyanine and flavonoid biosynthesis have the CRE-motif module of P-, A- and L-boxes close to the ATG. Moreover, phenylpropanoid genes with the P-A-L-boxes in their promoter are co-regulated under diverse stress treatments (Fig.5D). This supports our finding that the combination of the three boxes might form a conserved regulatory DNA-motif module, which is important for the concerted co-regulation of gene expression under stress.

Within the phenylpropanoid biosynthetic pathway the 4CL subfamily of the acyl:coenzymeA synthases (ACS) constitutes the branch point between lignin biosynthesis and the production of anthocyanines or flavonoids [40,39]. There are 13 different 4CL genes that have been identified based on protein sequence similarities [40]. Only six of them are stress responsive - all of which do have the P-A-L-box DNA-motif module in their promoter sequences (Fig.5C). Amongst them is the *bona fide* group of 4CL1 - 4CL4, which are known to be UV-B light stress responsive [40,41].

To get a better impression on how specific the regulation of the different 4CL genes is, we conducted a novel approach for the comparison of different microarray expression datasets.

The principle approach uses a variance based comparison of gene expression datasets. The variances in gene expression have been computed on the expression trajectories for both datasets independently [19]. Expression estimates of the abiotic stress and the hormone treatment experiments of the AtGenExpress project have been calculated and plotted (Fig.5C). The axis labels represent relative numbers that increase with the variability of a gene expression trajectory within each of the datasets.

There are 4CL genes that can clearly be associated with only one of the stresses, whereas others are not found to be regulated by any of the conditions at all. Only two of the 4CL genes are responding in both experimental datasets (Fig.5C). The six genes with the P-A-L-box DNA-motif module are all stress responsive, which is supportive of our previous findings.

These preliminary results indicate that combinatorial approaches and the integration of different information sources are beneficial for expression data analysis.

4. Conclusions

In the present work we discuss the value of microarray-unrelated biological information to increasing the quality of expression data analysis. In the two cases we presented here, both sides are profiting during the investigation. The identification of co-regulated genes in distantly related species is a crucial step for future developments. Our analysis has shown that the integration of an evolutionary context was beneficial for the evaluation of expression data. On the other hand, a high correlation between gene expression trajectories can be used as supportive information to infer species relatedness.

Many different algorithms exist, that are capable of identifying gene expression trajectories, which can be correlated with the presence of CREs. Only few approaches have been successful in the identification of CRE-modules that account for co-regulation of genes after a certain stimulus. Therefore, we are optimistic to find more of such modules the more suitable information on functional CREs is available. Moreover, we have demonstrated that the variance based analysis of gene expression constitutes a suitable means for the comparison of different datasets.

Finally, the synergy between both computational and experimental methods for gene expression analyses will help to complete our views of the Arabidopsis genome, and will be essential to our understanding of other eukaryote genomes, including the human genome sequence.

Acknowledgements

We thank Martin Strauch and Andreas Zell for their continuous support in the development of microarray analysis algorithms. This work was supported by the National Genome Research Network (NGFN II) of the Federal Ministry for Education and Research in Germany under contract number 0313323, by the Arabidopsis Functional Genomics Network (AFGN) of the DFG (HA2146/5-2) and by the DFG-funded AtGenExpress project.

References

1. A. G. Initiative, Nature 408, 796(Dec 2000).
2. J. C. Redman, B. J. Haas, G. Tanimoto and C. D. Town, Plant J 38, 545(May 2004).

3. M. Schmid, T. S. Davison, S. R. Henz, U. J. Pape, M. Demar, M. Vingron, B. Schlkopf, D. Weigel and J. U. Lohmann, Nat Genet 37, 501(May 2005).
4. J. Kilian, D. Whitehead, J. Horak, D. Wanke, S. Weinl, O. Batistic, C. D'Angelo, E. Bornberg-Bauer, J. Kudla and K. Harter, Plant J 50, 347(Apr 2007).
5. H. Goda, E. Sasaki, K. Akiyama, A. Maruyama-Nakashita, K. Nakabayashi, W. Li, M. Ogawa, Y. Yamauchi, J. Preston, K. Aoki, T. Kiba, S. Takatsuto, S. Fujioka, T. Asami, T. Nakano, H. Kato, T. Mizuno, H. Sakakibara, S. Ya-maguchi, E. Nambara, Y. Kamiya, H.Takahashi, M. Y. Hirai, T. Sakurai, K. Shinozaki, K. Saito, S. Yoshida and Y. Shimada, Plant J (Apr 2008).
6. Y. Yamazaki and P. Jaiswal, Plant Cell Physiol 46, 63(Jan 2005).
7. D. Ware, P. Jaiswal, J. Ni, X. Pan, K. Chang, K. Clark, L. Teytelman, S. Schmidt,W. Zhao, S. Cartinhour, S. McCouch and L. Stein, Nucleic Acids Res 30, 103(Jan 2002).
8. S. Avraham, C.-W. Tung, K. Ilic, P. Jaiswal, E. A. Kellogg, S. McCouch, A. Pujar, L. Reiser, S. Y. Rhee, M. M. Sachs, M. Schaeffer, L. Stein, P. Stevens, L. Vincent, F. Zapata and D. Ware, Nucleic Acids Res 36, D449(Jan 2008).
9. T. J. Close, S. I.Wanamaker, R. A. Caldo, S. M. Turner, D. A. Ashlock, J. A. Dickerson, R. A.Wing, G. J. Muehlbauer, A. Kleinhofs and R. P.Wise, Plant Physiol 134, 960(Mar 2004).
10. L. Shen, J. Gong, R. A. Caldo, D. Nettleton, D. Cook, R. P. Wise and J. A. Dickerson, Nucleic Acids Res 33, D614(Jan 2005).
11. J. Supper,M. Strauch, D.Wanke, K. Harter and A. Zell, BMC Bioinformatics 8, p. 334 (2007).
12. H. B. Nielsen, J. Mundy and H. Willenbrock, PLoS ONE 2, p. e676 (2007).
13. J. Zhou, X. Wang, Y. Jiao, Y. Qin, X. Liu, K. He, C. Chen, L. Ma, J. Wang, L. Xiong, Q. Zhang, L. Fan and X. W. Deng, Plant Mol Biol 63, 591(Mar 2007).
14. B. Kaplan, O. Davydov, H. Knight, Y. Galon, M. R. Knight, R. Fluhr and H. Fromm, Plant Cell 18, 2733(Oct 2006).
15. K. W. Berendzen, K. Stber, K. Harter and D. Wanke, BMC Bioinformatics 7, p. 522 (2006).
16. T. Obayashi, K. Kinoshita, K. Nakai, M. Shibaoka, S. Hayashi, M. Saeki, D. Shibata, K. Saito and H. Ohta, Nucleic Acids Res 35, D863(Jan 2007).
17. E. Mangelsen, J. Kilian, K. W. Berendzen, U. H. Kolukisaoglu, K. Harter, C. Jansson and D. Wanke, BMC Genomics 9, p. 194 (2008).
18. I. B. Jeffery, D. G. Higgins and A. C. Culhane, BMC Bioinformatics 7, p. 359 (2006).
19. C. Truntzer, C. Mercier, J. Estve, C. Gautier and P. Roy, BMC Bioinformatics 8, p. 90 (2007).
20. D.Wanke, J. Kilian, J. Supper, K.W. Berendzen, A. Zell and K. Harter, QPPQ: QUANTUM BIO-INFORMATICS - From Quantum Information to Bio-Informatics, Quantum Probability and White Noise Analysis Vol. 21 (World Scientific, 2008), ch. The Analysis of Gene Expression and Cis-Regulatory Elements in Large Microarray Expession Datasets., pp. 294 – 314.

21. H. K. Lee, A. K. Hsu, J. Sajdak, J. Qin and P. Pavlidis, Genome Res 14, 1085(Jun 2004).
22. D. Walther, R. Brunnemann and J. Selbig, PLoS Genet 3, p. e11(Feb 2007).
23. H. U. Kolukisaoglu, L. Bovet, M. Klein, T. Eggmann, M. Geisler, D. Wanke, E. Martinoia and B. Schulz, Planta 216, 107(Nov 2002).
24. M. J. Sanderson, M. F. Wojciechowski, J. M. Hu, T. S. Khan and S. G. Brady, Mol Biol Evol 17, 782(May 2000).
25. S. Stefanovic, D. W. Rice and J. D. Palmer, BMC Evol Biol 4, p. 35(Sep 2004).
26. L. Ma, C. Chen, X. Liu, Y. Jiao, N. Su, L. Li, X. Wang, M. Cao, N. Sun, X. Zhang, J. Bao, J. Li, S. Pedersen, L. Bolund, H. Zhao, L. Yuan, G. K.-S. Wong, J. Wang, X. W. Deng and J. Wang, Genome Res 15, 1274(Sep 2005).
27. K. M. Olsen, U. S. Lea, R. Slimestad, M. Verheul and C. Lillo, J Plant Physiol (Jan 2008).
28. M. M. Babu, L. M. Iyer, S. Balaji and L. Aravind, Nucleic Acids Res 34, 6505 (2006).
29. V. Brambilla, R. Battaglia, M. Colombo, S. Masiero, S. Bencivenga, M. M. Kater and L. Colombo, Plant Cell 19, 2544(Aug 2007).
30. R. Sibout, P. Sukumar, C. Hettiarachchi, M. Holm, G. K. Muday and C. S. Hardtke, PLoS Genet 2, p. e202(Nov 2006).
31. S. J. Douglas, G. Chuck, R. E. Dengler, L. Pelecanda and C. D. Riggs, Plant Cell 14, 547(Mar 2002).
32. I. Ciolkowski, D. Wanke, R. P. Birkenbihl and I. E. Somssich, Plant Mol Biol 68, 81(Sep 2008).
33. U. Hartmann, M. Sagasser, F. Mehrtens, R. Stracke and B. Weisshaar, Plant Mol Biol 57, 155(Jan 2005).
34. 34. A. Druka, G. Muehlbauer, I. Druka, R. Caldo, U. Baumann, N. Rostoks, A. Schreiber, R. Wise, T. Close, A. Kleinhofs, A. Graner, A. Schulman, P. Langridge, K. Sato, P. Hayes, J. McNicol, D. Marshall and R. Waugh, Funct Integr Genomics 6, 202(Jul 2006).
35. L. Zhang, K. Zuo, F. Zhang, Y. Cao, J.Wang, Y. Zhang, X. Sun and K. Tang, BMC Genomics 7, p. 323 (2006).
36. R. L. Hong, L. Hamaguchi, M. A. Busch and D. Weigel, Plant Cell 15, 1296(Jun 2003).
37. M. A. Koch, B. Weisshaar, J. Kroymann, B. Haubold and T. Mitchell-Olds, Mol Biol Evol 18, 1882(Oct 2001).
38. M. A. Koch, B. Haubold and T. Mitchell-Olds, Mol Biol Evol 17, 1483(Oct 2000).
39. E. Kombrink and K. Hahlbrock, Plant Physiol 81, 216(May 1986).
40. C. de Azevedo Souza, B. Barbazuk, S. G. Ralph, J. Bohlmann, B. Hamberger and C. J. Douglas, New Phytol (Jun 2008).
41. B. Hamberger and K. Hahlbrock, Proc Natl Acad Sci U S A 101, 2209(Feb 2004).

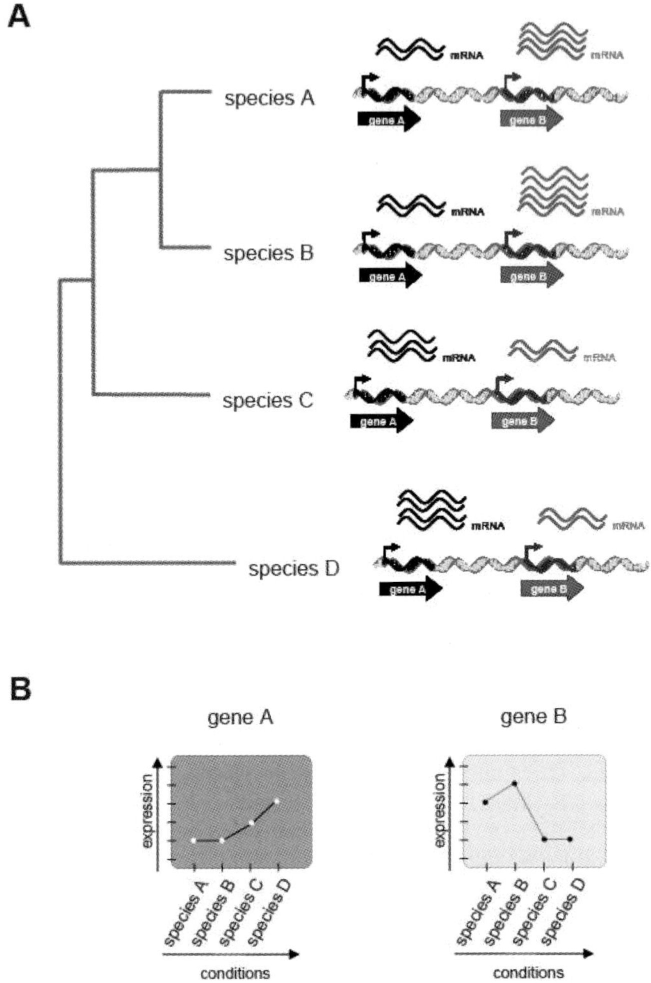

Figure 3. Scheme of phylogenetically conserved Gene Expression Trajectories. (A) The phylogenetic tree displayed on the left side gives an impression on the evolutionary relatedness of four species. The mRNA abundance of the two related genes A and B is changing with the evolutionary distance of the species. Hence, similarities in gene expression can be used as a measure of relatedness. The more similar the gene expression trajectories of orthologs are the closer is the relationship between species. (B) Schematic view of line graphs that display a gene wise expression trajectory with species as different conditions. Simple correlation coefficients can be used as a measure of species relatedness on the basis of gene expression similarities.

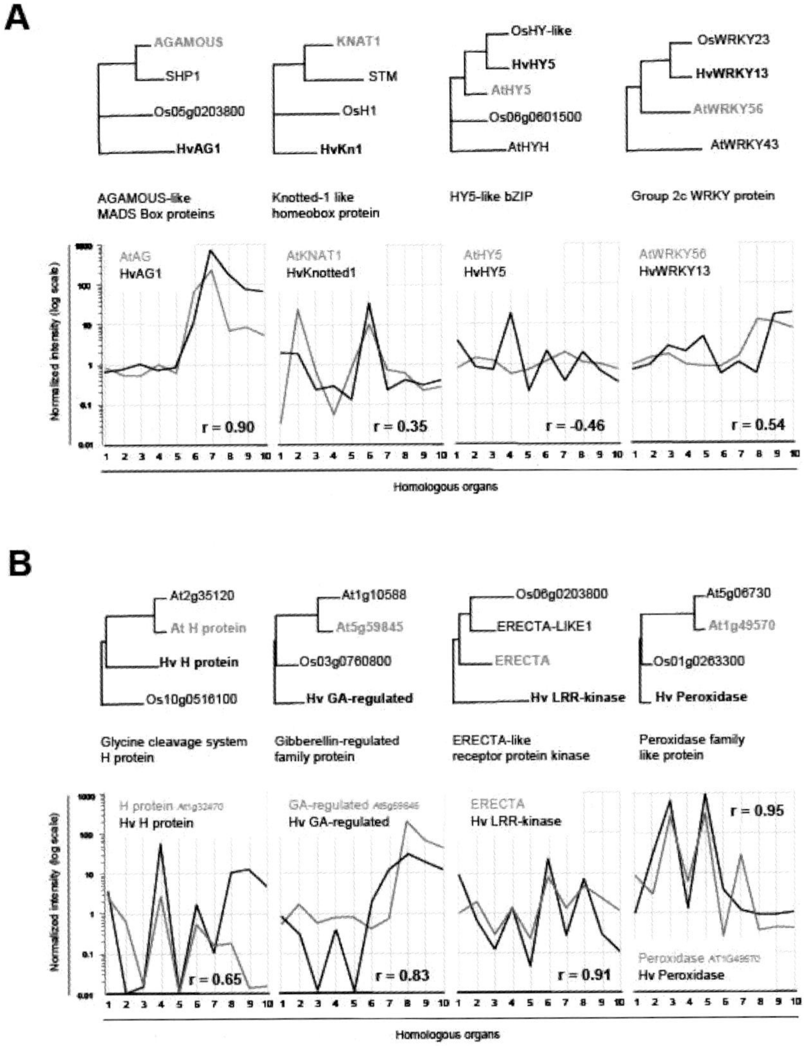

Figure 4. Selected expression profiles barley genes (Hv) and their designate orthologs in Arabidopsis (At). Cross-species comparison of orthologs of selected transcription factor genes (A) and genes involved in diverse metabolic pathways (B). Phylogenetic trees have been calculated on the basis of sequence similarities between Arabidopsis, rice and barley (Hv) proteins (top). For orthologous gene pairs, expression trajectories of ten homologous plant organs have been displayed using line graphs (bottom). Signal intensities of the developmental baseline experiments by Druka *et al.*[7] and Schmid *et al.*[3] are normalized to the mean expression values and are plotted in log-scale for different homologous plant organs (1-10). The respective organs of barley and Arabidopsis are: 1 (coleoptile/ cotyledon), 2 (mesocotyl/ hypocotyl), 3 (radicule/ seedling roots), 4 (leaf/ leaf), 5 (root/ root), 6 (inflorescence/ inflorescence), 7 (anthers/ stamen), 8 (caryopsis 5dpa/ seed stage 6), 9 (caryopsis 10dpa/ seed stage 8) and 10 (caryopsis 15dpa/ seed stage 10). Pearson correlation coefficients have been computed for the putative orthologs on the basis of the mean expression values.

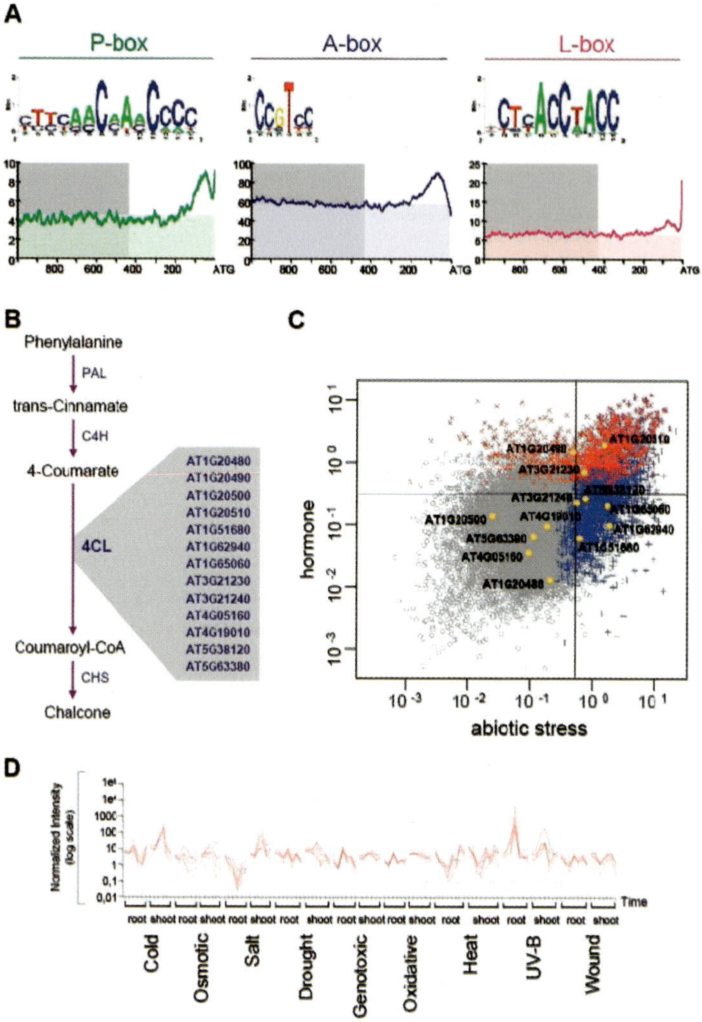

Figure 5. Identification of a *cis*-regulatory module by combined DNA-motif and microarray expression analysis. The boxes P, A and L are well known transcription factor binding sites (A), that have been identified from phenylpropanoid pathway genes (B). Frequency distribution curves of P-, A- and L-box consensi have been conducted with Motif Mapper [15] and rooted to the translation initiation codon ATG (A, bottom). The motif consensi have been displayed as WebLogos (http://weblogo.berkeley.edu/) for reasons of clarity (A, top). The distribution curves for each of the motifs exhibit distinct frequency disequilibria. Genes of the phenylpropanoid pathway (B) that contain a combination of P-A-L-boxes are co-regulated under different abiotic stress conditions (D). A per-gene Variance Analysis (VA) has been computed on the expression trajectories for the abiotic stress [4] and the hormone [5] treatment datasets independently (C). VA allows the direct comparison of differences in gene expression trajectories between two datasets. Differentially regulated genes are color-coded in blue (abiotic stress) or red (hormone). The 4CL gene family was plotted on top. This approach reveals distinct function for 4CL family members with and without the P-, A- and L-box consensi present in their promoters.

QBIC'08

March 12-16, 2008

INTERNATIONAL CONFERENCE IN QUANTUM BIO-INFORMATICS CENTER

Main Session
March 12, Wednesday 8:50-18:00
March 13, 14, and 15, 9:00-18:30
at Conference Room in Canal Hall

Noda Campus, Tokyo University of Science
Noda City, Chiba Japan

Satellite Session
March 16, Sunday 9:00-17:00
at Frontier Research Center for Computational Sciences(4F)

Poster Session
March 12, Wednesday 12:30-13:30
March 14, Friday 12:30-13:30
at Canal Hall

Welcome Party
March 12, Wednesday 18:30-
at Cafeteria(2F) in Canal Hall

Banquet
March 14, Friday 18:30 -
at Cafeteria(2F) in Canal Hall

Speakers(Tentative)
L. Accardi (Roma II University, Italy)
V. Belavkin (University of Nottingham, UK)
D. Chruscinski (Nicolaus Copernicus University, Poland)
K-H. Fichtner (Friedrich Schiller University Jena, Germany)
L. Fichtner (Friedrich Schiller University Jena, Germany)
W. Freudenberg (Brandenburg Techn. University Cottbus, Germany)
A. Fujiyama (National Institute of Informatics, Japan)
T. Hida (Emeritus Professor, Nagoya University, Japan)
W. Im (The University of Kansas, USA)
A. Khrennikov (University of Växjö, Sweden)
S. Kikuchi (National Institute of Agrobilogical Sciences, Japan)
S. Kondo (Nagoya University, Japan)
A. Kossakowski (Nicolaus Copernicus University, Poland)
A. Majewski (University of Gdańsk, Poland)
M. Michalski (Nicolaus Copernicus University, Poland)
I. Ojima (Kyoto University, Japan)
M. Regoli (Roma II University, Italy)
J. Skolnick (Georgia Institute of Technology, USA)
I. Volovich (Steklov Mathematical Institute, Russia)
D. Wanke (University of Tuebingen, Germany)
Some QBIC members in Tokyo University of Science
... and more

Purpose
The main aim of QBIC and the conference is to create a new paradigm synthesizing Quantum Information and Bio-Informatics based on efforts by active researchers traversing various fields of Mathematics, Physics, Information and Life Science.

Organizer
M. Ohya, Chief (Tokyo University of Science, Japan)
H. Sakata (Tokyo University of Science, Japan)
Y. Togawa (Tokyo University of Science, Japan)
I. Yamato (Tokyo University of Science, Japan)

Advisory Committee
L. Accardi (Roma II University, Italy)
W. Freudenberg (Brandenburg Techn. University Cottbus, Germany)
T. Hida (Emeritus Professor, Nagoya University, Japan)
H. Kamimura (Tokyo University of Science, Japan)
I. Ojima (Kyoto University, Japan)

Local Committee
K. Kuchitsu (Tokyo University of Science, Japan)
S. Miyazaki (Tokyo University of Science, Japan)
H. Takayanagi (Tokyo University of Science, Japan)
M. Takeda (Tokyo University of Science, Japan)
S. Tomizawa (Tokyo University of Science, Japan)
N. Watanabe (Tokyo University of Science, Japan)

Contacts
M. Taneda, Secretary
Quantum Bio-Informatics Center
Tokyo University of Science
Noda City, Chiba 278-8510 Japan
Tel:+81-4-7124-1501 ext.3350,3319,3329,5249
Fax:+81-4-7124-7130
E-mail:qbic@rs.noda.tus.ac.jp
URL http://www.noda.tus.ac.jp/qbic/

Sponsor
The QBIC (Quantum Bio-Informatics Center) project from 2006 to 2011 is sponsored by Japanese Ministry of Education and Tokyo University of Science